D0323401

COMPARATIVE NEUROBIOLOGY

WILEY SERIES IN NEUROBIOLOGY
R. Glenn Northcutt, Editor

DORSAL VENTRICULAR RIDGE: A TREATISE ON FOREBRAIN ORGANIZATION IN REPTILES AND BIRDS
by Philip S. Ulinski

COEXISTENCE OF NEUROACTIVE SUBSTANCES IN NEURONS
edited by Victoria Chan-Palay and Sanford L. Palay

COMPARATIVE NEUROBIOLOGY: MODES OF COMMUNICATION IN THE NERVOUS SYSTEM
edited by Melvin J. Cohen and Felix Strumwasser

COMPARATIVE NEUROBIOLOGY

Modes of Communication in the Nervous System

Edited by

Melvin J. Cohen
Department of Biology
Yale University

and

Felix Strumwasser
Department of Physiology
Boston University School of Medicine

A Wiley-Interscience Publication

John Wiley & Sons

New York • Chichester • Brisbane • Toronto • Singapore

Library of Congress Cataloging in Publication Data:

Main entry under title:

Comparative neurobiology.

(Wiley series in neurobiology)
"A Wiley-Interscience publication."
Based on a symposium held at the National Academy of Sciences Woods Hole Study Center on Sept. 9 and 10, 1983.
Includes index.
1. Neurophysiology—Congresses. 2. Neurobiology—Congresses. I. Cohen, Melvin J. II. Strumwasser, Felix. III. Series. [DNLM: 1. Neurophysiology—congresses. WL 102 C7365 1983]

QP351.C57 1985 591.1'88 85-9324
ISBN 0-471-87853-7

Printed in the United States of America

10 9 8 7 6 5 4 3 2 1

SERIES PREFACE

Neuroscience is a rapidly expanding interdisciplinary field that is yielding significant insights into the organization and function of nervous systems. An outgrowth of several more traditional disciplines—Animal Behavior, Comparative Biology, Cybernetics, Neuroanatomy, Neurochemistry, Neurophysiology, and Physiological Psychology—Neuroscience arose because of many reasons, but central to the focus of Neuroscience is the growing realization that no single approach or discipline can fully explain how nervous systems are organized; how they come into being ontogenetically, as well as phylogenetically; how a specific nervous system works and what operational principles are applicable to most, if not all, nervous systems.

From subcellular organelles and processes to entire networks mediating behavior, the complexity and diversity exhibited by nervous systems is staggering. The goal of Neuroscience is to understand how these complex and diverse systems work as devices for information processing, control, and communication. Unlike artificial devices that are man-made and thus specifically designed to solve limited, well-defined sets of problems, nervous systems are the result of a historical process called evolution. Thus their analysis is further confounded by the fact that they have arisen opportunistically and without optimal design. Understanding how they have arisen, how they have adapted to solve problems that are virtually unlimited, is a challenge that can not be met by a single discipline. Although individual neuroscientists will continue to focus on specific questions related to a particular facet of neural organization or a single species, achievement of the goals of Neuroscience demands an eclectic approach that is rapidly becoming its hallmark.

The Wiley Series in Neurobiology reflects this eclecticism, and the Series will present work ranging from subcellular to behavioral topics, from specialized monographs to contributions spanning several disciplines. As a forum in which nervous systems are viewed and analyzed from widely different perspectives, it is hoped that these offerings will not only provide information to researchers in all disciplines of Neuroscience, but will also provide further stimulation for an eclectic approach to the evolution, organization, and function of nervous systems.

R. GLENN NORTHCUTT

Ann Arbor, Michigan

PREFACE

The vigor of contemporary studies on the nervous system owes much to the diversity of preparations employed in asking fundamental questions ranging from development through neuroethology. The literature is permeated with studies carried out on preparations drawn from the entire spectrum of the animal kingdom. Thus work spanning a range from coelenterates to birds, as exemplified in this volume, rests comfortably alongside investigations of the cat spinal cord and monkey brain. Indeed, the very term *neurobiology*, commonly used to describe contemporary neural studies, stems from this broad comparative approach to basic questions of nervous system function. Theodore Holmes Bullock, perhaps more than any other individual, has illustrated by constant example the power and beauty of the comparative approach to investigations of neural function. It is to him that this volume is dedicated.

The essence of Ted Bullock's work stems from his basic grounding in zoology. The zoologist's keen interest in animals and immersion in the diversity of the animal kingdom underlie the free range in selection of preparations that is so characteristic of Bullock's work. He combines an appreciation of diversity in animal form with a keen interest in the variety of behavior. For Ted, an important aspect of the future of comparative neuroscience is the determination of the critical differences in the structure and function of nervous systems that are related to broad differences in capabilities, that is, what accounts for the great jump in cognitive abilities between ape and man? "We will really be on the road to understanding how the brain achieves the functions for which it evolved when we ask two questions: (i) What are the neural correlates relevant to known behavioral differences among animals (such as thoughtfulness or tameness), and (ii) what are the behavioral correlates relevant to known neural differences (such as lamination of the tectum and size of the cerebellum)?"*

*All quotations in this Preface are from T. H. Bullock, Comparative neuroscience holds promise for quiet revolution. *Science* **225**:473–478 (1984).

Ted Bullock

Modes of communication at all levels, between the organism and its environment and also between excitable cells processing this incoming information, are another important theme in this field. Interaction between the animal and its environment involves asking questions about the properties of the sensory information available to particular organisms. Here again, Ted has asked questions of novel systems. The infrared receptors in the pit organs of vipers, electroreceptors in fish, and hearing in porpoises have been his concern in this realm.

Interest in the processing of this sensory information led Ted further into examining the propagation of signals and their transmission from one cell to the next. Although he pioneered approaches to the study of classic synaptic potentials, as seen by his introduction of the squid giant synapse to physiology in 1946, we find again a breaking away from the conventional concepts currently in vogue. Work on the slow potentials of the cardiac ganglion led to the exploration and cataloging of the diverse mechanisms by which cells can communicate. Thus we have the classic paper* on the *quiet revolution* in cellular communication, calling attention to the diverse forms of slow electrical signals, ranging from pacemaker potentials to nonspiking interneurons. Here, again, the work is characterized by

*T. H. Bullock, The neuron doctrine and electrophysiology. *Science* **129:**997–1002.

vision extending beyond the concentration of activity in single units to the realization that we may have to ask different questions and invoke a variety of functions to move from the single unit to properties of large populations of excitable cells.

Ted is often a spokesman for neuroscience, as well as a spur to neuroscientists. He can ably address the concept of selfish DNA, turning it around to at least please the neuroscientist. "Everything else in the body has evolved to maintain and reproduce the behavior machine—that is, to enable animals to act. The common conclusion that behavior and metabolism are no more than means to the end of reproducing and disseminating DNA deserves reformulation from the perspective of the animals that evolution has produced."

Again, he goads us toward the future and looks forward to conceptual change: "One might well regard comparative neuroscience as Mission Impossible. . . . The most certain predictions are that our view of the nervous system today will appear naive tomorrow and that a vigorous comparative neuroscience can accelerate this hoped for obsolescence."

Perhaps we are most grateful to Ted for the environment of excitement and exchange that has always characterized his laboratory. The sheer energy and intellectual ferment created by a stream of young investigators from all over the world gave rise to an outflow of ideas and research that continue to this day. In thanks for these gifts of energy and vision some of Ted's students, friends, and colleagues gathered in Woods Hole in September 1983. Their presentations and discussions provide the framework for this volume in tribute to Theodore H. Bullock.

MELVIN J. COHEN
FELIX STRUMWASSER

New Haven, Connecticut
Boston, Massachusetts
May 1985

ACKNOWLEDGMENT

The meeting upon which this volume is based was held at the National Academy of Sciences Woods Hole Study Center on September 9 and 10, 1983. Financial support for this meeting was generously provided by Electrobiology, Incorporated (EBI) of New Jersey, and by a second donor who wishes to remain anonymous.

We express our appreciation to Mr. Robert N. Smith, manager of the National Academy of Sciences Woods Hole Study Center, and his staff for their gracious and expert support during the course of this symposium.

Our thanks also to Catherine B. Cohen for her able administrative efforts during the course of the meeting.

M.J.C.
F.S.

CONTENTS

Part One

DEVELOPMENT, REGENERATION, AND PLASTICITY

Commentary

NICHOLAS C. SPITZER

Biology Department
University of California, San Diego
La Jolla, California

The six following chapters address many of the major issues of development, regeneration, and plasticity and illustrate principles exemplified in Ted Bullock's work. The remarkable breadth of scientific problems investigated by his students and colleagues mirrors his own contributions. The authors have all focused on central problems and are concerned with deriving the general rules underlying their experimental results. Each extends conventional modes of thinking about these areas of investigation. In each case there has been a judicious choice of experimental preparation, ranging from arthropods to amphibia, birds, and mammals, selected for the special features leading it to yield the answers to a particular set of questions. In each instance there has been the imaginative application of physiological and anatomical techniques.

Understanding the basis of the guidance of growing axons, which contributes ultimately to the formation of specific synaptic connections, is one of the principal goals of developmental neurobiologists. The problem seems formidable in higher organisms, partly as a result of the large numbers of neurons. In recent years, several of the cues guiding axons have been identified in some of the numerically simpler and more stereotyped invertebrate nervous systems. Palka discusses his results from studies of sensory neurons in the fruitfly wing. Previous work has suggested that filopodia grow out randomly from neurons differentiating in a particular sequence and are guided in particular directions by contacts with other cellular or extracellular elements. In the fly wing, however, filopodial extension occurs from groups of neurons nearly synchronously, and exhibits a polarity that may be intrinsic to each neuron; axons appear to follow paths formed by physical channels, the wing veins. The flattened and complex geometry of the wing is certainly different from the cylindrical insect legs and antennae examined by others, and promotes a more general hypothesis for the mechanism of axon guidance.

The metamorphosis of homometabolous insects provides an opportunity to study development in an organism that is larger and more experimentally accessible and manipulable than the embryo. Truman and his colleagues have examined the endocrinology of this process in the moth and the changes that ensue in its CNS; it is clear that these phenomena can be viewed either as an example of development or of plasticity of reorganization. At the morphological level, neurons that lose their original target muscles often lose some dendrites and grow others when they innervate new targets. Parallel changes are seen in the function of some reflexes, which are rapidly activated at the time of ecdysis. Larval reflexes may persist even when neurons have already achieved most of their adult morphology, indicating an intriguing separation of function and structure, at least at the light microscopic level. Finally, cell death of neurons occurs during the transition from larva to adult; here, as in other developing systems (see Chapter 5 by Knudsen), there exists a critical period, in this case, a time during which the normal hormonal stimuli must act to trigger the process. The knowledge that the control of neuronal morphology, synaptic connectivity, and cell number occurs under the influence of three identified hormones will allow detailed analysis of the mechanisms involved.

Although the frog neuromuscular junction has been extensively studied with respect to its short-term function and early development, the findings of Grinnell and his collaborators have more recently revealed relationships involved in longer-term maintenance and plasticity and have focused attention on several unresolved issues. The length of a nerve terminal is proportional to the diameter of the innervated muscle fiber; for fibers of similar size, the amount of release along a terminal is inversely proportional to its length. The nature of the signals from the nerve or muscle that govern these steady-state relationships is unknown, but impulse activity seems unlikely to play a major role; characteristics of individual motoneurons and the size of their motor unit appear to be involved in regulating synaptic strength. Competition exists not only at the level of developmental exclusion of other inputs but also at the level of reduction of effectiveness of two endplates terminating on the same muscle fiber. Plasticity in synaptic strength can be demonstrated by experimental manipulations. The differences in long-term efficacy seem to be entirely presynaptic, and stronger synapses may be the consequence of regulation of the intraterminal calcium concentration at higher levels. The basis of possible intrinsic differences between motoneurons, as well as the reason for stable polyneuronal innervation of some muscle fibers, are only two of the many unanswered questions that await investigation.

The elaboration and maintenance of processes by a neuron create its most distinctive phenotype, and Cohen addresses the mechanisms by which this is controlled. The location of sprouting processes and the extent of growth in an identified, axotomized neuron in the cricket are described. Axotomy close to the cell body induces sprouting from dendrites, to a degree reciprocal with the extent of axonal sprouting. Axotomy distant to the cell body elicits sprouting only from the axon. These and other results suggest that newly synthesized membrane is preferentially inserted into the axon unless this is prevented—by the dying back of the axon, for example. Furthermore, the membrane properties of the cell body as well as some of its organelles are altered by section of the axon close to the cell soma. Transection of the lamprey spinal cord results in production of steady currents that enter the cut end of the proximal stump. The influx of ions should alter the ionic composition of the tissue locally. Axons sprout and regenerate across the site of the lesion. The responses of both preparations may be mediated by an influx of cations, in particular, calcium, affecting the cytoskeleton, neurite outgrowth, and electrical excitability. In support of this hypothesis, experimental application of steady electrical current of appropriate polarity to the transected lamprey spinal cord can either enhance or retard morphological regeneration.

The ultimate purpose of the development of specific neuronal connections is the generation of behavior. Knudsen analyzes some of the requirements for the development of sound localization in the owl, in which this essential capability is extraordinarily well developed. Attenuation of the sound arriving at one ear still permits detection of patterned stimuli, unlike the monocular occlusion performed in studies of the visual system. The animal's orientation to auditory stimuli is then compared to its response to visual stimuli. At early stages, there is sufficient plasticity to allow matching of abnormal auditory cues with locations in space. When owls are monaurally occluded prior to eight weeks of age, they correct sound localization errors fairly quickly, whereas older animals never do so. The sensory experience of the birds appears to be the decisive factor in bringing this sensitive period to a close, although the age of the animal may be important as well. The readjustment of the sound localization accuracy of animals monaurally occluded past the end of the sensitive period occurs until the age of sexual maturation, defining a critical period for consolidation of the association of auditory cues and locations in space. These behavioral results are paralleled by neurophysiological observations made by recording from the optic tectum. Occlusion past the end of the critical period results in alignment of auditory and visual receptive fields while the ear plug is in place, and in stable misalignment after it is removed.

The sources of variation in the cortical maps of higher primates and the extent to which these maps can be altered by experience in the adult are reviewed by Merzenich. He discusses experiments in which he and his colleagues have recorded from the monkey sensory cortex and mapped the receptive fields of the skin of the hand in substantial detail. There is variability in the detailed topography, superimposed on a constant, orderly preservation of neighborhood relationships. Use-dependent alteration of these somatosensory cortical maps is illustrated by amputation of several digits of adults, after which there is expansion of the receptive fields of the remaining digits. Similar changes ensue following peripheral nerve transection and ligation. The involvement of more normal experience as a determinant of routinely observed variability is supported by observations of changes in receptive field sizes following chronic tactile stimulation. These findings stand in contrast with the widespread view of the limited capabilities for reorganization of the CNS. The function of the system appears to play a role in defining connections at times long after initial development.

These studies direct attention to several future avenues of investigation. Ted Bullock stimulated thought about the variety of methods of information coding in the nervous system, both with and without impulses. This theme is extended by the demonstration that electrical activity and ion fluxes can carry still other kinds of information on slower time scales (see Chapter 4 by Cohen). Are there further, at present unidentified forms of long-term signaling that involve release of energy stored in ion concentration gradients? Since such signals are probably smaller and slower than those with which we are now familiar, sensitive assays will be required to detect them. Some processes described in these chapters appear to involve interactions mediated by agents other than electrical activity, suggesting roles for growth or differentiation factors. Purification of these factors and determination of their mechanisms of action are major challenges. Analysis of this kind of chemical signaling has already proven more demanding than the investigation of neurotransmitter biochemistry: the agents studied to date are effective at such low concentrations that detection is a problem. Future prospects in this regard will be enhanced by application of some of the tools of molecular biology (see chapters by Strumwasser, Walters et al., and Bloom). Genetic studies of organisms such as the fruitfly, nematode, and mouse have identified essential genes; these can now be isolated, sequenced, and the polypeptides they encode can be determined. One may also expect to see increasing application of tissue culture to investigate phenomena *in vitro,* where the environment can be more easily manipulated and relevant parameters identified in the analysis of mechanisms. The increasing extent to which it is possible to observe development, regeneration, and plasticity ocurring in the dish encourages this direction. Finally, it seems likely that we are now substantially closer to having a sufficient understanding of these processes to begin to influence them *in vivo.* One may look forward to successful efforts to achieve exogenous guidance of embryonic axonal outgrowth, enhancement of the duration of sensitive or critical periods, and an increased extent of regeneration and plasticity in the adult nervous system.

Chapter One

FACTORS INFLUENCING NEURAL DIFFERENTIATION IN THE PERIPHERY

JOHN PALKA

Department of Zoology
University of Washington
Seattle, Washington

How is a nervous system constructed from its components? How are all its various cells correctly laid out and interconnected to process information and generate behavior? What mechanisms ensure that each complex individual develops much like others of its species? Conversely, what is the nature of the variability in neural elements and circuits upon which natural selection can act during the course of evolution?

Questions like these are not new. They have been formulated in various guises at least since the time of Darwin and are a restatement of the interest of ethologists in the evolution of behavior, of developmentalists in the relationship of the genotype to the phenotype, of sensory physiologists in the many exquisite adaptations of sensory systems to specific tasks. In fact, one might even argue that they underlie our age-old philosophical quest to understand who we are and our place in the universal scheme of things.

Although I cannot provide answers to these grand questions, I can relate some observations made in our laboratory during the past few years on neurogenesis in the wing of the fruitfly, *Drosophila.* Small and humble though a fly's wing may be, its study quickly brings us face to face with the grand questions, and in this chapter I propose to set forth some observations and speculations resulting from my encounter with the problems of neurogenesis in the context of evolution.

A MODEL SYSTEM—THE WING OF THE FLY

The wing of an adult insect is a very thin, often transparent, sheet of cuticle reinforced by hollow struts, the veins (Fig. 1). It is an object of great beauty (Who can resist delighting in a butterfly's wings?) and utility: insects are the most varied group of animals, and it is hard to imagine that their spectacular diversification is not related to their ability to fly. The presumed importance of wings in insect evolution is reflected in the names we give to the various orders: Orthoptera (crickets and grasshoppers) have straight wings, Coleoptera (beetles) have wings that act like covers, Lepidoptera (butterflies and moths) have wings with scales on them, Hymenoptera (bees and wasps) have membranous wings, and so forth. Diptera, flies and their relatives, have two wings instead of the four found in other orders.

In spite of their ability to propel and steer their owners, insect wings have no intrinsic muscles but rather are moved by muscles in the thorax by way of the complex articulation of the wing hinge. This, of course, means that a wing also lacks motor neurons. It does, however, carry an array of sensory neurons, most though not all of them associated with cuticular elements having an important role in transduction. Each miniature sense

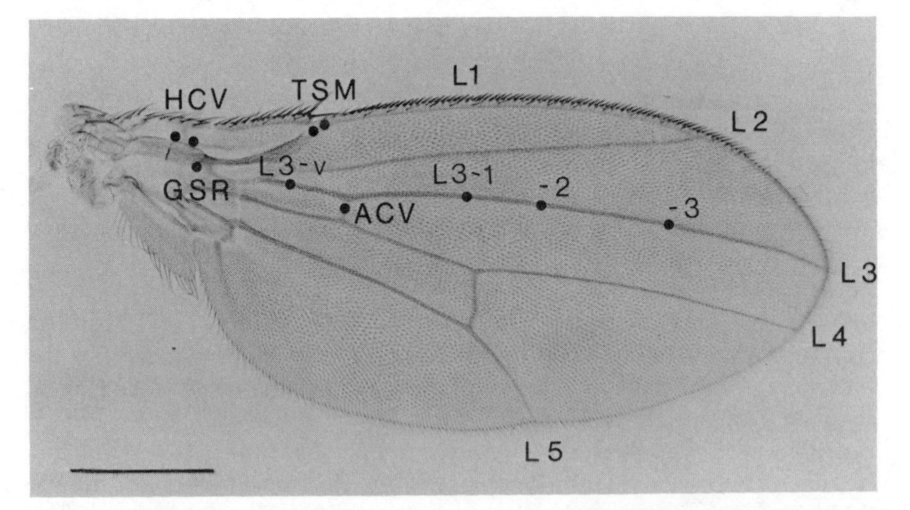

FIGURE 1. Distribution of sensilla on the wing of *Drosophila*. Bristles, both mechano- and chemosensory, line the anterior edge of the wing (up in the photograph). All 10 large campaniform sensilla are indicated, named according to their location on specific veins: HCV, humeral cross-vein sensilla; GSR, giant sensillum of the radius; TSM, twin sensillum of the margin; ACV, anterior cross-vein sensillum; L3-v, ventral sensillum of 3rd vein; L3-1, 2, 3, dorsal sensilla of 3rd vein. The longitudinal veins are numbered L-1 to L-5. Scale = 0.5 mm.

organ consisting of specialized cuticle, the two cells that secrete it, one or a few nerve cells, and a sheath cell is called a *sensillum*. Two types of sensilla are found on the wing of *Drosophila*: bristles with long central shafts, and campaniform sensilla with rounded domes (Fig. 2).

All of the sensory neurons of the wing send their axons to the central nervous system (CNS) where they ramify and form synapses. The axons of bristles and of campaniform sensilla form very different projections, which presumably reflect the particular sensory modalities they subserve and the uses to which the information they supply to the CNS is put (Fig. 3).

The various sensilla have very regular locations on the wings: the bristles are confined to the wing's leading edge, and the campaniform sensilla occur in characteristic places along some of the veins (Fig. 1). Their axons follow consistent paths and ultimately assemble into two nerves before leaving the wing on the way to the CNS. We can think of the wing as a double-sided circuit board, with components (sensilla) laid out in the *X-Y* plane and joined together by conduction lines (axons).

There are at least four characteristics of this biological circuit board that are notable for their consistency from individual to individual within the species:

1. The *number* and *location* of the components.
2. Their *qualitative nature*, seen in the external morphology and the central projections.
3. The peripheral *paths* followed by the axons.
4. The *direction* in which axons grow along given paths.

Let us consider this consistency from two different viewpoints: what factors bring it about, and how much variability is available for modifying the board by natural selection?

GENETIC INFLUENCES ON WING SENSILLA

It is no surprise to find that the number, location, and nature of the sensilla are under rather close genetic control. They are, after all, used by taxonomists for species identification, even though in that context they are thought of simply as "characters" and no special attention is paid to their sensory function. Let me give two examples of easily demonstrated genetic effects.

The map of Figure 1 shows two campaniform sensilla near the anterior, bristle-studded edge which are labeled TSM (twin sensilla of the margin). In commonly used wild type strains, such as Canton-S and Oregon R, one finds two identical sensilla in this location in more than 95% of individuals (Fig. 4*A*). In the strain called Sevelen, however, cases of single sensilla are

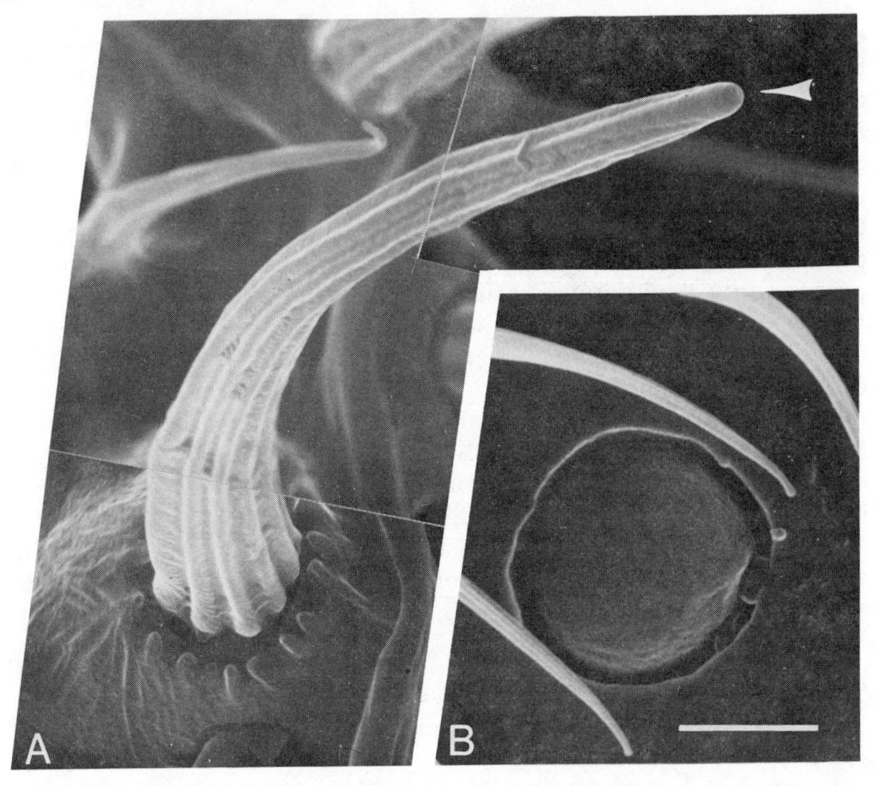

FIGURE 2. Scanning electron micrographs of wing sensilla. (*A*) Mixed-modality chemo/mechanosensory bristle. The dendrite of one associated neuron attaches near the base and responds to bristle displacement; the dendrites of four other neurons grow through the lumen to the tip where a pore through the cuticle (arrow) gives them access to molecules in the environment. (*B*) Campaniform sensillum with its dome nestled within a socket. The dendrite of the single associated neuron attaches to the inner surface of the dome and responds to cuticle deformation. Scale = 2.5 μm.

quite common. Following just five generations of selecting and inbreeding flies with single TSM sensilla on both wings, we have obtained lines where this phenotype occurs in more than 90% of individuals (Fig. 4*B*). We have not yet recognized any other consistent phenotypic effect of this selection process, and the phenotype has remained stable for several generations in the absence of selection. Preliminary experiments, carried out with the help of L. Sandler and M. Prout of the Department of Genetics of the University of Washington, indicate that this is a polygenic effect, influenced by factors on at least chromosomes II and III. Evidently, then, the number of sensilla, or the presence or absence of a sensillum in a specific location, is genetically regulated and subject to selection.

In rare cases (less than 1%) the more proximal of the two TSMs develops not as a campaniform sensillum but as a bristle (Fig. 4*C*). We have not yet

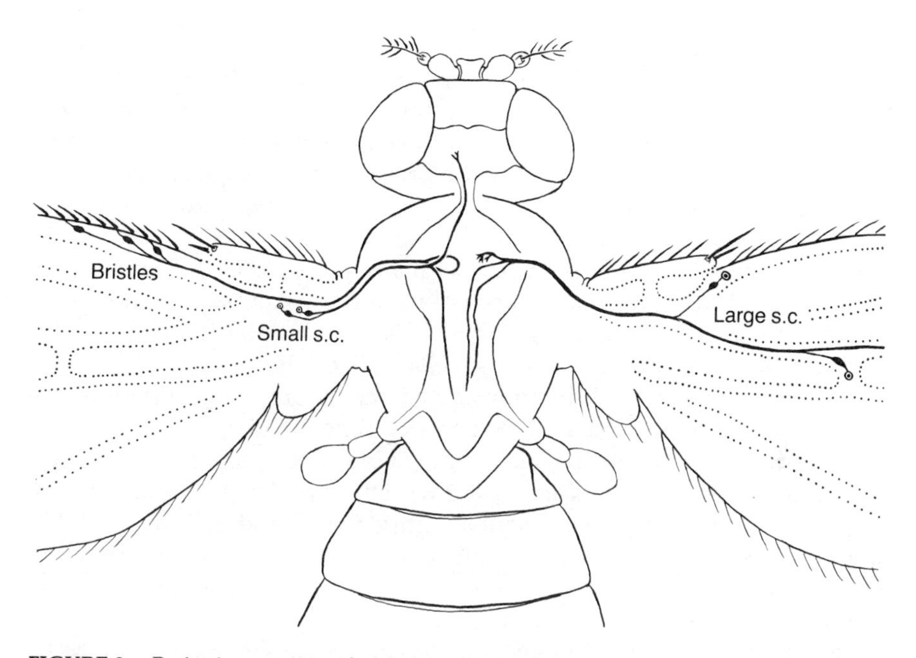

FIGURE 3. Projection patterns of wing sensilla. Bristle axons (left) project to a discrete area of the mesothoracic neuromere of the CNS; axons of large campaniform sensilla (those described in this chapter and shown on the right of the figure) branch in the mesothorax and form two ventral bundles projecting posteriorly to the metathorax; axons of small campaniform sensilla located proximally on the wing (left) form a bifurcating tract dorsally in the CNS and reach both the metathorax and the subesophageal ganglion in the head. Thus each class of sensilla forms a discrete and distinctive projection pattern.

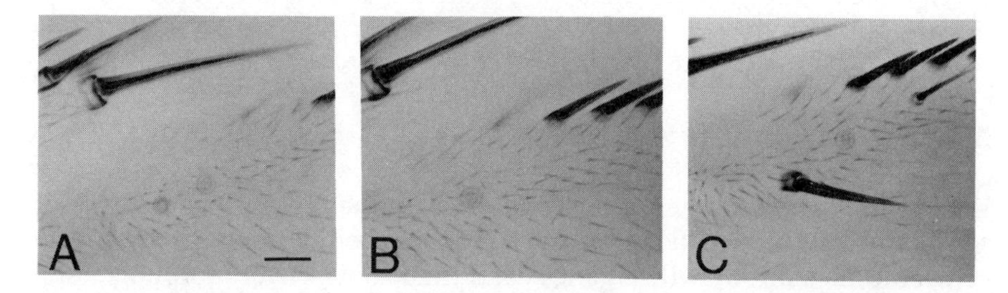

FIGURE 4. Variation in the complement of wing sensilla. (*A*) In most wild type individuals of *D. melanogaster*, there are two nearly identical sensilla near the edge of the wing (TSM, Fig. 1). (*B*) Some individuals show only a single sensillum here. The frequency of this phenotype can be greatly increased by artificial selection. (C) Very rarely, the more proximal sensillum develops as a bristle. This is a common phenotype in at least one other species of *Drosophila*, *D. busckii*. Scale = 20 μm.

tried breeding for this phenotype. However, we have found it with high frequency (more than 50%) in a species from another subgenus of *Drosophila, D. busckii,* though not in a number of other species both closely and distantly related to *D. melanogaster* (Throckmorton, 1975). It seems very likely, therefore, that not only the presence or absence of a sensillum in a given location, but also its nature, is genetically regulated and selectable.

When a bristle forms in the place normally occupied by a campaniform sensillum in *D. melanogaster,* the axon of its nerve cell projects to the bristle area of the CNS (D. Wigston and R. Ellison, unpublished). Thus the phenotypes of the cuticle-secreting cells and of the neuron of a given sensillum are usually tightly coupled, as seems only appropriate if the sensillum is to perform a specific sensory function. However, the (unknown) coupling mechanism(s) can fail. For example, uncoupling has been described in certain mutants associated with the so-called bithorax complex of genes which influence segment identity (Palka and Schubiger, 1980; Ghysen, Janson, and Santamaria, 1983). The latter authors offer an interesting genetic model to explain the uncoupling phenotype, but it would take us too far afield to discuss it here.

There is an extensive older literature on the genetics of bristle patterns in *Drosophila* (see Sondhi, 1963, for a detailed review). These are largely formal, not mechanistic, analyses, and, as Waddington (1962) has aptly put it, such studies "should be regarded as facing us with a problem and not as providing us with an explanation." Nonetheless, from the neurobiological point of view, the genetic variants that alter the components of our biological circuit board are important in two ways: (1) they provide useful experimental material, as illustrated later; and (2) they indicate features that are susceptible to selection and hence are profitable to examine from an evolutionary point of view.

A RANDOM SEARCH MODEL FOR ESTABLISHING PATHWAYS

We have seen so far that genetic influences on the presence and nature of sensilla in specific locations on the wing are readily demonstrated. The actual mechanisms by which genes exert their control over morphogenesis are quite unknown, but let us close our eyes to this problem and assume that the positioning of components on our biological circuit board is accomplished by reliable mechanisms acting ultimately under genetic control. Having put the components in their places, we need to ask how they become wired up—how the connections between them are established.

Let me introduce the steps we have taken toward shedding light on this issue by adapting to the wing of *Drosophila* a model proposed for neurogenesis in the legs and CNS of grasshoppers. In its starkest and most explicit form (Bentley and Keshishian, 1982), this model proposes that a differentiating neuron sends out filopodia randomly in all directions; that neuronal filopodia adhere better to other neurons than they do to non-neural cells; and that the cell's axon will develop where the filopodia adhere best. Consequently, all that is required to establish regular peripheral pathways is to ensure that neurons differentiate in the proper places and in the proper sequence, as illustrated for the wing in Figure 5. Here we suppose that Cell 1 differentiates first. Some of its filopodia encounter a nearby, preexisting larval nerve (l.n.), the axon develops along the trajectory of the successful filopodia, and the most proximal portion of the nerve path in the wing is established. Cell 2 differentiates next. Its only neural neighbor is Cell 1, so the exploring filopodia will adhere to it most strongly. The axon will form accordingly, and the next segment of the pathway will have been established. The cycle repeats for the last cell in line, Cell 3. Once the pathway is established, other cells can use it, given the additional assumption that the earlier axons can impart correct polarity to the latter ones.

This is an attractive hypothesis, providing for the self-assembly of neurons into pathways with only a few reasonable assumptions. We have already seen that peripheral sensory neurons do, in fact, develop in specific locations. We show in this chapter that the sequence in which they differentiate is likewise specific. In our material we cannot see a halo of filopodia emanating from the soma prior to axonogenesis, but this has been described in other cases (Keshishian and Bentley, 1983). There is abundant evidence that growing axons tend to fasciculate. Finally, we have some experimental evidence that later-differentiating axons do obtain polarity information for the correct direction of growth from something associated with preestablished peripheral pathways, quite possibly the older axons (see below).

For all these reasons, the random search model seems an excellent starting point for thinking about the formation of peripheral neural pathways. We do not actually believe that it adequately accounts for neurogenesis in the *Drosophila* wing, but it serves an extremely useful heuristic function.

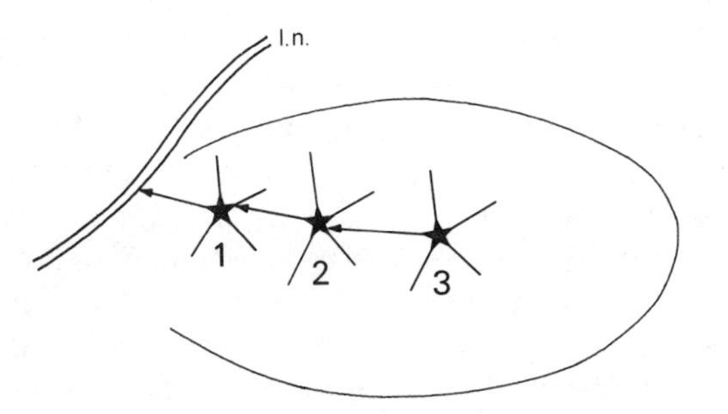

FIGURE 5. Diagram illustrating the most extreme version of the hypothesis that peripheral neural pathways are established by random filopodial contact between neurons differentiating in specific locations and in a specific sequence. See text for a more detailed explanation.

This is reflected in the discussion that follows, which is organized around the specification of factors that might restrict the filopodial exploration process, making it more efficient and reliable than a purely random search would be.

EARLY NEURAL DIFFERENTIATION

Like the entire body wall of *Drosophila,* the wing develops from an imaginal disc, an invaginated sac of cells that do not participate in forming the larva but are held in reserve for making adult structures at the time of metamorphosis. The timing of adult differentiation is conventionally measured starting with the beginning of pupariation, when the third-stage larva stops burrowing in the food, crawls up onto a dry surface, everts its spiracles, attaches to the substrate, and shortens. Within 20 minutes, the white, flexible larval cuticle begins to harden and tan, forming a protective case called the puparium. Zero hours after pupariation (AP) is defined as the short time after attachment and shortening but before tanning. Twenty-four hours later (at 25°C), the eversion of the wing from an invaginated sac into an evaginated appendage, and the differentiation of sensory neurons within it, are complete.

The sequence of neural differentiation is shown in Figure 6 (based on Murray, Schubiger, and Palka, 1984). We recognize nerve cells on the basis of their affinity for antibody to horseradish peroxidase (Jan and Jan, 1982), a property that they acquire prior to axonogenesis and maintain into adulthood. On this criterion, the neurons of six campaniform sensilla and four bristles have already begun to differentiate prior to 0 hr AP, and one of the TSM neurons has started to grow its axon. During the first hour AP, axonogenesis begins in three other neurons. A few hours later other cell bodies begin to stain, and their axons soon begin to grow. The full sequence is set out in Figure 6. Figure 7 illustrates axonogenesis in specific campaniform neurons.

The large population of bristle neurons accumulates over an extended period of time. Axonogenesis starts only at 12 hr AP, when the population of cell bodies is approximately complete.

The arrows in Figure 6 show the direction of initial axon growth for the six campaniform neurons born before 0 hr AP. A number of details in this pattern suggest that random search for pre-positioned neural targets is not a full explanation for the process of circuit assembly in the wing. The axons of the two HCV cells (humeral cross-vein sensilla), for example, grow parallel to the edge of the wing, not toward the nearby giant sensillum of the radius (GSR) neuron. They seem to ignore this conspicuous target, and we cannot find any other neural target that would account for their course. The axon of the first TSM neuron presents a similar puzzle. Initially it grows directly inward from the wing margin, but soon angles proximally.

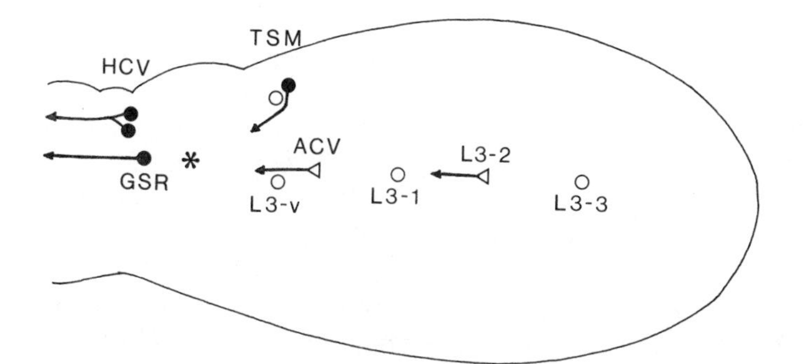

FIGURE 6. Summary of the sequence of differentiation of neurons in the wing. The diagram represents a wing six hours after pupariation. The neurons indicated by closed circles were born before zero hours, while the wing was still an imaginal disc, and acquired immunoreactivity within the first hour. Those indicated by open triangles differentiated a few hours later. At six hours, both groups have prominent axons which follow well-defined courses (cf. Fig. 7). The neurons marked by open circles have yet to be born. The HCV axons shun the GSR and grow parallel to its axon; they have no detectable neural target at this time. The TSM and ACV axons will meet at the point marked by the asterisk. No neural target is located there, and their behavior is unaltered if all of the proximal wing, including the GSR, is removed.

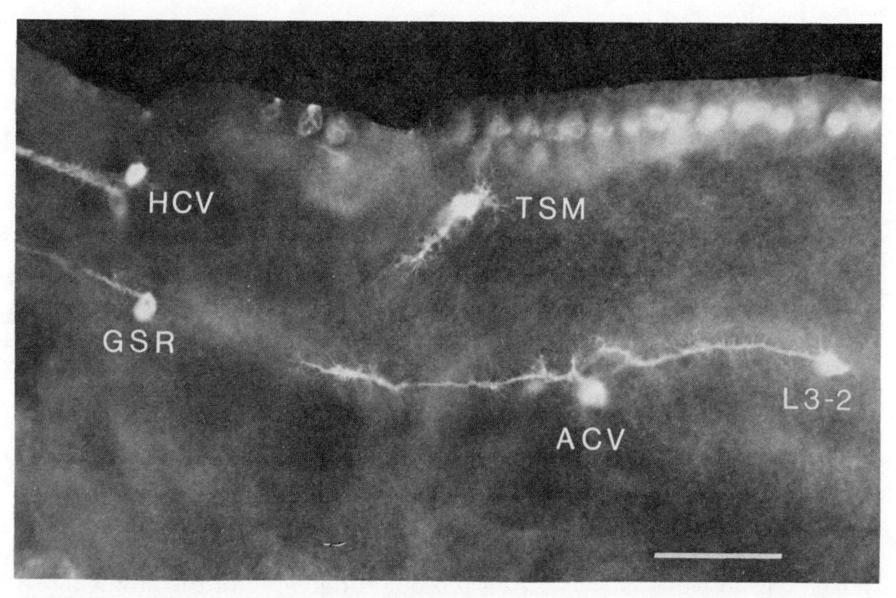

FIGURE 7. Some of the neurons revealed in a 6-hr wing by the anti-HRP technique. Compare with the diagram and legend of Figure 6. Scale = 25 μm.

There is no neuron present at any time at the point where it changes direction. The axon then heads for the rendezvous point marked by the asterisk in Figure 6, where it meets the axon of the anterior cross-vein sensillum (ACV) neuron. Again, no neural target exists at this point, now or later. The TSM and ACV axons head for this distant meeting point rather than toward each other or toward a common neural target cell.

Wherever we look in the normally developing wing, we see signs that axon growth is guided by more factors than just the random search of filopodia for strategically positioned nearby neurons. In the following sections, we consider what some of these other factors might be.

VEINS AND OTHER SPACES

In the normal adult wing all sensilla are located over veins, and their axons run within veins. This close relationship between sensilla, axons, and veins is not obligatory, however. There are a number of mutants that produce supernumerary sensilla, both bristles and campaniform sensilla, and these can form anywhere on the wing: on veins that carry sensilla in wild type flies, on veins that are normally free of sensilla, and between veins (Fig. 8).

During the first few hours after pupariation, while the wing's first axons are growing, the epithelial sheets are joined except for a system of internal spaces. We do not yet understand the three-dimensional arrangement of these spaces, but some of them certainly seem to be the forerunners of the definitive adult veins (Fig. 9). In whole-mounted preparations, the growing axons lie within the best-defined channels, and their filopodia often seem to outline them. Thus there is a possibility that physical channels contribute to the efficient establishment of neural pathways, as has been suggested for certain cases in vertebrates (Singer, Nordlander, and Egar, 1979; Silver and Sidman, 1980; Krayanek and Goldberg, 1981).

Later, between about 12 and 24 hours AP, the already-joined epithelia first separate, so that a large, continuous open space forms between them, and then gradually rejoin (Waddington, 1940; Anderson, 1984). The axons of the ectopic cells shown in Figure 8 start to grow during this time, and it seems reasonable to suggest that the absence of defined channels encourages wide-ranging filopodial exploration and the establishment of multiple neural pathways. As far as we can tell, axons in all locations, between as well as within veins, survive the subsequent rejoining of the epithelia, so that the precise association of veins and axons in wild type flies cannot be ascribed to the retraction or death of errant branches.

Thus there is a correlation between the presence of discrete physical channels and discrete neural pathways early in development, and between open spaces and multiple pathways during a curious developmental inter-

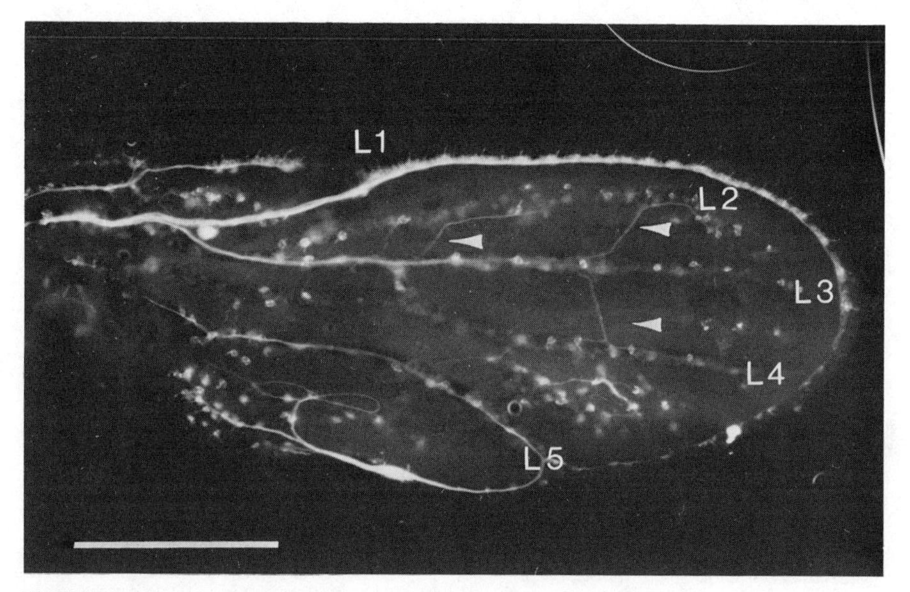

FIGURE 8. Pupal wing of a *Hairy wing* mutant. There are many supernumerary sensory neurons. Their axons can apparently grow anywhere in the wing, and the patterns they form vary widely among individuals. A regular feature is bundles of axons crossing from veins L2 and L4 to L3 (arrows), where the normal cells and axons are found in their usual places. Scale = 200 μm.

lude. We are now starting to examine this correlation in detail (see the postscript to this chapter).

POLARITY INFORMATION

The presence of physical channels in differentiating wings may well help to position growing axons in the *X–Y* plane, reducing the amount of exploration required of the filopodia. The process of establishing the correct axonal

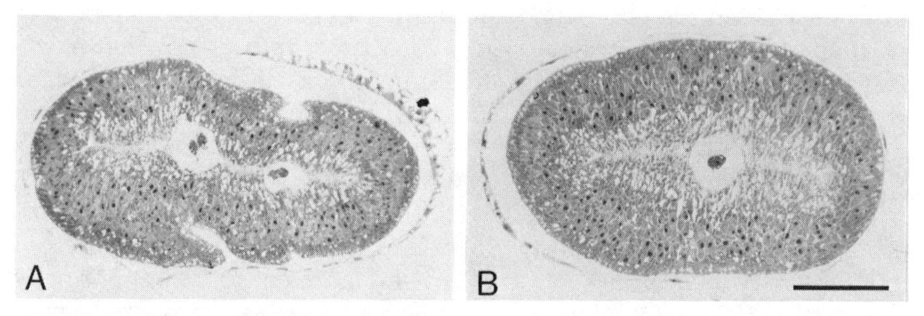

FIGURE 9. Cross sections through a 4-hr wing (*A*) proximal, (B) distal. Note the well-defined channels bounded by columnar epithelium and containing blood cells. These channels may help to guide growing axons. Scale = 50 μm.

trajectory would be made still more efficient if the filopodia had cues for correct polarity prior to reaching the next, sometimes rather remote, "guidepost" neurons.

We can imagine at least two possible sources of such polarity information: (1) molecules in the environment, perhaps incorporated in cell surfaces (Goodman et al., 1982), basal laminae (Sanes, 1983), or an extracellular matrix (Nardi, 1983); (2) intrinsic factors within the neurons themselves, perhaps cytoskeletal elements whose distribution reflects the plane of the previous cell division (e.g., Solomon, 1979; Kirschner, 1982). Here I summarize some experimental data bearing on this issue.

It is possible to cut the developing wing into fragments along lines that separate neighboring neurons before any axon outgrowth has occurred, to keep the fragments in organ culture for a day, and then to examine axonogenesis in the absence of putative target cells. Thus far, no cut separating neighboring neurons in wild type wings has produced disoriented axon growth (Blair and Palka, 1985). Since correct polarity is regularly achieved in the absence of all known neuronal targets, we are forced to conclude that filopodial contact with these targets is not essential for the establishment of correct polarity. We would not, of course, argue that it is not a contributing factor under normal circumstances, only that it is not essential for most cells.

A parallel experiment done using genetic techniques has yielded comparable results. It is possible to create wings that are genetic mosaics in which most of the wing is wild type and has the normal set of sensilla and neurons, but a patch is mutant and lacks sensilla and neurons. Thus some neurons are deprived of their targets. In addition, in some genotypes, certain of the sensilla simply fail to develop; here, too, some neurons lack their normal targets. The result for the majority of cells is just like that of the surgical fragmentation experiment—axons grow in the correct direction even if their normal targets are missing (Schubiger and Palka, 1985). However, interesting exceptions to this generalization do occur, and we are currently trying to define and understand them better.

Both the surgical and the genetic experiments argue for polarity cues provided by sources other than the normal target neurons, but neither discriminates between extrinsic cues, associated with a growing axon's environment, and factors intrinsic to the neuron itself. Intuitively most of us tend to think in terms of extrinsic cues, but there are several observations in the *Drosophila* wing that make us pause and not exclude intrinsic polarity factors too hastily.

The axons of ectopic neurons in *Hairy wing* flies seem not to recognize any global polarity cues (Palka, Schubiger, and Ellison, 1983). If the cell body of such a neuron lies in vein L2 or L4, adjacent to L3 but normally devoid of sensilla, the axon crosses over to L3, joins the normal nerve there, and grows with the correct polarity toward the CNS (Fig. 8). In more remote veins, such as L5, crossing over to L3 is observed only rarely. The axons tend to remain within the vein but exhibit no consistent polarity—about half grow outward and half grow inward. If the cell body lies between veins, the axon can grow in any direction. We see no signs of a general distal-to-proximal gradient in the wing that effectively guides the ectopic axons toward the CNS, as might have been expected from the experiments of Nardi (1983) on moth wings.

In contrast, two observations are consistent with the idea that intrinsic cell polarity may be an important determinant of the direction of initial axon outgrowth, an idea applied to neurogenesis in insect wings at least as early as the study of Henke and Rönsch (1951). The more general observation is that the intervein ectopic cells of *Hairy wing* not only produce axons growing in many different directions, but they have a variable polarity, defined for each cell by an arrow drawn through the dendrite and soma, even before the axons can clearly be distinguished. Ectopic neurons associated with veins may point either inward or outward, just as their axons are seen to grow in both directions. Thus there is a correlation between axonal behavior and cell polarity on a popularity basis (J. Palka, unpublished).

This correlation can be analyzed more precisely by studying the normal neurons occurring along vein L3 of wild type flies. At 6 hr AP, for example, all of these neurons point toward the base of the wing *except* the single neuron on the ventral surface, the L3-v, which regularly points outward. Under normal circumstances, of course, its axon does grow into the CNS. But if, in a genetic experiment, one of the earlier-differentiating, more distal neurons on the dorsal surface is eliminated, the axon of the L3-v frequently grows *outward*, presumably along the epithelium. Thus in the absence of a putative guiding axon the L3-v axon's polarity correlates best with the polarity of its own soma and not with that of potential cues associated with the vein epithelium (J. Palka, M. Murray, and M. Schubiger, unpublished; see the postscript to this chapter).

We certainly do not take this as proof that intrinsic cell polarity determines the polarity of axon outgrowth, and that nonneural extrinsic cues are not important, only as a reminder that our observations up to the present time are more easily reconciled with this hypothesis than with the extreme alternative, that axon outgrowth is unpolarized until a polarity is imposed by external cues from the epithelium. However, our attention at present is focused on a third alternative, that extrinsic guidance cues are important, but are specifically associated with the epithelial paths that axons normally follow rather than being distributed throughout the wing (see the postscript to this chapter).

AN INTERIM MODEL

Our search for factors that might constrain random searching by filopodia, and thereby make the process of linking sensory components distributed on the wing more efficient, has produced three leading candidates: veins or other physical channels, localized pathway cues, and intrinsic neuron polarity. Surprisingly, we have not found equally suggestive evidence for the operation of a general epithelial factor for establishing axon polarity.

At present, therefore, we can summarize our observations in the form of the following suggestions of factors influencing peripheral neurogenesis in *Drosophila*.

1. The construction of the wing, like the rest of the body, is ultimately controlled by genes, though how this is done is no clearer here than in any other experimental system. However, we can show by breeding experiments that the number, location, and nature of sensilla are genetically regulated in such a way that they respond to selection pressures.
2. Most of the axons of the wing in wild type flies grow within veins; indeed, the filopodia often give the impression of outlining the lumen of the vein. Thus the veins appear to define normal axon paths.
3. Our techniques do not reveal a halo of filopodia emanating from the sensory cell body before axonogenesis. Rather, filopodia seem to form mainly in association with the growth cone and the axon it leaves behind. Thus they seem to be concerned more with the details of axon growth than with establishing its initial polarity.
4. Judging by the behavior of ectopic axons, regions of the wing (whether within or between veins) that are not normally transversed by axons do not supply useful polarity information for axon growth.
5. Even along the paths that axons normally take, we have not found clear evidence for potent extrinsic polarity cues. It is difficult to distinguish extrinsic from intrinsic cues experimentally, but in at least one case, the L3-v neuron, it seems possible that intrinsic factors are important.

A possible working hypothesis, therefore, is that in the wing of *Drosophila* axons start out in the right direction primarily because of factors intrinsic to each neuron, and they follow standard paths formed by physical channels. These constraints are so strong that random filopodial exploration generally plays a relatively minor role in the establishment of neural pathways. Likewise, the precise sequence of differentiation of neurons is not critical in most places. However, exploration and timing assume greater roles whenever intrinsic polarity (as in the case of the L3-v neuron) and/or physical channeling do not provide adequate information for reliable circuit assembly, and in some places or circumstances they may well be the dominant factors. Polarity information derived from nonneural cells remains an attractive candidate factor, but our current evidence argues against assigning it a dominant role.

Neither channeling nor intrinsic neuronal polarity have been given prominence in the many elegant studies on peripheral neurogenesis in the antennae, legs, and cerci of grasshoppers and crickets (Bentley and Caudy, 1983; Edwards, 1982; Goodman et al., 1982). It seems entirely possible that in these tubular appendages with a single lumen the balance of relevant factors may be different than in the large, flat wings whose multiple internal channels may serve several purposes simultaneously.

THE COMPARATIVE APPROACH

I want, finally, to return to some rudimentary evolutionary considerations relating to neurogenesis in the wing. It is difficult to think about the adaptive significance of specific sensilla on the wing and to try to understand the meaning of any modifications in their number, nature, or location that we might find or select for, because we do not know what specific physiological, let alone behavioral, functions the sensilla serve. Do the chemosensory bristles function in feeding, courtship, or some other behavior? Do the mechanosensory bristles play a role in flight, grooming, both, or neither? Flies with supernumerary sensilla, such as *Hairy wing* and the genetically quite distinct *hairy*, do not fly well, their deficit being roughly proportional to the number of extra sensilla (C. Miles, unpublished). However, it would be premature, to say the least, to conclude that the extra sensilla are the cause of flightlessness. And even if they were, this would not enable us to make confident statements about the function of the normal sensilla. Stronger evidence comes from physiological studies which indicate that the total sensory feedback from the wings has an effect on the rhythm of the flight motor output (see Heide, 1983). However, this function cannot as yet be attributed to specific sensilla. Many questions remain that are appealing to sensory physiologists, and naturally we are taking up the challenge, especially since electrophysiological recordings from the campaniform sensilla have proven to be feasible.

Since we know so little about sensillar function at the present time, though, we are seeking to interpret the results of our selection experiments along quite different lines. We find that certain sensilla, such as the proximal (but not the distal) TSM, are somewhat variable in location, number, and/or qualitative nature, whereas others are extremely stable. Why should this be? In what ways might the relatively variable sensilla differ from the extremely stable ones? If a particular sensillum is variable in *D. melanogaster*, will it also vary across species? If yes, can we formulate and test hypotheses that would account for a linkage between intra- and interspecific variability? Can these hypotheses be directly related to developmental events?

It is too early to commit our current hypotheses to print, but I consider the approach to be a promising one. In contrast to the limited data on sensillar function, there is an enormous literature on the ecology, behavior, and evolution of many drosophilid species (e.g., Ashburner, Carson, and Thompson, 1981–), which helps us to choose with reasonable confidence among closely or distantly related species, primitive or derived species, or species with distinctive, well-described behaviors, all according to the demands of the question or experiment. Furthermore, this literature is laden with tantalizing observations and ideas for the peripatetic neurobiologist—for example, the huge and beautiful picture-winged flies in Hawaii, courting and jousting with their wings! I look forward to a delightful period of true Bullockian neurobiology, studying different animals both because they are intrinsically fascinating and because comparisons among them can be expected to enrich our understanding of the structure, function, development, and evolution of nervous systems in the natural world.

POSTSCRIPT

In the 1½ years since this essay was written, we have been able to test some of the working hypotheses suggested herein. In particular, we have been able to produce dorsal and ventral wing fragments that develop without physical channels, and thus to test whether these channels are essential to axon guidance (Blair et al., 1985). Like guidepost cells, they prove *not* to be essential. Furthermore, axons continue to grow along their normal paths even in the simultaneous absence of channels and of guidepost cells. The most extreme case is the L3-v neuron, the only one found in the central region of the ventral wing epithelium. In ventral fragments, it is seen as a single neuron in a sea of epithelium, yet its axon grows with the correct axial alignment (along the longitudinal axis of the wing, rather than at some arbitrary angle). It exhibits some uncertainty with respect to polarity, sometimes growing distally rather than proximally, which is consistent with its behavior in genetically mosaic wings lacking putative guidepost cells (see the section on "Polarity Information"). Nevertheless, information that specifies the axon's pathway, and in most cases also its polarity, is present in this most reduced system of a single neuron and its epithelial environment. We are now trying to establish whether any extrinsic cues used by the neuron are generally distributed in the epithelium, or whether the pathway is preestablished in the properties of the cells of the vein epithelium even though the vein as a physical channel does not exist.

Acknowledgments

I am grateful to my colleagues Seth Blair, Marjorie Murray, and Margrit Schubiger for permission to refer to unpublished work and for comments on the manuscript. Dr. Murray contributed the material for Figures 7 and 9; Steve Hart and Eric Cole made the SEMs of Figure 2. Supported by grants NS-07778 from the NIH and 82-04088 from the NSF.

REFERENCES

Anderson, H. (1984) The development of sensory nerves within the wing of *Drosophila melanogaster*. Wilhelm Roux's Arch. **193:**226–233.

Ashburner, M., H. L. Carson, and J. N. Thompson, (1981–) *The Genetics and Biology of Drosophila*, Vol. 3a–e, Academic Press, New York.

Bentley, D. and M. Caudy (1983) Navigational substrates for peripheral pioneer growth cones: Limb-axis polarity cues, limb-segment boundaries, and guidepost neurons. *Cold Spring Harbor Symp. Quant. Biol.* **48:**573–585.

Bentley, D. and H. Keshishian (1982) Pioneer neurons and pathways in insect appendages. *Trends NeuroSci.* **5:**354–358.

Blair, S. S., M. A. Murray, and J. Palka (1985) Axon guidance in cultured fragments of the *Drosophila* wing. *Nature* **315:**406–409.

Blair, S. S. and J. Palka (1985) Axon outgrowth in cultured wing discs of *Drosophila. Dev. Biol.*, **108:**411–419.

Edwards. J. S. (1982) Pioneer fibers: The case for guidance in the embryonic nervous system of the cricket. In *Neuronal Development*, N. C. Spitzer, ed. Plenum, New York, pp. 255–266.

Ghysen, A., R. Janson, and P. Santamaria (1983) Segmental determination of sensory neurons in *Drosophila. Dev. Biol.* **99:**7–26.

Goodman, C. S., J. A. Raper, R. K. Ho, and S. Chang (1982) Pathfinding by neuronal growth cones in grasshopper embryos. In *Developmental Order: Its Origin and Regulation*, S. Subtelny and P. B. Green, eds. Alan R. Liss, New York, pp. 275–316.

Heide, G. (1983) Neural mechanisms of flight control in Diptera. In *Biona-Report 2, Akad. Wiss. Mainz*, W. Nachtigall, (ed.) G. Fischer, Stuttgart, New York, pp. 35–52.

Henke, K. and G. Rönsch (1951) Über Bildungsgleichheiten in der Entwicklung epidermaler Organe und die Entstehung des Nervensystems im Flügel der Insekten. *Naturwiss.* **38:**335–336.

Ho, R. K. and C. S. Goodman 1982. Peripheral pathways are pioneered by an array of central and peripheral neurons in grasshopper embryos. *Nature* **297:**404–406.

Jan, L. Y. and Y. N. Jan (1982) Antibodies to horseradish peroxidase as specific neuronal markers in *Drosophila* and in grasshopper embryos. *Proc. Natl. Acad. Sci. USA* **79:**2700–2704.

Keshishian, H. and D. Bentley (1983) Embryogenesis of peripheral nerve pathways in grasshopper legs. I. The initial pathway to the CNS. *Dev. Biol.* **96:**89–102.

Kirschner, M. (1982) Microtubules and their role in cell, tissue and organismal polarity. In *Developmental Order: Its Origin and Regulation*, S. Subtelny and P. B. Green, eds., Alan R. Liss, New York, pp. 117–132.

Krayanek, S. and S. Goldberg (1981) Oriented extracellular channels and axonal guidance in embryonic chick retina. *Dev. Biol.* **84:**41–50.

Murray, M. A., M. Schubiger, and J. Palka. Neuron differentiation and axon growth in the developing wing of *Drosophila melanogaster*. *Dev. Biol.* **104:**259–273.

Nardi, J. B. (1983) Neuronal pathfinding in developing wings of the moth *Manduca sexta*. *Dev. Biol.* **95:**163–174.

Palka, J. and M. Schubiger (1980) Formation of central patterns by receptor cell axons in *Drosophila*. In *Development and Neurobiology and Drosophila*, O. Siddiqi, P. Babu, L. M. Hall, and J. C. Hall, eds. Plenum, New York and London, pp. 223–246.

Palka, J., M. Schubiger, and R. L. Ellison (1983) The polarity of axon growth in the wings of *Drosophila melanogaster*. *Dev. Biol.* **98:**481–492.

Sanes, J. R. (1983) Roles of extracellular matrix in neural development. *Ann. Rev. Physiol.* **45:**581–600.

Schubiger, M. and J. Palka (1985) Genetic suppression of putative guidepost cells: Effect on establishment of nerve pathways in *Drosophila* wings. *Dev. Biol.*, **108**:399–410.

Silver, J. and R. Sidman (1980) A mechanism for the guidance and topographic patterning of retinal ganglion cell axons. *J. Comp. Neurol.* **189**:101–111.

Singer, M., R. H. Nordlander, and M. Egar, (1979) Axonal guidance during embryogenesis and regeneration in the spinal cord of the newt: The blueprint hypothesis of neuronal pathway patterning. *J. Comp. Neurol.* **185**:1–22.

Solomon, F. (1979) Detailed neurite morphologies of sister neuroblastoma cells are related. *Cell* **16**:165–169.

Sondhi, K. C. (1963) The biological foundations of animal patterns. *Quart. Rev. Biol.* **38**:289–327.

Throckmorton, L. H. (1975) The phylogeny, ecology and geography of *Drosophila*. In *Handbook of Genetics, Vol. 3, Invertebrates of Genetic Interest*, R. C. King, ed. Plenum, New York, pp. 421–469.

Waddington, C. H. (1940) The genetic control of wing development in *Drosophila*. *J. Genet.* **41**:75–137.

Waddington, C. H. (1962) *New Patterns in Genetics and Development*, Columbia University Press, New York.

Chapter Two

DEVELOPMENTAL PLASTICITY DURING THE METAMORPHOSIS OF AN INSECT NERVOUS SYSTEM

JAMES W. TRUMAN, JANIS C. WEEKS, and RICHARD B. LEVINE

Department of Zoology
University of Washington
Seattle, Washington

The development of the central nervous system (CNS) does not draw to a close with the birth or hatching of an individual. Indeed, the CNS continues to change both qualitatively and quantitatively as postembryonic life proceeds. These postembryonic changes are particularly pronounced in animals that undergo metamorphosis, such as the amphibia and the holometabolous insects. In these animals the nervous system must change to accommodate the profound alterations in body form and behavioral repertoire that accompany the diffent life stages.

The cellular changes that go on in the CNS during metamorphosis are not qualitatively different from those seen during embryogenesis. Embryonic development of the nervous system involves such processes as cell birth, cell death, and sequential changes in the form and function of particular cells (e.g., Goodman and Bate, 1981). Metamorphosis may be viewed as a second embryonic period, but the emphasis on each developmental process varies depending on the region of the nervous system that is considered. In certain areas of the insect brain, such as the optic and antennal lobes, cell generation is a dominant theme, as thousands of new neurons are added for the processing of information from new adult sensory structures (Edwards, 1969). By contrast, in the lower segmental ganglia, the addition of new cells has a lesser role, and most of the changes occur through changes in cell function and selective degeneration (Truman et al., 1985).

From a behavioral point of view, metamorphosis has the unique attraction that this second embryonic state is interposed between two actively behaving stages, the larva and the adult, which exhibit radically different body form and behavior. This raises the question of how the neuronal circuitry of the larva is related to that of the adult. Furthermore, in holometabolous insects, this reorganization of the nervous system takes place during the pupal stage, which may exhibit its own stage-specific behaviors. Thus the transition from larval to adult circuits must occur while still allowing the ongoing performance of normal pupal behaviors.

A dominant force controlling cellular fates during metamorphosis is the ensemble of developmental hormones that are present at this time. As we discuss, their actions on the nervous system appear to constitute the major thrust for development, although in some cases (e.g., development of the optic lobes (Edwards, 1969) and the antennal lobes (Schneiderman et al., 1982)), the interaction between the periphery and the CNS is essential for normal differentiation. The focus of this chapter is on the plastic changes that go on in the CNS during metamorphosis of the moth *Manduca sexta,* the chemical signals that direct these changes, and the little that we know about other factors that may be involved in influencing these processes.

METAMORPHOSIS AND ITS REGULATION IN INSECTS

In the moth, *Manduca sexta,* growth occurs during the larval stage (Fig. 1). During this time, the caterpillar goes through a series of four molts, each involving the retraction of the epidermis from the old cuticle, followed by the secretion of a new cuticle. At the end of each molt, the insect then sheds the remains of the old skin, an event termed *ecdysis*. The larval growth period is followed by metamorphosis, during which the insect transforms into the pupal stage and then into the adult. The pupal stage is relatively quiescent since the cuticular sheaths in which the adult appendages will form are plastered to the surface of the head and body and are not capable of movement. Thus the pupa shows only behaviors involving abdominal movements. The moth that emerges 18 days later has well-formed wings, walking legs, genitalia, and other structures that are unique to the adult.

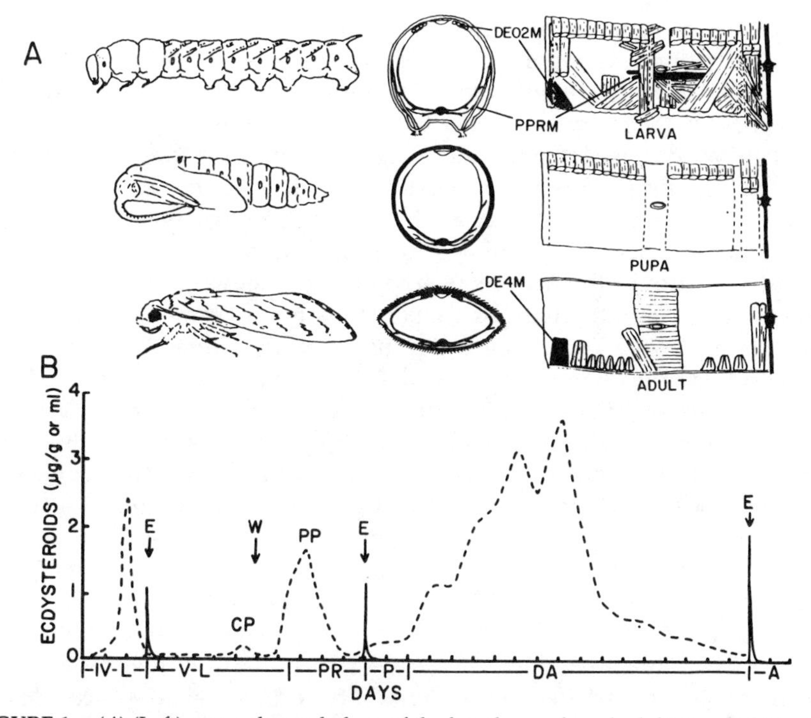

FIGURE 1. (*A*) (Left) external morphology of the larval, pupal, and adult stages of *Manduca sexta*. (Middle) cross section of the abdomen showing the positions of the target muscles for MN-1 and PPR. Muscle designations are as described in the text. The ventral ganglion and the projections of the doral and ventral segmental nerves are shown in black. (Right) an internal view of the musculature on the left side of the segment, including in black the muscles that are innervated by motoneurons MN-1 and PPR; the internal muscles have been cut away posteriorly. The ventral nerve cord and ganglion appear at the far right. (*B*) Representation of ecdysteroid (in 20-hydroxyecdysone equivalents) and eclosion hormone blood titers during the late stages of the life history of *Manduca sexta*. Dashed line, ecdysteroid titer; solid line, eclosion hormone titer. The commitment peak (CP) and prepupal peak (PP) of ecdysteroids are indicated. The eclosion hormone titer indicates relative changes with time. IV-L, fourth larval instar; V-L, fifth larval instar; PR, prepupa; P, pupa; DA, developing adult; A, adult; E, ecdysis; W, onset of wandering behavior. Ecdysteroid data modified from Bollenbacher et al., 1981.

The change in body form during metamorphosis is accompanied by a massive reorganization of the musculature (Finlayson, 1956; Fig. 1*A*). We have focused on the changes that occur in the mid-abdominal segments A3 to A6. In the larva these segments each bear about 50 muscle groups, including the *internal muscles* that span each segment and are primarily longitudinal in their orientation and the *external muscles* that show various orientations and are generally shorter than the internal groups. During the transition from the larva to the pupa, all of the external muscles degenerate, leaving the pupa with only some of the internal muscle groups. These larval internal muscles persist and participate in behavior throughout the pupal-adult transition while the new external muscles differentiate. After adult ecdysis, the internal muscles then degenerate, leaving the insect with just its new adult external muscles. The only muscles that persist through the life of the insect are the spiracular closer muscles and the weak muscle strands associated with the segmental stretch receptor organs. Thus in the transition from larva to the adult there is an almost complete changeover in the musculature of the abdomen. Interestingly, although the muscles are all replaced, their motoneurons are not (Taylor and Truman, 1974; Cassady and Camhi, 1976). The changes that occur in individual motoneurons to accommodate the changes in musculature are discussed later.

The cellular events that occur during metamorphosis are regulated by three hormones (reviewed in Riddiford, 1980). The ecdysteroids, most notably 20-hydroxyecdysone (20-HE), are steroid hormones that initiate the molting process and induce the epidermis to make a new cuticle. A second hormone, the juvenile hormone, determines the type of cuticle produced. In addition to the epidermis, these two hormones also govern changes in internal organs such as the CNS. During larval life, ecdysteroids always appear in the presence of juvenile hormone so that at each molt the insect makes an essentially identical, but larger larval stage. During the last larval instar, however, the juvenile hormone titer drops, and the next appearance of ecdysteroids occurs in the absence of juvenile hormone. This small pulse of ecdysteroids, the "commitment pulse" (Fig. 1*B*), commits the insect to pupal differentiation (Riddiford, 1976) and causes the larva to stop feeding and seek out a suitable pupation site (Dominick and Truman, 1980). This is then followed two days later by a large ecdysteroid surge (the "prepupal peak"), which causes the pupal stage to be formed. After pupal ecdysis, ecdysteroids reappear to bring about adult differentiation, and the adult emerges about 18 days later. A third hormone involved in the molting process is a peptide hormone, eclosion hormone, which is released at the end of each molt to trigger the ecdysis behavior (Fig. 1*B*; Truman, 1980). In the case of metamorphic molts, such as from the larva to the pupa or from the pupa to the adult, the time of ecdysis signals an abrupt switch in the behavior of the insect. Eclosion hormone triggers these behavioral changes as well (see later discussion).

Although neuronal modifications in the CNS are correlated with the pronounced changes in overt morphology and musculature, the two are not necessarily causally related. Obviously there are interactions between the CNS and the periphery, but as we discuss later, most of the changes in the abdominal CNS appear to result from the direct action of the developmental hormones on this tissue. The coordinated development of the CNS and the periphery then results from the parallel responses of the two areas to the same hormonal cues.

PLASTICITY IN NEURONAL MORPHOLOGY DURING METAMORPHOSIS

Dismantling of Larval Circuits

Logically, the transition from the larva to the adult is a two-step process with first the dismantling of the neural circuits for larval behavior and then the construction of the new circuits to be used in the adult stage. In the motor system the end of larval life involves the death of all but a few of the abdominal muscles, necessitating the reorganization of the larval motoneurons in anticipation of the subsequent adult musculature and behaviors. We have examined the fates of a number of identified motoneurons throughout this time period. The few motoneurons that innervate muscles that do not die at pupation remain essentially unchanged morphologically (Levine and Truman, in preparation; Weeks and Truman, unpublished). In contrast, motoneurons that experience the death of their larval target muscles show two different fates: some simply die along with their target muscle, whereas others persist and become respecified both functionally and morphologically to innervate new muscles that are generated during adult development. The next sections describe an example of each of these two latter classes.

During the larval-pupal transformation, the insect loses its abdominal prolegs which were the principal locomotory appendages of the caterpillar. This loss involves the death of all of the proleg muscles, including the principal planta retractor muscle (PPRM), an external muscle that causes retraction of the tip of the abdominal proleg. PPRM begins degeneration by three days after the initiation of metamorphosis, and the muscle is gone by pupal ecdysis (Fig. 2; Weeks and Truman, 1983 and in preparation). Over the same period of time, the motoneuron that innervates this muscle, PPR, undergoes a marked reduction in its dendritic arbor, representing a loss of about 40% of the neuropilar area that the cell covered in the larva. Two days after pupal ecdysis, PPR dies (Fig. 2).

The temporal correlation between the death of PPRM and loss of dendrites and subsequent death of PPR suggested that a causal relationship might exist between the two. Alternatively, the two events might coincide because they are in response to a common endocrine signal, for example, the prepupal peak of ecdysteroids. In fact, if this peak is prevented by placing a blood-tight ligature between the abdomen and the thorax (the site of ecdysteroid secretion), then PPRM does not die, and PPR's dendritic reduction and death do not occur (Weeks and Truman, in preparation). Importantly, infusion of steroid into such ligated abdomens, so as to replace the prepupal peak, results in the degeneration of PPRM and the appropriate dendritic loss in PPR. Although these experiments implicate ecdysteroids in regulating these events, they cannot distinguish whether

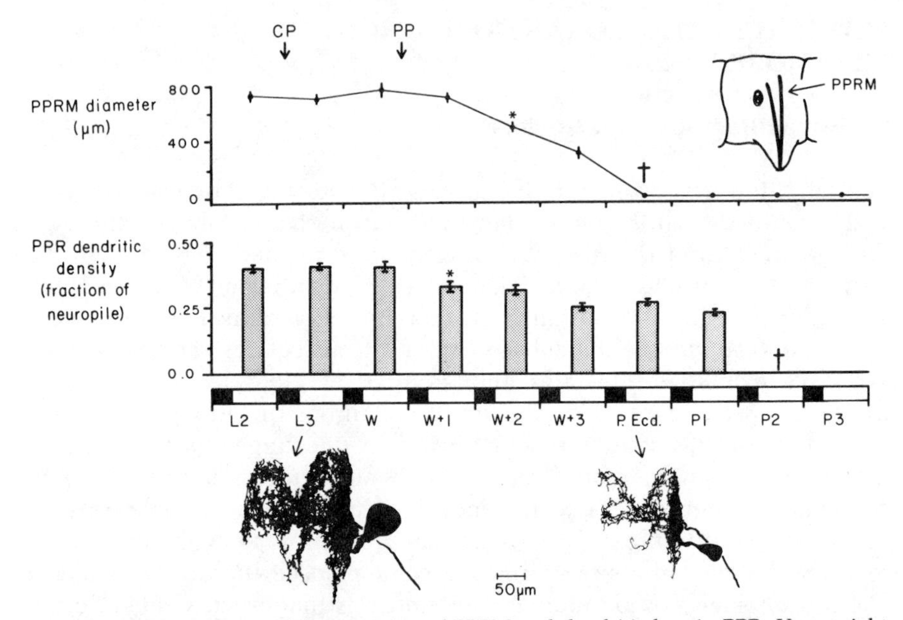

FIGURE 2. Time course of degeneration of PPRM and dendritic loss in PPR. Upper right inset is an external view of an abdominal segment showing the location of the two fibers that make up the PPRM. Upper graph: degeneration of PPRM as followed by measuring the sum of the muscle fiber diameters in fresh, unfixed tissue. Each point is the mean ± SEM of at least 20 muscles. Asterisk indicates the first day, showing a significant loss in fiber diameter ($p < .0005$; one tailed T-test). Lower graph: progression of dendritic loss in PPR as determined in $CoCl_2$ fills of the cell. Quantification was restricted to the dendritic arbor that extended into the hemineuropile that was contralateral to the cell body. Density was computed as the fraction of the hemineuropile when viewed ventrally that was covered by the processes of PPR. Each point is the mean ± SEM of five neurons. Asterisk indicates the first day of significant dendritic loss ($p < .0025$; one tailed T-test). At bottom are drawings of PPR stained by intracellular iontophoresis of $CoCl_2$.

the two effects are induced by a common signal, or whether the death of the muscle causes degeneration of the motoneuron. These possibilities were distinguished by examining the effect on PPR of the chronic removal of its target muscle. This was accomplished by surgically removing PPRM from one proleg and then ligating the abdomen to prevent the prepupal ecdysteroid peak. The structure of the PPR that had innervated the ablated muscle was examined seven days later. These neurons remained alive and retained the normal larval amount of branching, indicating that loss of the target muscle alone is not sufficient to bring about a reduction in dendritic density (Weeks and Truman, 1983, and in preparation). These results suggest that PPR loses dendrites not in response to the death of its target muscle, but rather in response to the prepupal ecdysteroid peak, which at the same time triggers the programmed death of PPRM. We have not yet determined whether ecdysteroids act directly on PPR, or indirectly by means of some other central action.

It should be noted that even though at certain times motoneurons may lack a target muscle, they may nevertheless continue to participate in patterned behavior. For example, one characteristic input onto PPR is seen during larval ecdysis, when the prolegs are retracted rhythmically as the abdomen shows peristaltic ecdysis waves. The same motor pattern is also used at pupal ecdysis, and PPR continues to fire in its proper phase of the cycle, even though it has lost substantial dendritic area (Fig. 2), and the proleg and its muscles are no longer present (Truman and Weeks, 1983; Weeks and Truman, in preparation). This suggests that motoneurons might be involved in central circuits independently of their output function onto muscles, a possibility that has been observed in other arthropods (Heitler, 1978) and which is currently under investigation in *Manduca*.

Construction of Adult Neural Circuits

An understanding of how morphological changes in neurons relate to their incorporation into new adult circuits has been provided by studies of the motoneuron MN-1 (Levine and Truman, 1982). MN-1 innervates the dorsal oblique 2 muscle (DEO2M) in the larva (Fig. 1*A*). This muscle dies shortly after pupal ecdysis, but MN-1, unlike PPR, survives the death of its target and eventually innervates a new muscle (dorsal external 4 muscle, DE4M; Fig. 1*A*) that differentiates during adult development. In the larva, the motoneuron's major dendritic field is ipsilateral to its target muscle (Fig. 3). During the transition to the pupa, the larval dendritic tree of this cell is reduced, but small processes near the soma begin to sprout (Fig. 3). With the initiation of adult development, these new dendritic processes extend caudally so as to eventually cover most of the previously unoccupied half of the neuropil. During this same time, the old larval portion of the cell also reelaborates over its portion of the neuropil. Thus, in the transition from larva to adult, the dendritic field of MN-1 comes to encompass both hemineuropiles, rather than one.

The significance of this morphological change in MN-1 during the transition from larva to adult was revealed by studying the inputs to the cell from a pair of bilaterally arranged, segmental stretch receptors (SRs) (Levine and Truman, 1982). In the larval stage the stretch receptor neuron ipsilateral to MN-1's target muscle excites the cell through a monosynaptic contact on the larval portion of the dendritic tree. The contralateral SR inhibits the neuron through a polysynaptic pathway involving at least one interneuron (Fig. 3). Thus the right and left MN-1s receive antagonistic inputs as the caterpillar bends to the right or to the left. The same relationship of the right and left SRs to MN-1 is seen in the pupa. By contrast, in the adult, both the ipsilateral and contralateral SRs, which persist through metamorphosis, directly excite MN-1. The receptor ipsilateral to the new target of MN-1 retains its contact onto the larval portion of the cell, whereas the contralateral SR makes monosynaptic contact with the new, adult-specific

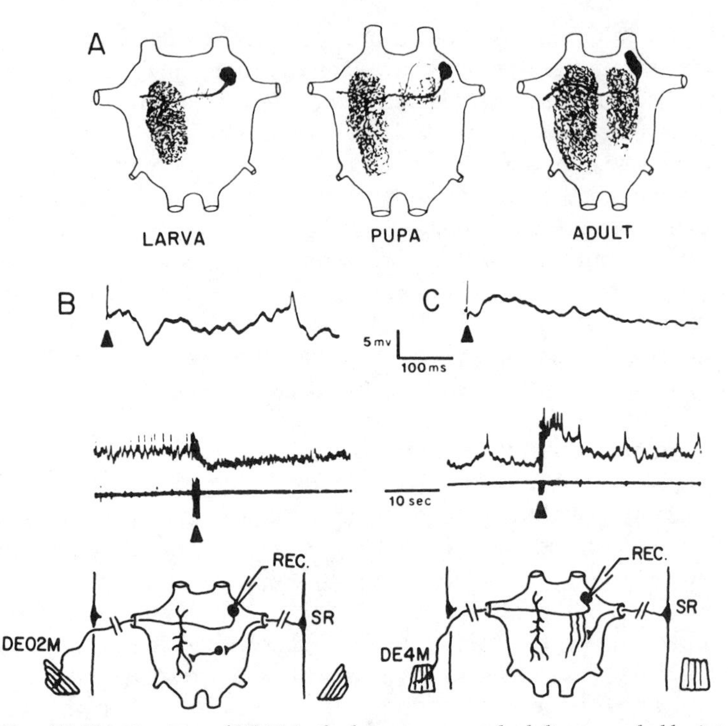

FIGURE 3. (*A*) The structure of MN-1 in the larva, pupa, and adult as revealed by intracellular dye injection. (*B*) the interaction between a larval MN-1 and the dorsal stretch receptor (SR) contralateral to the motoneuron's target muscle. The SR inhibits the larval MN-1 through a polysynaptic pathway, with single stimuli (arrow head) evoking a long-latency IPSP (top), and bursts of stimuli (arrow head and artifacts on extracellular trace) resulting in the inhibition of MN-1 activity (middle). (Bottom) shows the proposed relationship between the contralateral SR and MN-1. (*C*) the interaction between the adult MN-1 and the same stretch receptor. Single stimuli elicit a short latency EPSP (top), and multiple stimuli cause excitation (middle). The proposed new relationship between the adult MN-1 and the SR is shown at the bottom. REC., recording site for intracellular records. Muscle designations are as in the text. (Modified from Levine and Truman, 1982.)

dendritic field. Thus the right and left MN-1s of a given segment receive the same input from each stretch receptor, rather than receiving antagonistic inputs as in the larva. Not only do the two homologues in the adult share common input from the segmental SRs, but also they seem to share other inputs. Simultaneous recordings from the two MN-1s in the adult stage show that the neurons typically fire in synchrony, a condition that was not seen in the pupa or larva. Since the cells are not electrically coupled, this synchrony of activity suggests that both neurons receive numerous common synaptic inputs.

The significance of the change in the inputs shared by the right and left MN-1s as the insect transforms from larva to adult can be seen by considering the behavior of the respective stages. The abdomen of the larva is cylindrical in cross section and extremely flexible (Fig. 1*A*). During its movements the caterpillar frequently bends to the right or left, and the mid-dorsal muscles innervated by the two MN-1s function as antagonists. This lateral flexibility is preserved in the pupal stage and through adult development. In the adult, however, the mid-abdominal segments are dorsoventrally flattened and no longer can be flexed to the right or to the left (Fig. 1*A*); movements are primarily confined to the dorsoventral plane. During these motions the two muscles innervated by the adult MN-1s act as synergists. Consequently, these neurons that were primarily antagonists in the larva become synergists in the adult, a situation accomplished by the selective growth of both motoneurons, so as to give both members of the pair access to similar synaptic inputs. These morphological changes shown by the MN-1 pair are shared by many other abdominal motoneurons that are unilateral in the larva and then become bilateral as they acquire their new adult muscles. That this may be related to the mechanical constraints imposed on the abdomen during metamorphosis is suggesed by the observation that the thoracic homologues of MN-1 do not change from unilateral to bilateral during this time (Truman, unpublished).

It is of interest that, in the examples thus far examined, the functional alteration of neural circuits occurring in the abdomen during metamorphosis involves major restructuring of the motoneurons. Comparison of the central arbor of the SR neurons before and after metamorphosis showed essentially no change in the cell's central projections, whereas MN-1 undergoes extensive change. Whether other neurons that are presynaptic to MN-1 are also as conservative morphologically as the SR neurons during metamorphosis needs to be established. Specifically, the metamorphic fates of abdominal interneurons remain largely unexplored.

The basic stimulus that causes a cell to produce its adult-specific arbor is most likely hormonal in nature. Through larval life, the short processes in MN-1 that will give rise to the future adult dendritic tree are apparently dormant. They begin to grow when the pupal transition is triggered by the appearance of ecdysteroids. The majority of the growth of the new dendritic field in MN-1 occurs during adult development, which is promoted by the high ecdysteroid titer throughout this time (Fig. 1*B*). If ecdysteroids are withheld, as occurs naturally when the insect enters a facultative pupal diapause, then the neuron remains in an immature condition for months until ecdysteroids reappear to initiate development. This action of ecdysteroids requires the absence of juvenile hormone. Indeed, if the latter hormone is applied a day after pupal ecdysis, the adult differentiation of the neuron is prevented, even in the presence of ecdysteroids (Truman, unpublished). It should be noted that the effects of the juvenile hormone treatment are not confined to the CNS, and such insects become second pupae or pupal-adult intermediates. Therefore in neither case can we yet say that the response of the cell is a direct response to the respective hormones, since these experiments have all been done with systemic applications of hormone *in vivo*.

Although ecdysteroids directly or indirectly stimulate dendrite growth,

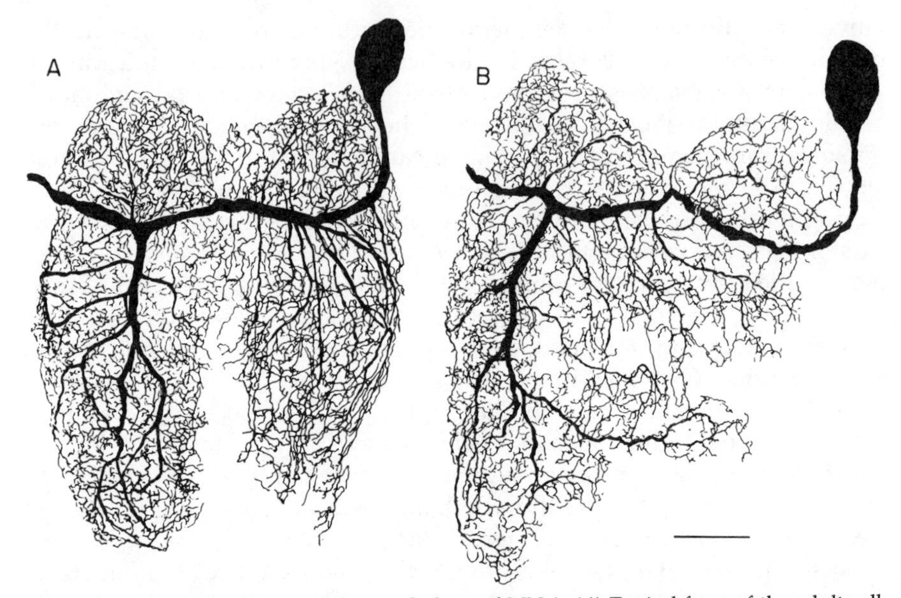

FIGURE 4. Examples of the adult morphology of MN-1. (*A*) Typical form of the adult cell; (*B*) example in which posterior sprouts failed to grow from the neurite ipsilateral to the cell body, but compensation occurred by the sprouting of branches from the larval portion of the dendritic tree. Camera lucida drawings were made of motoneurons stained with $CoCl_2$. Bar equals 50 μm.

the extent of growth may be regulated by local interactions in the ganglion. This conclusion is based on a few atypical cells, illustrated in Figure 4. In this example the neurite ipsilateral to the cell body failed to extend processes caudally into the neuropil. Apparently, to compensate for this lack, some of the neurites from the larval portion of the dendritic tree sprouted processes that went across the midline to invade some of the unoccupied territory. The reason that the ipsilateral region of the cell failed to produce processes is unknown, but the growth response of the remainder of the cell is interesting in that it suggests that the previously unoccupied area of neuropile became "attractive" to the motoneuron and thereby stimulated the neurites to grow into an area in which they would normally never go.

FUNCTIONAL PLASTICITY IN THE CNS

In addition to the long-term changes that involve rather substantial changes in the morphology of central components, the metamorphosing nervous system also shows relatively rapid functional changes that are not associated with marked anatomical changes in the cells involved. These functional changes are controlled by the peptide, eclosion hormone. The ecdysial behaviors that occur at the end of each molt are triggered by this

8000-dalton peptide. It acts directly on the CNS to trigger the ecdysis motor patterns (Truman, 1978; Truman and Weeks, 1983), and during metamorphosis it also turns on behaviors that are characteristic of the new stage that the insect has just entered. The actions of eclosion hormone are relatively rapid as compared to those of the steroids previously discussed. In *Manduca* the behavioral responses to the peptide occur 40 min to 2 hr after hormone appears in the blood.

The activation of stage-specific behaviors at the time of adult ecdysis has been studied in a number of giant silkmoths and in *Manduca* (Blest, 1960; Kammer and Kinnamon, 1979; Truman, 1976). In these insects circuits that are constructed during adult development appear to become repressed just before adult ecdysis (in the so-called pharate adult stage). Indeed, if the surrounding pupal cuticle is peeled from around the animal a few hours before its normal time of emergence, the newly freed moth shows a very limited behavioral repertoire. Importantly, in response to the release of eclosion hormone at its normal time of ecdysis, the peeled animal performs a pantomime ecdysis behavior and then "turns-on" the appropriate adult behavior patterns (Truman, 1976). Injection of eclosion hormone at slightly earlier times brings about premature ecdysis and activation of these adult behaviors.

A partial cellular analysis of the action of eclosion hormone has been carried out for two simple reflexes. The most conclusive data showing the involvement of eclosion hormone are seen for the pupal gin-trap reflex. This reflex is specific to the pupal stage and involves pairs of cuticular pits arranged on the lateral margins of segments 5 through 7 of the pupal abdomen. Stimulation of sensory hairs in the pit results in the reflexive contraction of ipsilateral longitudinal muscles of the next anterior segment, thereby pulling the pit under a shelf of cuticle and pinching the object in the trap. This reflex is thought to serve as a defense against small insects that might attack the relatively helpless pupa.

The reflex is mediated through a group of about 20 trap sensory neurons, an integanglionic interneuron, and a defined set of motoneurons (Bate, 1973; Levine and Truman, 1983; Fig. 5). The reflex circuitry appears to be built during the larval-pupal transition from preexisting larval neurons, but the reflex does not become functional until the time of pupal ecdysis. Its activation comes about through the action of eclosion hormone; for example, treatment of intact prepupae or semi-intact deafferented preparations with the peptide leads to the rapid activation of the reflex pathway, as well as to the activation of the pupal ecdysis motor pattern (Levine and Truman, 1983).

The site of action of the peptide in turning on the gin-trap reflex was inferred as follows. Treatment with eclosion hormone did not change the physiological properties of the sensory neurons and motoneurons involved in the reflex; for example, in the prepupa, even prior to hormone exposure, the gin-trap afferents responded to deflections of the gin trap

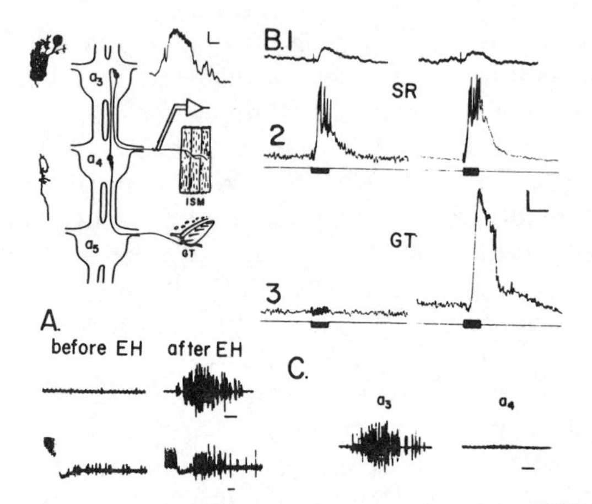

FIGURE 5. Activation of the gin-trap reflex by eclosion hormone (EH). Insert (upper left) shows a representation of the proposed circuitry underlying the reflex and the location of extracellular electrodes recording motoneuron activity. ISM, intersegmental muscles; GT, gin-trap. On the left are drawings (top) of the central arborization of an intersegmental muscle motoneuron and (bottom) the terminal arborization of a gin-trap sensory neuron. On the upper right is an intracellular record from the cell body of one such motoneuron showing the response to tactile stimulation of the gin-trap (Cal: 2.5 mV, 50 msec). (*A*) Extracellular records of the response of ISM motoneurons to tactile stimularion of the gin-trap receptors (top) and electrical stimulation of the gin-trap nerve (bottom) before and 1 hr after exposure to EH (Cal: 50 msec). (*B*) The same type of preparation as in (*A*) except that recordings are intracellularly from the cell body of the motoneuron. Records on the left are before EH addition, those on the right are after EH exposure and subsequent ecdysis. Before EH addition, a single electrical stimulus to the ipsilateral stretch receptor evokes an excitatory potential in an ISM motoneuron (*B*1), and a train of stimuli (stimulus marker) evokes a burst of action potentials (*B*2). A train of stimuli to the gin-trap nerve evokes no response (*B*3). After ecdysis, the motoneuron response to stretch receptor stimulation is unchanged (*B*1, *B*2), while stimuli to the gin-trap nerve evoke a large depolarization and a high-frequency burst of action potentials (*B*3) (Cal: top, 5 mV, 100 msec; middle and bottom, 5 mV, 400 msec.) (C) Selective exposure of individual ganglia to EH. Recordings from the lateral nerve of a_4 as in (*A*). Left, exposure of a_3 to EH activates the entire reflex pathway, as shown by the motoneuron response to tactile stimulation of the receptors 1 hr after EH exposure. Right, exposure of a_4 in another preparation to EH does not activate the pathway (Cal: 50 msec). (From Levine and Truman, 1983.)

hairs by generating action potentials that propagated into the CNS. Likewise, in this stage, the motoneurons drove the muscles and received numerous synaptic inputs of normal strength from non-gin-trap pathways (Fig. 5*B*). These observations suggested that the blockade of the gin-trap reflex prior to exposure to eclosion hormone was a selective block of the pathway and not due to a general inhibition within the CNS. The normal functioning of the sensory and motor elements in the pathway prior to peptide exposure suggested that the interneuron was involved in the activation of the pathway.

Intracellular studies of the interneuron have not yet been carried out, but insight into a possible site of action of the peptide was obtained by exposing single ganglia of the intact nerve cord to the hormone. This approach takes advantage of the fact that the interneuron is interganglionic and receives contacts from the sensory neurons in one ganglion while synapsing upon motoneurons in another (Fig. 5). Addition of peptide to the ganglion containing the sensory neuron-interneuron synapses had no effect on the pathway, whereas treatment of the ganglion containing the interneuron-motoneuron synapses activated the entire pathway. Also, in these single ganglion exposure experiments, the reflex pathway was activated without the simultaneous turning on of the ecdysis motor pattern. Thus the peptide appears to act directly on the elements of the gin-trap circuit to turn it on, and its primary site of action appears to be associated with the synapse between the gin-trap interneuron and the longitudinal muscle motoneurons.

Another simple reflex that is altered rapidly by eclosion hormone at adult ecdysis is the reflex previously described involving the stretch receptors and MN-1. In the "pharate" adult, just a few hours prior to ecdysis, the insect is still covered with pupal cuticle and shows the lateral flexibility characteristic of the larval and pupal stages. Interestingly, when trains of stimuli are applied to the contralateral stretch receptor in the pharate adult, MN-1 is inhibited, just as in the earlier stages (Levine and Truman, 1982). Thus, just prior to adult ecdysis. MN-1 shows larval-like responses, even though the cell has had its adult morphology for a number of days.

The seeming conflict between the adult morphology of the cell and the larval-like manner in which it interacts with the stretch receptor was resolved by examination of the response of MN-1 to single impulses from the SR contralateral to its target muscle (Levine and Truman, 1982). The synaptic response was biphasic, with an initial rapid excitatory phase followed by inhibition (Fig. 6). With repetitive stimuli, the inhibition predominated and the neuron was inhibited. Our interpretation of the biphasic response of MN-1 to the contralateral SR stimulation is given in Figure 6. The polysynaptic pathway that mediates the inhibition of MN-1 is seen in both the larva and the pupa. We have assumed that this inhibitory pathway persists through adult differentiation while the monosynaptic contacts between the contralateral SR and the adult-specific regions of MN-1 are forming. Thus, in the pharate adult, stimulation of the contralateral SR activates both the old inhibitory pathway and the new, excitatory pathway. The excitation precedes the inhibition because the former is monosynaptic, whereas the latter is polysynaptic. The inhibitory component then abruptly disappears within 30 minutes after the adult ecdysis and remains absent for the remainder of adult life.

From its demonstrated action of switching on other new behaviors at ecdysis, it is likely that eclosion hormone is responsible for turning off the inhibitory pathway in this circuit. A direct test of this hypothesis should be possible by determining if the peptide can alter the circuit in the dish as one sees it *in vivo*.

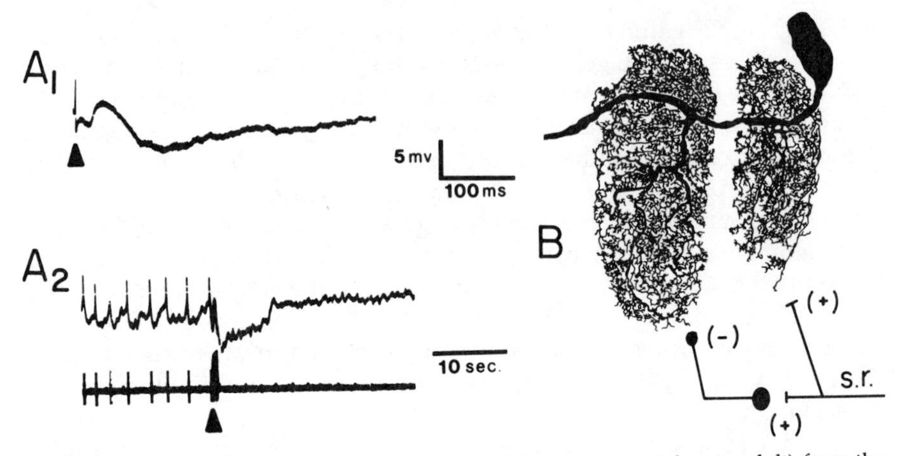

FIGURE 6. Effects on MN-1 on the day before adult emergence (pharate adult) from the stimulation of the stretch receptor (SR) contralateral to the neuron's target muscle. (A_1) A single stimulus to the dorsal SR (arrow) evokes a biphasic response in the motoneuron; a short-latency EPSP followed by an IPSP. (A_2) A burst of stimuli to the SR (arrow) causes inhibition of MN-1; top, intracellular record from MN-1; bottom, extracellular record showing MN-1 spikes in the dorsal nerve and stimulus artifacts at the arrow. (*B*) a model to account for the biphasic response of MN-1 to the stimulation of the SR. See text for further details. (Modified from Levine and Truman, 1982.)

PLASTICITY IN CELL SURVIVAL

A third type of plasticity in the nervous system has to do with cell survival. The attainment of final cell numbers seems to be regulated in most cases by the process of cell death rather than by a strict control of cell proliferation. Commonly, cell death occurs during development to match cell numbers between different parts of the CNS or between the CNS and the periphery (Hamburger and Oppenheim, 1982). In other instances cells may be produced that have a transient function, but after that function is completed the cells are no longer required, and they die (e.g., Bate et al., 1981). A third context for cell death occurs when cells are produced as a by-product of cell lineages to obtain sibling cells of a particular phenotype (see, e.g., Horvitz et al., 1982).

In the metamorphosing insect nervous system, cell death occurs in two contexts. The developing optic lobes are a site of active cell proliferation as interneurons are generated to accommodate the forming compound eyes. Optic lobe neurons are overproduced, and, as connections are made between the developing retinular cell axons and optic lobe neurons, the cells that do not receive appropriate connections apparently degenerate (Nordlander and Edwards, 1968). By contrast, in the abdominal ganglia cell death also occurs during metamorphosis, but many of the cells that die are mature motoneurons that have functional synaptic connections with other cells and target muscles up until the time of their death.

The death of mature neurons occurs at two principal times during the metamorphosis of the CNS of *Manduca*: immediately following pupal ecdysis, when cells such as PPR die (Fig. 2; Weeks and Truman, in preparation), and then after adult ecdysis (Truman, 1983). A both times degenerating cells include both interneurons and motoneurons. Following adult ecdysis, neurons die according to a stereotyped program, with individual cells dying at their own characteristic times. The end result is that at three days after ecdysis the abdominal ganglia of the adult have approximately half the number of nerve cells that were present at ecdysis (Taylor and Truman, 1974; Truman, 1983).

The first cells to die at adulthood are the interneurons, which begin to degenerate at the time of adult ecdysis. By about 12 hours after ecdysis the number of cells undergoing degeneration reaches its peak, and by about 30 hours, the process is essentially complete. During this 30-hr span, ganglion A4 loses almost 50% of its interneurons. The functions of the cells that die have not yet been established, but presumably some are involved in the patterning of the behaviors characteristic of the pupal stage and of the specialized behaviors shown at adult ecdysis. Also, as mentioned before, many of the adult motor patterns that arise during adult development are thought to be repressed or altered (e.g., the SR reflex described) by inhibitory interneurons. Presumably, these interneurons also die after the emergence of the adult. The loss of interneurons is accompanied by the death of about half of the motoneurons, all of which innervate muscles that degenerate after adult emergence. The stereotyped cell body location of many of these motoneurons has allowed them to be followed as individuals through the degeneration process (Fig. 7). Each identified cell has its own characteristic time of death, with certain cells dying as early as eight hours after ecdysis and others not starting to degenerate until about 36 hours.

Most of the motoneurons that die after adult ecdysis innervated the internal muscles that persisted through metamorphosis from the larval stage. Interestingly, the dendritic arbors of these conserved larval motoneurons do not change significantly as the insect transforms from the larval to the adult stage (Levine and Truman, in preparation). Furthermore, the proprioceptive inputs onto these cells appear to remain stable throughout metamorphosis. This is in marked contrast to the external muscle motoneurons, which gain new external muscle targets (e.g., MN-1) and which show changes in both anatomy and patterns of connectivity.

The trigger for the neuronal degeneration appears to be hormonal. This was first indicated by the fact that ganglia implanted into the body cavity of insects early in metamorphosis showed interneuron and motoneuron death in synchrony with that of the host (Truman and Schwartz, 1984). The nature of the humoral signal triggering degeneration was further analyzed by experiments in which the abdomens of pharate adults were isolated from the head and thorax at various times prior to adult ecdysis. When isolated within a day of ecdysis, the treatment had little effect on the timing

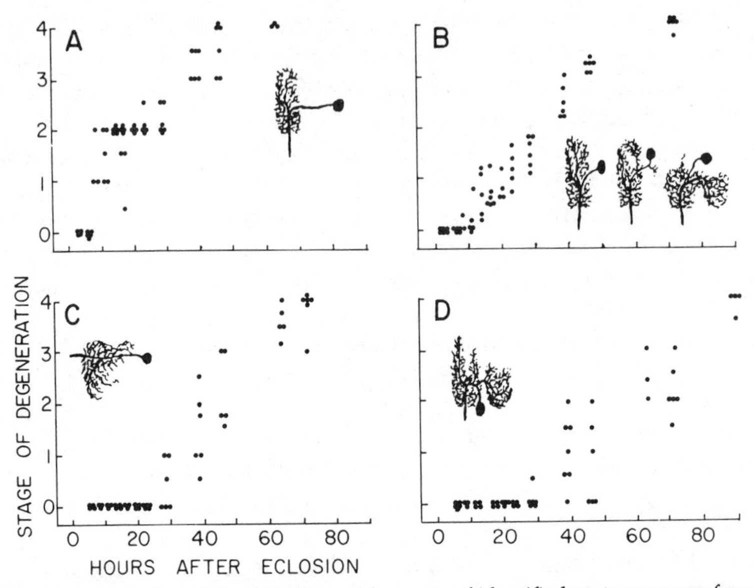

FIGURE 7. Time course of degeneration of four sets of identified motoneurones from adult ecdysis in *Manduca*. (*A*) MN-11; (*B*) MNs 7, 9, and 10; (*C*) MN-2; (*D*) MN-12. The stages of degeneration are based on the appearance of the cell body of the respective neuron in histological sections of the abdominal ganglia. The stages range from a healthy cell (0) to a highly degenerate cell body composed of shrunken collection of membranes (4). Each dot represents the condition of the respective cells in a single ganglion. (Based on Truman, 1983.)

of cell death. However, abdomens isolated earlier in development showed a progressive advancement in the time of cell degeneration. This result suggested that a factor from the head or thorax normally prevented the onset of degeneration and that abdomen isolation resulted in its early disappearance and hence the early death of the cells. Because the ecdysteroids are produced by glands in the thorax, they seemed to be likely candidates; furthermore, the ecdysteroid titer is normally declining at the time of adult ecdysis, and abdomen isolation would serve to accelerate this decline.

The involvement of ecdysteroids was tested directly by injecting steroid into both isolated abdomens and intact animals (Truman and Schwartz, 1984). When various dosages of steroid were injected into *Manduca* late in development, neuronal death was delayed to an extent proportional to the amount of steroid injected. Continuous maintenance of high steroid levels by the infusion of 20-hydroxyecdysone prevented the degeneration for the duration of the treatment. Based on this evidence, we concluded that the decline in the ecdysteroid titer at the end of adult development is the normal signal that triggers the death of the interneurons and motoneurons.

A decline in the ecdysteroid titer is also involved in the cell death that occurs at pupation. In the case of PPR, although it is the rising phase of the prepupal ecdysteroid peak that initiates dendritic loss, the cell does not become irrevocably committed to die until it experiences the subsequent fall in ecdysteroids that precedes pupal ecdysis. As with the cell death in the adults, constant administration of ecdysteroids to prepupae beginning prior to pupal ecdysis prolongs the life of PPR (Weeks and Truman, in preparation).

A number of aspects of this steroid-regulated cell death are of interest. Treatment with ecdysteroids late in adult development blocks the death of both the neurons and the internal muscles (Truman and Schwartz, 1984). When a given dosage of steroid is applied at various times during the last day of development, one can identify a critical time before which the steroid treatment blocks the death but after which the treatment is without effect. Presumably, this critical time indicates the time at which the cells have commenced processes that will result in their degeneration. Importantly, the critical period for the muscles occurs about 16 hours before that for the motoneurons that innervate them. Thus, by appropriately timed steroid treatment, one can produce moths that show normal emergence and muscle death, but whose neurons continue to live. Thus the death of the neurons in the adult is not the simple result of the previous degeneration of their target muscles, as was also indicated by the case of the proleg muscle and motoneuron described previously.

Not only do the muscles and motoneurons have different critical periods after which they can no longer be saved by steroid treatment, but also among the motoneurons themselves the cells vary with respect to their critical times (Truman and Schwartz, 1984). The sequence of critical times within a group of cells parallels the normal sequence in which the cells die. By application of steroid at various times around the time of ecdysis, one can interrupt the program of motoneuron and interneuron degeneration at many stages. This observation is contrary to a cascade hypothesis in which the death of a small number of cells then results in the death of other neurons, which then cause others to die, and so on. Also, from preliminary experiments at both pupal and adult ecdysis in which intracellular recordings were made from motoneurons at various times during degeneration, it is clear that the cells continued to receive many synaptic inputs well into the lytic process (Truman and Levine, 1980; Weeks and Truman, unpublished). These observations are consistent with a hypothesis that the neurons degenerate in direct response to an ecdysteroid signal, rather than in response to a withdrawal of intercellular communication.

CONCLUSIONS

Metamorphosis involves a variety of plastic changes in the CNS, and these changes are orchestrated through the hormonal signals that play on the

CNS. On the cellular level, the most dramatic changes that occur relate to the reorganization of existing larval neural circuits to take on new functions in the adult stage. This has been most extensively studied in the abdominal motor system, although similar changes have also been demonstrated in thoracic motoneurons (Cassady and Camhi, 1976). Neurons that retain the same target muscle into adulthood tend to remain relatively stable in their morphology, whereas those that change muscles exhibit a much more plastic morphology. As typified by motoneuron PPR, these cells often show a loss of dendrites at the time that their target muscle dies. Motoneurons such as MN-1 that then obtain a new muscle show a reelaboration and expansion of their dendritic trees. These new dendritic fields are related to the new synaptic connections made by the cells in the adult, as compared to the larval, stage.

This restructuring is under the control of the hormone 20-hydroxyecdysone, but the exact mode of action of the steroid is not known. One possibility is that the particular cells are genetically preprogrammed to grow in a particular configuration and that the steroid triggers this growth program. Another view is that the pattern of growth of the cell is entirely controlled by factors extrinsic to the cell and that the ecdysteroids are responsible for the production of these factors by other cells and/or the ability of the neuron to respond to these factors. Most likely, the truth will lie somewhere in between these two extremes.

In addition to the growth of neural processes and the establishment of new connections which occur over relatively long periods of time, there are also rapid functional changes associated with the activation of these pathways at the times of ecdysis. This activation is necessitated by the fact that metamorphosis in insects occurs in the context of the periodic molting of the exoskeleton. Thus while the insect is undergoing the transition to the adult, it is encased in the cuticle of the pupal stage, and its behavior remains appropriate for a pupa. When this external covering is cast off at ecdysis, the insect must rapidly assume the behavior appropriate for its new morphology. These functional changes are triggered by the action of the peptide eclosion hormone on the CNS. The results from the study of the gin-trap circuit indicate that the peptide is targeted to specific sites in the neural circuit to activate that pathway.

A third type of cellular plasticity during metamorphosis involves cell death in the CNS. In the transition from the larva to the mature adult, the abdominal nervous system loses more than 50% of its neurons. Some of the loss occurs at the larval-pupal transition, but most of the cells die after the emergence of the adult. This situation is in marked contrast to the timing of death in the thoracic ganglia, in which essentially all of the degeneration that is going to occur takes place at the end of larval life, and only a few cells subsequently die in the adult (Truman, unpublished). The postponement of cell death in the abdomen until adulthood probably relates to the fact that the abdomen is the only region of the body that continues muscular activity throughout metamorphosis. Thus the larval portion of the abdominal motor system is retained during pupation to mediate pupal behaviors, while the remainder of the CNS is free to construct its adult circuits. After metamorphosis is complete, these pupal behavior circuits, and probably also those that are used for adult ecdysis behavior, can be disposed of. This phase of cell death is also under hormonal control and is regulated by the withdrawal of ecdysteroids at the end of adult development. Thus endocrine-mediated changes in neuronal morphology, synaptic connectivity, and cell number all contribute to the reorganization of the CNS during metamorphosis.

Acknowledgments

Unpublished work reported in this paper was supported by grants from NSF, NIH, and the McKnight Foundation.

REFERENCES

Bate, C. M. (1973) The mechanism of the pupal gin trap. III. Interneurones and the origin of the closure mechanism. *J. Exp. Biol.* **59**:121–135.

Bate, M., C. S. Goodman, and N. C. Spitzer (1981) Embryonic development of identified neurons: segment-specific differences in the H cell homologues. *J. Neurosci.* **1**:103–106.

Blest, A. D. (1960) The evolution, ontogeny, and quantitative control of settling movements of some new world saturniid moths, with some comments on distance communication by honey bees. *Behaviour* **16**:188–253.

Bollenbacher, W. E., S. L. Smith, W. Goodman, and L. I. Gilbert (1981) Ecdysteroid titer during larval-pupal-adult development of the tobacco hornworm, *Manduca sexta. Gen. Comp. Endocrinol.* **44**:302–306.

Cassady, G. B. and J. M. Camhi (1976) Metamorphosis of flight motor neurons in the moth *Manduca sexta. J. Comp. Physiol.* **112**:143–158.

Dominick, O. S. and J. W. Truman (1980) Central effects of ecdysterone controlling wandering behavior in the caterpillar, *Manduca sexta. Soc. Neurosci. Abstr.* **6**:862.

Edwards, J. S. (1969) Postembryonic development and regeneration of the insect nervous system. *Adv. Insect Physiol.* **6**:97–137.

Finlayson, L. H. (1956) Normal and induced degeneration of abdominal muscles during metamorphosis in the Lepidoptera. *Quart. J. Microsc. Sci.* **97**:215–233.

Goodman, C. S. and M. Bate (1981) Neuronal development in the grasshopper. *Trends NeuroSci.* **4**:163–169.

Hamburger, V. and R. W. Oppenheim (1982) Naturally occurring neuronal death in vertebrates. *Neurosci. Comment.* **1**:39–55.

Heitler, W. J. (1978) Coupled motoneurons are part of the crayfish swimmeret central oscillator. *Nature* **275**:231–233.

Horvitz, H. R., H. M. Ellis, and P. W. Sternberg (1982) Programmed cell death in nematode development. *Neurosci. Comment.* **1**:56–65.

Kammer, A. E. and S. C. Kinnamon (1979) Maturation of the flight motor pattern without movement in *Manduca sexta. J. Comp. Physiol.* **130**:29–37.

Levine, R. B. and J. W. Truman (1982) Metamorphosis of the insect nervous system: changes in the morphology and synaptic interactions of identified cells. *Nature* **299**:250–252.

Levine, R. B. and J. W. Truman (1983) Peptide activation of a simple circuit. *Brain Res.* **279**:335–338.

Nordlander, R. H. and J. S. Edwards (1968) Morphological cell death in the postembryonic development of the insect optic lobes. *Nature* **218**:780–781.

Riddiford, L. M. (1976) Hormonal control of insect epidermal cell commitment *in vitro*. *Nature* **259**:115–117.

Riddiford. L. M. (1980) Insect endocrinology—action of hormones at the cellular level. *Ann. Rev. Physiol.* **42**:511–528.

Schneiderman, A. M., S. G. Matsumoto, and J. G. Hildebrand (1982) Trans-sexually grafted antennae influence development of sexually dimorphic neurons in the moth brain. *Nature* **298**:844–846.

Taylor, H. M. and J. W. Truman (1974) Metamorphosis of the abdominal ganglia of the tobacco hornworm, *Manduca sexta*. Changes in populations of identified motor neurons. *J. Comp. Physiol.* **90**:367–388.

Truman, J. W. (1976) Development and hormonal release of adult behavior patterns in silkmoths. *J. Comp. Physiol.* **107**:39–48.

Truman, J. W. (1978) Hormonal release of stereotyped motor programmes from the isolated nervous system of the Cecropia silkmoth. *J. Exp. Biol.* **74**:151–174.

Truman. J. W. (1980) Eclosion hormone: its role in coordinating ecdysial events in insects. In *Insect Biology in the Future* M. Locke and D. S. Smith, eds. Academic Press, New York, pp. 385–401.

Truman, J. W. (1983) Programmed cell death in the nervous system of an adult insect. *J. Comp. Neurol.* **216**:445–452.

Truman, J. W. and R. B. Levine (1980) Programmed cell death in the nervous system of an insect: histological and physiological aspects. *Soc. Neurosci. Abstr.* **6**:668.

Truman, J. W. and L. M. Schwartz (1984) Steroid regulation of neuronal death in a moth nervous system. *J. Neurosci.* **4**:274–280.

Truman, J. W. and J. C. Weeks (1983) Hormonal control of the development and release of rhythmic ecdysis behaviours in insects. *Symp. Soc. Exp. Biol.* **37**:223–241.

Truman, J. W., R. B. Levine, and J. C. Weeks (1985) Reorganization of the nervous system during metamorphosis of the moth *Manduca sexta*. In *Metamorphosis*, M. Balls and M. Bownes, eds., British Soc. Dev. Biol. Symp. 8. Cambridge University Press, Cambridge, in press.

Weeks, J. C. and J. W. Truman (1983) Cellular interactions between a muscle and its motor neuron are not involved in their endocrine-mediated deaths during metamorphosis in the tobacco hornworm *Manduac sexta*. *Soc. Neurosci. Abstr.* **9**:605.

Chapter Three

PLASTICITY AND THE REGULATION OF SYNAPTIC PROPERTIES OF ANURAN NEUROMUSCULAR JUNCTIONS

A. D. GRINNELL, B. M. NUDELL, and L. O. TRUSSELL

Jerry Lewis Neuromuscular Research Center
UCLA School of Medicine
Los Angeles, California

A. A. HERRERA

Department of Biology
University of Southern California
Los Angeles, California

P. A. PAWSON and A. J. D'ALONZO

Jerry Lewis Neuromuscular Research Center
UCLA School of Medicine
Los Angeles, California

There is probably no synapse more thoroughly studied or better understood than the frog neuromuscular junction. Mechanisms of transmitter release, postsynaptic receptor/channel mechanisms, and ultrastructural correlates of function have all been worked out initially, and most thoroughly, at this junction (Katz, 1966, 1969; Heuser and Reese 1973; Heuser et al., 1979). Similarly, studies of the development and specificity of neural connections and developmental changes in connectivity associated with neuronal cell death and synapse elimination have focused especially on this synapse (Grinnell and Herrera, 1981; Bennett, 1983; Van Essen, 1982). However, there are several aspects of connectivity and function that have been little investigated at any synapse, including the neuromuscular junction. Prominent among these are questions concerning the quantitative match between neuron and target: the size and number of terminals formed, the amount of release per terminal (or bouton), the placement of synapses on the target cells, competitive interaction between sources of innervation, and plastic changes in synaptic properties due to experience or other factors.

Although there is a tendency to treat the mature neuromuscular junction in twitch muscles as a stable, high safety margin final relay between integrative centers in the CNS and muscle, these synapses are, on the contrary, highly diverse and plastic (Grinnell and Herrera, 1981). In fact, they provide excellent material for study of mechanisms of plasticity and the intercellular interactions involved in regulation of synaptic number, position, size, and strength. The innervation of frog twitch muscle fibers, for example, can differ enormously in ways that are little understood. Fibers of some muscles have only one endplate per fiber; those of other fast twitch muscles may have anywhere from one to five. What is the reason for the additional junctions, which normally are so far apart that their synaptic potentials do not sum? Junctions within a given muscle may differ by more than tenfold in size and twentyfold in endplate potential (EPP) amplitude. Can it be predicted whether a synapse will be large or small, strong or weak? The mean quantal content of junctions in the *Rana pipiens* cutaneous pectoris muscle is approximately 2.4 times greater than that of sartorius junctions (Grinnell and Herrera, 1980). In the latter muscle, a sizeable fraction of the synapses are so weak that a single stimulus does not release enough transmitter to excite the muscle fiber. In fact, perhaps 20% of the sartorius fibers have no synapses that are suprathreshold upon single stimulation. In another frog muscle, the extensor 1. dig. IV of the toe, mean quantal content is reportedly much smaller still, with approximately one-tenth the release/unit length of sartorius terminals (Weakly and Yao, 1983). Moreover, frog neuromuscular junctions have been shown to grow and retract in response to seasonal or hormonal changes (Wernig et al., 1980; Grinnell and Herrera, 1981), to compete with and sometimes suppress or displace other junctions over considerable distances along the muscle fiber (Grinnell et al., 1979; Mark 1980; Grinnell and Herrera, 1981), and to sprout in response to the presence of denervated muscle fibers

(Morrison-Graham, 1983b). Furthermore, a reduction in the axonal arborization of a motoneuron can sharply increase transmitter release from its terminals (Herrera and Grinnell, 1980), and denervation of the contralateral homologous muscle can result in terminal sprouting (Rotshenker, 1979, 1982) and an increase in transmitter release (Herrera and Grinnell, 1981). Thus frog neuromuscular junctions exhibit long-term plasticity in structure and function that, if understood, might help considerably in understanding comparable phenomena in the central nervous system (CNS). It is our intention to show that at least some of these forms of plasticity are indicative of regulatory processes that are beginning to become amenable to analysis.

THE REGULATION OF TERMINAL LENGTH

It has long been recognized that as muscle fibers grow, the motor nerve terminals on them grow longer (Kuno et al., 1971; Bennett and Pettigrew, 1975; Harris and Ribchester, 1979; Bennett 1983). In fact, the correlation between muscle fiber diameter and total terminal branch length in frog muscle is quite good. Figure 1 shows examples of this correlation for frog cutaneous pectoris (c.p.) and sartorius muscles. (Unless otherwise specified, all of our experiments have been done with *Rana pipiens*.) Note that fibers from frogs of different sizes fit nicely into the same relationship. The somewhat smaller slope in this relationship for sartorius fibers may in part result from the addition of endplates during growth (Bennett and Pettigrew, 1975), which probably reduces the stimulus for growth of individual terminals (Nudell and Grinnell, 1983). The scatter of data for large diameter fibers suggests that the correlation begins to break down in very large fibers. However, at least part of this scatter is artifactual, since large fibers commonly are elliptical or triangular in shape, making visual determination of diameter inaccurate in whole mounts. For that reason, we normally measure muscle fiber input resistance (R_0), which can be accurately converted into effective muscle fiber diameter (Katz, 1948; Trussell, 1983a). Apparently, muscle fibers, in proportion to their size, provide a signal to the nerve terminals that end on them, telling the terminals how large to grow. It is of interest to ask what this signal is, how a given nerve terminal knows the diameter of the fiber it innervates, whether the terminals formed by a given axon are all on fibers of the same diameter, and, if not, how the local decision is made that matches terminal size to muscle fiber size. Since the synaptic current necessary to evoke a comparable postsynaptic depolarization varies with R_0 in just the way terminal length does, it is possible that the terminal somehow is able to sense quantitatively the postsynaptic effects of the synaptic current, enlarging to compensate for changes in R_0, so that, to a first approximation, excitatory postsynaptic potentials (EPPs) of adequate safety margin are produced.

Although there is quite good correlation between terminal and fiber

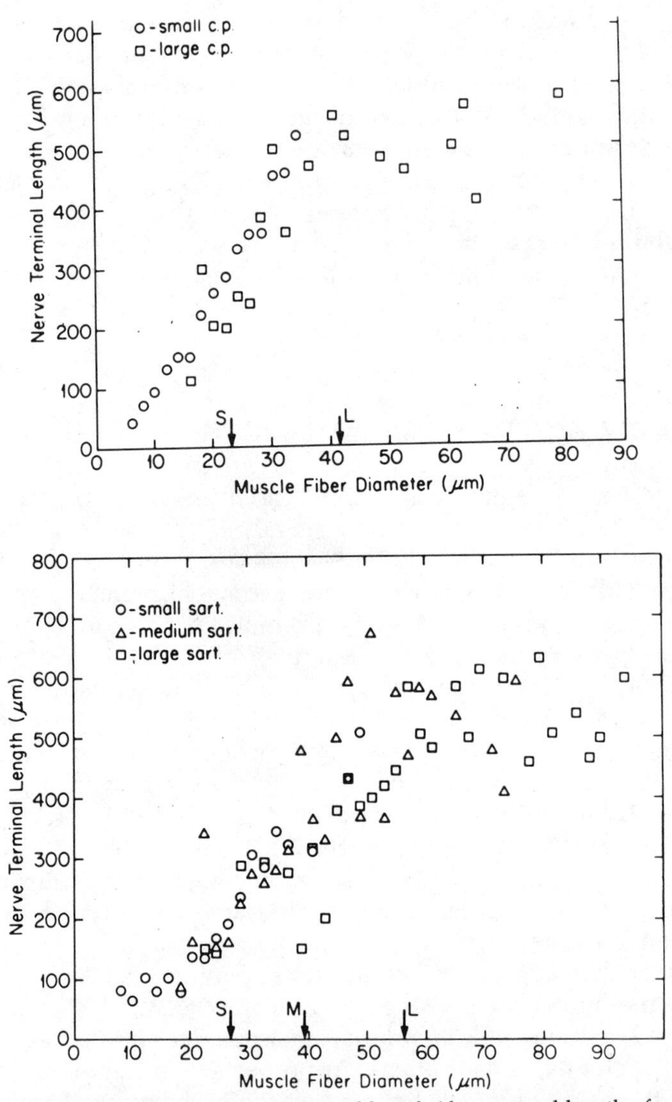

FIGURE 1. Correlation between nerve terminal length (the summed length of a terminal's branches, judged by the presence of cholinesterase strain) and the apparent diameter of the muscle fiber it innervates, for cutaneous pectoris (top) and sartorius (bottom) muscles. Each point is the mean of all fibers within a 2-μm bin, and data are subdivided by the size range of the muscles used. For sartorius muscles, the small examples (S) weighed 21–23 mg and were 18 mm long (78 fibers); medium muscles (M) weighed 56–79 mg and were 25–32 mm long (126 fibers); large muscles (L) weighed 254–466 mg and were 40–45 mm long (78 fibers). The c.p. muscles were from the same animal, with small and medium-sized animals' muscles (grouped as S) weighing up to 62 mg (175 fibers), large muscles weighing between 133–325 mg (75 fibers). Means for different sized muscles are indicated by arrows.

size, there still can be significant variability. For any given R_0, terminal size (most easily quantified in frogs by measuring the summed length of all its branches) can vary by nearly 100%. It turns out, however, that much of the variability in terminal length for a group of muscle fibers of the same size, in the same muscle, follows certain rules that imply interesting nerve-muscle interactions. If one looks not only at the length of terminals on such a group of fibers, but also at the amount of transmitter released by each terminal, it can be seen that the greater the amount of transmitter released per unit length, the shorter the terminal (Fig. 2). The implication is that the signal from each muscle fiber, telling the terminal how large the muscle fiber is and how long to grow, is modulated in inverse proportion to the strength of the input. Large fibers tend to have larger, stronger synapses; but on any group of fibers of similar size, there is an inverse relationship between terminal length and strength (Nudell and Grinnell, 1982).

Interestingly, in a muscle in which most fibers acquire two endplates, the *Xenopus* pectoralis muscle, it has been shown that there is a strong tendency for both junctions to be of the same length. They are much more similar than are randomly compared junctions on fibers of the same input resistance from the same muscle (Fig. 3*A*; Nudell and Grinnell, 1983). Thus the terminal size-regulating signal from a muscle fiber appears to be expressed relatively uniformly along its length. Surprisingly, however, there is no corresponding tendency for the two junctions on a given fiber to be of more similar strength than any two junctions on different fibers (Fig. 3*B*).

Table 1 shows sample data for two sets of fibers of similar R_0 in two muscles. Although most pairs of terminals were quite similar in total branch length, they could differ dramatically in transmitter release levels. However, it would not be accurate to conclude that there is no relationship between terminal length and the amount of transmitter released by the junctions of doubly innervated fibers. On the contrary, there is again a good inverse correlation, if one takes into account the *total* terminal branch length and the *summed* transmitter output of both junctions. Figure 4 shows this relationship for eight fiber groups, including the two in Table 1 (4*E*′ and *G*′). As in the case of the c.p., larger muscle fibers tend to have larger terminals, presumably reflecting a positive feedback stimulation of terminal growth proportional to fiber size. And again, as Figure 4 shows, within each group of similar sized fibers the amount or effect of this postulated signal is reduced in proportion to the synaptic input. But now each muscle fiber appears to be integrating the combined inputs from both its terminals.

There are many unknowns in this relationship. The nature of the signal produced by the muscle cell that governs terminal growth is unclear, although the existence of such a signal has been postulated by many workers (see, for example, Purves, 1976; Jansen et al., 1978; Kuffler et al., 1980; Grinnell and Herrera, 1981), and it may be similar or identical to the substance(s) produced by denervated or paralyzed muscle fibers that cause

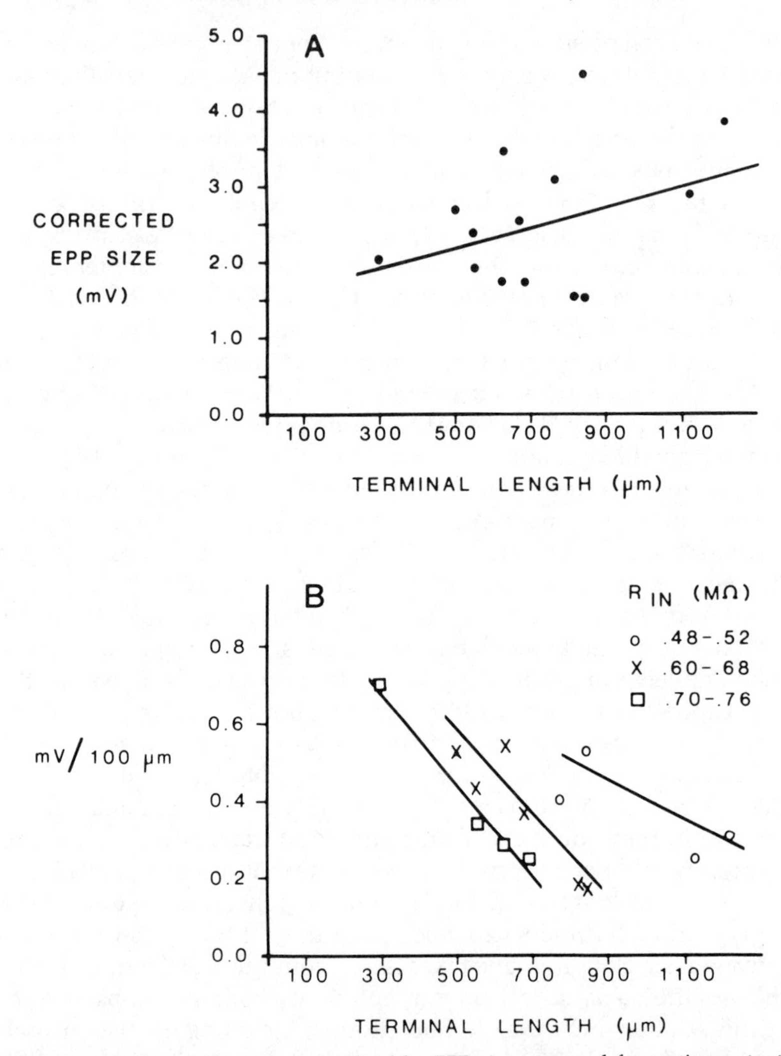

FIGURE 2. (*A*) Transmitter release, judged by EPP size corrected for resting potential (to −90 mV), nonlinear summation, and input resistance, plotted against terminal length for 14 fibers in one curarized c.p. muscle (correlation coefficient .4). (*B*) The same data plotted as release per unit terminal length, with junctions grouped on the basis of muscle fiber input resistance. (From Nudell and Grinnell, 1982.)

sprouting of intact terminals and promote reinnervation of fibers (Hoffman, 1951; Brown et al., 1981), or the putative trophic factors that promote survival and differentiation of embryonic motoneurons (Berg, 1982; Dribin and Barrett, 1980; Bennett, 1983). Moreover, it is not known how a nerve terminal can down-regulate the expression of this sprouting, or growth-promoting, substance and how its effect comes to be equally distributed along the fiber instead of locally expressed in proportion to synaptic strength.

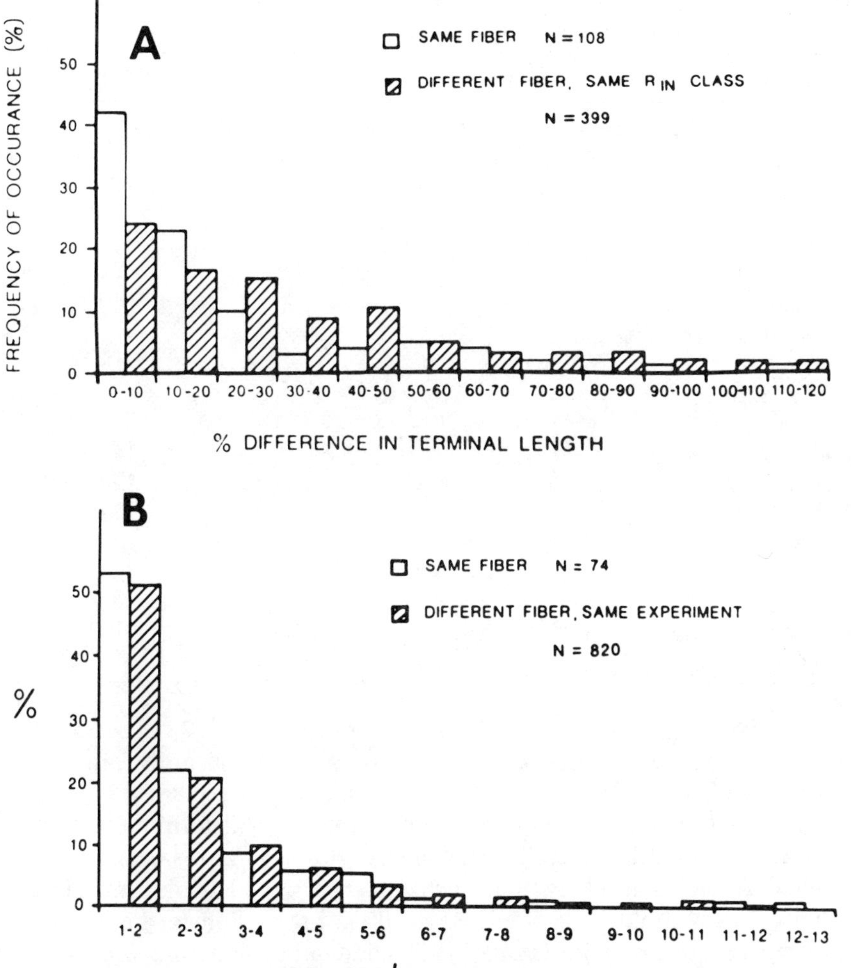

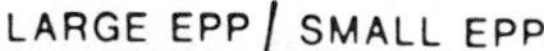

FIGURE 3. Histograms showing the degree of similarity in length (*A*) and EPP amplitude (*B*) for pairs of terminals on the same muscle fiber (clear bars), and for random comparison of terminals on different doubly innervated fibers (hatched bars) in the same preparation. Length comparisons were restricted to fibers of equivalent R_o. EPP amplitudes were corrected for resting potential and compared among all dually innervated fibers studied. Note the much greater similarity in length between terminals on the same as opposed to different fibers, and the lack of greater similarity in EPP amplitudes on the same muscle fibers. (Data from Nudell and Grinnell, 1983.)

An obvious candidate for the effect of synaptic input is muscle fiber activity, which would be felt equally throughout the fiber and which would be consistent with the known effect of muscle fiber activity in preventing or reversing the effects of partial denervation on nerve-sprouting and muscle membrane changes in mammalian muscle (Lømo, 1976). However, directly evoked muscle fiber activity is reportedly much less effective in reversing

TABLE 1. Length-Strength Relationships for Terminals in Two Groups of *Xenopus* Pectoralis Fibers of Similar Size

Fiber/Endplate	Terminal Length (mm)	Total Length (mm)	EPP (mV)	R_0
	Muscle E			
1 medial	400	1010	9.8	0.59
lateral	610		17.2	
2 medial	684	1357	1.7	0.54
lateral	673		9.5	
3 medial	673	1378	8.6	0.62
lateral	705		5.1	
4 medial	842	1684	1.2	0.54
lateral	842		1.9	
	Muscle G			
1 medial	630	1265	14	0.42
lateral	635		12.6	
2 medial	810	1750	6.4	0.34
lateral	940		15.2	
3 medial	918	1815	17.6	0.4
lateral	897		3.1	
4 medial	1175	2350	11	0.36
lateral	1175		3	

or preventing denervation changes in frogs (Anthony and Tonge, 1980; Sayers and Tonge, 1983). Moreover, it is difficult to explain the observed inverse relationship between transmitter release and terminal length on fiber action potential or contractile activity, for several reasons.

First, it seems unlikely that each synapse would evoke muscle fiber activity, a threshold phenomenon, in direct proportion to the amount of transmitter it releases. Junctions with a wide range of quantal outputs, and therefore of safety margins, would all be equally successful in evoking muscle fiber activity *in vivo* when an action potential invades the terminal (Nudell and Grinnell, 1982). Second, even though the *Xenopus* pectoralis is innervated by about 50 motoneurons (Haimann et al., 1981), we have found that in more than 50% of the dually innervated fibers in this muscle both endplates are formed by the same neuron; hence addition of the second input would not affect muscle fiber activity (Nudell and Grinnell, 1983). Finally, recent observations indicate that junctions on the same pectoralis muscle fiber that are formed by the same motoneuron and which therefore share the same activity pattern can differ dramatically in synaptic strength (Nudell, unpublished observation). On the basis of these arguments, it seems possible that a fiber's synaptic input may act through the transmitter itself, or some postsynaptic effect proportional to the integrated synaptic current (incuding nonquantal release), to help regulate the

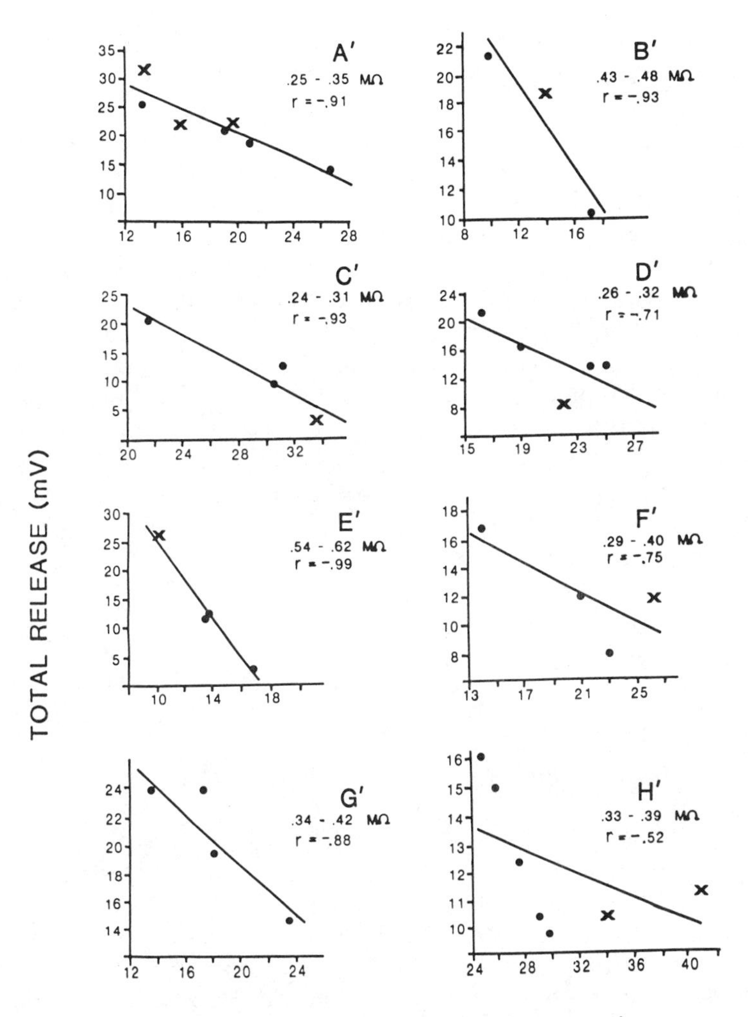

FIGURE 4. Inverse correlation between summed release from both terminals on a fiber and the summed length of both terminals for eight different groups of fibers of similar R_o. Release was judged by normalizing EPPs for resting potential and the mean R_o value of the group. Correlation coefficients for the lines of best fit are shown for each group. Circles represent fibers in which both terminals were within 20% of the same length, crosses represent data from fibers with terminals differing by more than 20% in length. (From Nudell and Grinnell, 1983.)

terminal-growth-promoting influence of the muscle fiber. It must be recognized, however, that any given neuron, either by virtue of its activity pattern or some other trophic influence, may help to determine the size of the fibers it innervates (Nudell and Grinnell, 1982; Trussell, 1983; Trussell and Grinnell, 1984). This of course would have a profound influence on the size of its terminals.

COMPETITION BETWEEN SYNAPTIC INPUTS AND DETERMINATION OF SYNAPTIC NUMBER AND LOCATION

When more than one axon terminal innervates a muscle fiber, the two inputs can compete with each other. This happens at most synapses during the developmental period of synapse elimination, when one input somehow suppresses or displaces all others (Brown et al., 1976; Purves and Lichtman, 1980; Morrison-Graham, 1983a; Van Essen, 1982). Even when they are far apart, two endplates on the same fiber tend to mutually reduce each other's effectiveness (Grinnell et al., 1979). The modulation by synaptic input of growth regulatory feedback from the muscle fiber to nerve terminals represents a mechanism that can, at least in part, account for this competitive interaction. If a given axon terminal shares a fiber with a strong second input, it will not grow as long or (presumably) release as much transmitter as it would if it were coupled with a weaker input (Nudell and Grinnell, 1983).

To this form of mutual competition must be added another obvious form—a localized exclusion of other inputs. As muscle fibers grow, for example, in the frog sartorius, new endplates are added, but these are always separated by a certain minimum distance (Bennett and Pettigrew, 1975). A minimum separation between endplates is also seen in the *Xenopus* pectoralis muscle (Nudell and Grinnell, 1983). Hence it appears that the presence of one endplate inhibits the formation of another very close to the first, even when the fiber is receptive to additional innervation at greater distances.

Together, these two forms of competition go far toward explaining the normal pattern of single and double innervation in the *Xenopus* pectoralis muscle (Nudell and Grinnell, 1983). During early postmetamorphic growth most fibers of this muscle acquire a second endplate, so that by the time the muscle can easily be studied physiologically about 95% of the fibers are doubly innervated. Those fibers that remain singly innervated tend to be smaller than fibers with two junctions, but a few are equally large. In these the single synaptic input tends to be extremely strong and located near the middle of the fiber. The junctions on dually innervated fibers are located to either side of the middle, usually separated by 20–40% of the muscle fiber length. Moreover, the summed strength of both junctions on dually innervated fibers is often less than the single input in singly innervated fibers (Haimann et al., 1981; Nudell and Grinnell, 1983). One can imagine that, among the whole population of fibers in the muscle, those that would best resist double innervation as the animal grows are those with synapses so strong that they effectively suppress production of growth- and innervation-promoting substances by the muscle fiber and those that have synapses placed near the middle of the fiber, where the localized inhibition of innervation by another axon is expressed for a considerable distance toward both ends.

It is appropriate to ask why there should be a mechanism producing multiple junctions on frog twitch fibers. This question was posed initially by Katz and Kuffler (1941), who speculated that the function of multiple endplates might be to ensure more synchronous activation and contraction of fibers. However, this appears not to be the case when tested experimentally (Nudell and Grinnell, 1983; and unpublished observations), nor are the endplates normally sufficiently close to one another for significant summation of EPPs. An obvious result of having inputs from two or more different axons is that any given fiber can contribute to more than one motor unit, theoretically giving finer control of contraction. This may be an important function of multiple innervation. However, the observation that in the *Xenopus* pectoralis a high percentage of dual junctions are formed by the same axon (Nudell and Grinnell, 1983) suggests that this is not the main driving force for the phenomenon. Instead, it seems likely that many frog twitch muscles retain a trophic need for multiple inputs, proportional in some way to their size. The unexpectedly high incidence of mononeuronal dual innervation may reflect a superimposed mechanism that selects for inputs that are simultaneously active. (Pursuing this reasoning, it would be of interest to know whether different inputs that survive together at polyneuronally innervated junctions tend to be active simultaneously or have similar activity patterns.)

REGULATION OF SYNAPTIC STRENGTH

The correlation between terminal length and amount of transmitter release is surprisingly poor. Since there appear to be clear-cut rules governing the length to which terminals will grow, this raises the possibility that length is regulated, at least to a certain degree, independent of release per unit length or the total output of a terminal. It is clear that one major determinant of transmitter output is the motoneuron of origin (Grinnell and Trussell, 1983). Motor units in the sartorius muscle vary widely in size: twitches of different motor units range from about 0.1% to more than 40% of the twitch tension elicited by direct supramaximal stimulation of the muscle or of the whole nerve. When these motor units are tetanized, the range of tetanus tensions is much narrower (only about fiftyfold), mainly because the smallest units evoke tetanus tensions up to 100 or more times their twitch tension (Fig. 5). Since the tetanus/twitch ratio for directly stimulated muscle fibers or the whole muscle is between 2 and 3, it is clear that most of the fibers belonging to a given small motor unit receive only subthreshold synaptic input from the motor axon. The largest motor units, on the other hand, have tetanus/twitch ratios of only 2–3, indicating that most or all of their muscle fibers receive suprathreshold synaptic excitation. Another way of showing this is to change the calcium concentration in the Ringer above and below the normal 1.8 m*M*. As Figure 6 shows, the large motor

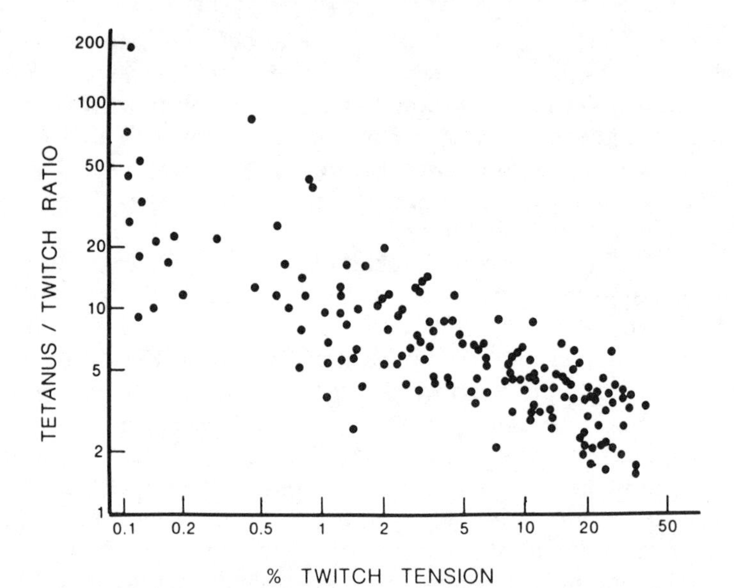

FIGURE 5. Tetanus/twitch ratios as a function of twitch size for individual motor units in the frog sartorius muscle. Twitch size is expressed as the percentage of the twitch tension produced by supermaximal direct stimulation of the muscle. The tetanus frequency was 50 per second. (Data from Grinnell and Trussell, 1983.)

units not only have stronger synapses, more resistant to lowering of calcium concentration, but they also tend to be much more uniform in synaptic strength than are junctions in the smaller motor units (Grinnell and Trussell, 1983). Since the range of fiber sizes among strong and weak motor units appears similar in the sartorius, these results imply that synaptic strength and transmitter release are determined in large part by the motoneuron forming the synapse.

This conclusion is borne out by study of individual junctions within identified motor units of the cutaneous pectoris (c.p.) muscle (Trussell 1983a,b; Trussell and Grinnell, 1985). Single axons have been isolated, the twitch tensions of their motor units measured, and the muscle then blocked with curare to permit study of as many junctions as can be found innervated by that axon. Motor units were observed with twitch tensions ranging from about 1 to 40% of the total muscle twitch tension, a narrower range of motor unit sizes than in the sartorius. Among these motor units, there was an approximately linear correlation between mean transmitter release and motor unit twitch size (Fig. 7). Small twitch motor units, with weaker synaptic strength, also appear to have slower axonal conduction velocity, suggesting overall that such motoneurons are small and metabolically weak (Burke, 1981; Luff and Proske, 1976; Trussell and Grinnell, 1985; see material to follow).

Many c.p. fibers, especially those of relatively large diameter, are poly-

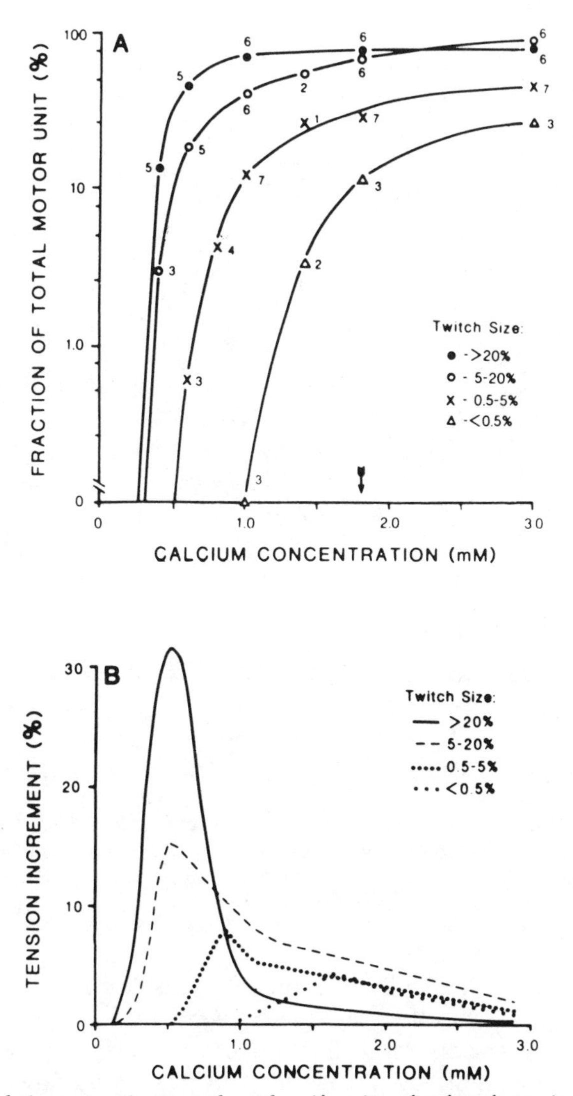

FIGURE 6. Relative synaptic strength and uniformity of safety factor in motor units of different size in sartorius muscles. (*A*) For motor units of different twitch size (as percent of total muscle twitch tension), the fraction of the total twitch produced by contraction of all the fibers in the motor unit is plotted against the calcium concentration in the Ringer. The arrow indicates the normal concentration, 1.8 m*M* calcium. Note that stronger motor units show higher safety factors at any given calcium concentration, and greater uniformity in this value. (Numbers beside each point represent the numbers of motor units studied.) Curves were fitted by *n*th-order regressions to obtain equations of best fit. (*B*) For different twitch size categories, the calculated tension increment for each 0.2 m*M* increase in calcium concentration is shown. See Grinnell and Trussell, 1983, for complete explanation of how ordinate values were obtained. (From Grinnell and Trussell, 1983.)

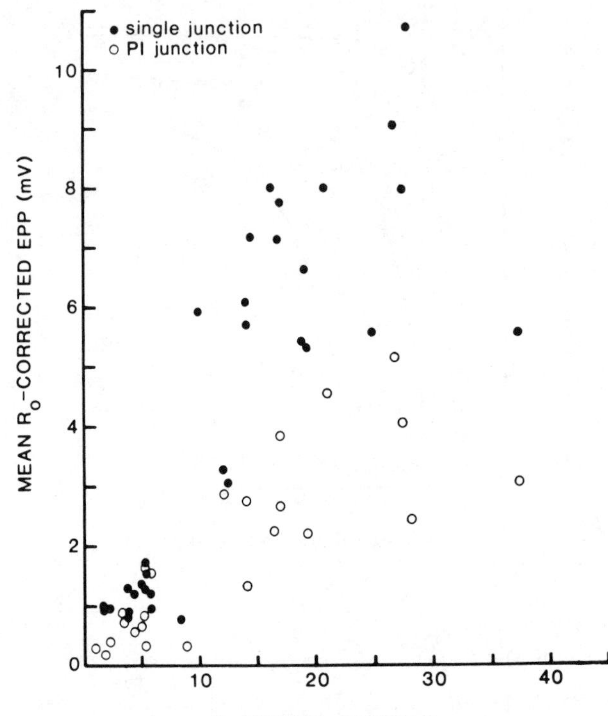

FIGURE 7. Dependence of transmitter release on motor unit size in isolated motor units of the frog c.p. muscle. Mean values for the amount of release from terminals of isolated axons are plotted against motor unit twitch size for junctions singly innervated by the isolated axon (closed circles) and in which the isolated axon provided an input to a polyneuronally innervated junction (open circles). Release is judged by EPP amplitude corrected for resting potential and nonlinear summation. (Data from Trussell, 1983a.)

neuronally innervated. In fact, when one first locates inputs from an isolated motor axon and then stimulates the whole nerve to look for other inputs, an unexpectedly large fraction of endplates are polyneuronally innervated—a mean of 36% (Trussell, 1983a; Trussell and Grinnell, 1985) compared with a value of about 16% obtained with the traditional method of looking for EPP amplitude steps in curarized preparations (Rotshenker and McMahan, 1976; Rotshenker, 1979;1982; Weakly and Yao, 1983). The explanation for the difference is that many components are so small that they would not be detectable riding on a larger lower threshold input. In fact, the 36% figure is certainly an underestimate, missing many of the cases in which the input from the identified axon is much larger than inputs from other axons (Trussell, 1983a; Trussell and Grinnell, 1985).

Figure 7 compares the relationship between transmitter release and motor unit size for singly innervated and polyneuronally innervated inputs. Although synaptic inputs to polyneuronally innervated endplates are uniformly weaker than singly innervating inputs from the same twitch size motor unit, their strength is still sharply dependent on the motoneuron of origin and the size of its motor unit. Since the total terminal length at polyneuronally innervated endplates apparently is quite comparable to that of singly innervated junctions on fibers of the same size, the reduced synaptic effectiveness for any given component may reflect mainly a smaller terminal. In any case, the efficacy of synaptic transmission in the neuromuscular junctions of the frog, and presumably also in the CNS, appears to depend both on the intrinsic characteristics of each neuron and competitive interactions between different neurons for synaptic space and influence.

Interestingly, motor units of different twitch strengths also show a systematic difference in mean muscle fiber diameter, ranging from means of 21–33 μm for weak units to 41–65 μm in strong units. Figure 8 shows the distribution of diameters in muscle fibers innervated by identified "strong" (> 14% of total muscle twitch tension) or "weak" (< 9% of total muscle twitch tension) motor unit axons. Again, although there is a wide range of fiber sizes in both groups of motor units, there is a clear-cut difference in means, suggesting that the muscle fiber size is strongly influenced by the synaptic input to it. This suggestion is reinforced by the observation (Fig. 8) that fibers polyneuronally innervated by two strong twitch motor units are as large in diameter as fibers singly innervated by one strong motor unit, whereas fibers belonging to both a small twitch motor unit and a strong twitch motor unit have intermediate diameters (Trussell, 1983a; Trussell and Grinnell, 1985).

In the sartorius muscle, the wide range in motor unit sizes can be attributed in large part to differences in the number of fibers innervated by different motor axons (Grinnell and Trussell, 1983). In the c.p., however, it is clear that much, or conceivably all, of the difference in motor unit twitch size is due to differences in muscle fiber size (Trussell, 1983a; Trussell and Grinnell, 1985). The considerable overlap of diameters of singly innervated fibers in strong and weak motor units (Fig. 8) is somewhat misleading. In any given muscle, there was little overlap. Nevertheless, some fibers of intermediate size could belong to either strong or weak motor units, presumably with terminals of similar length but with quite different EPP amplitudes (Trussell, 1983a; Trussell and Grinnell, 1985).

PLASTICITY IN SYNAPTIC EFFICACY

The "intrinsic" strength of a mature motoneuron's synapses is not a fixed, unalterable quantity. There is reason to believe that mean levels of transmitter release, as well as terminal length, change normally with seasonal or hormonal changes (see Wernig et al., 1980; Grinnell and Herrera, 1981). Moreover, we have now described two experimental manipulations that

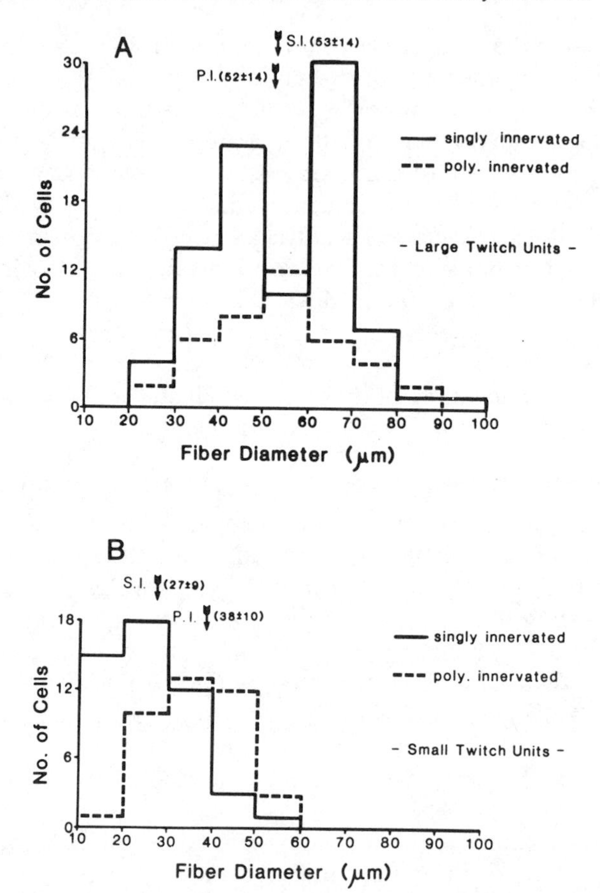

FIGURE 8. Histograms showing the distributions of fiber diameters in identified "strong" (*A*) and "weak" (*B*) motor units. Solid lines indicate fibers with singly innervated junctions; dashed lines represent polyneuronally innervated (P.I.) junctions receiving an input from an isolated axon. Arrows indicate means. Note the difference in fiber diameter distribution for strong and weak motor units and the intermediate diameter of fibers innervated by both "weak" axons and (usually stronger) unidentified inputs. (Data from Trussell, 1983a.)

can sharply alter transmitter release. In the first of these a reduction in the size of the motoneuron's axonal arborization results in a large increase in mean quantal output and safety margin (Herrera and Grinnell, 1980). In these experiments the nerve to a sartorius muscle was crushed and the lateral half of the muscle was removed. The full complement of axons regenerated, selectively reinnervating old endplates, but each apparently formed about one-half the normal number of synapses. As Table 2 shows, the result was a more than twofold increase in EPP quantal content and release per unit length, compared with reinnervated whole sartorius muscles. Our interpretation of this finding is that each motoneuron, as a result of its developmental history, matures with the capability of maintaining a

TABLE 2. Effects of Reduction in Motor Unit Size on Mean EPP Quantal Content $\bar{m}$ and Release/Unit Length in 0.3 mM Ca^{2+}, 1 mM Mg^{2+} Ringer (Mean ± s.d.)

	Quantal Content	$\bar{m}$/100 μm
Reinnervated half sartorius	6.30 ± 0.84 (n = 34)	1.89 ± 0.43 (n = 21)
Reinnervated whole sartorius	2.79 ± 0.59 (n = 19)	0.46 ± 0.09 (n = 16)
Unoperated normal	2.07 ± 0.36 (n = 25)	0.45 ± 0.09 (n = 29)

certain amount of release from its complement of terminals. When the number of synapses it must maintain is decreased, it continues to supply its normal amounts of transmitter-release components to the smaller number of terminals, with the result that each, on the average, is correspondently more effective. This is apparently a transient state, for 6 to 12 months after reinnervation, the mean synaptic strength is back to normal, and not significantly different than in reinnervated whole muscles (Herrera and Grinnell, 1985). Thus even mature motoneurons are sensitive to changes in connectivity or target size and readjust their metabolic rate accordingly. Interestingly, the junctions in two- to six-month reinnervated half-muscles also show much less polyneuronal innervation than junctions in reinnervated whole muscles, suggesting that stronger synaptic inputs may be more effective in excluding or eliminating second inputs (Herrera and Grinnell, 1985).

A second interesting form of plasticity is the enhancement of transmitter release from "normal" terminals when the contralateral homologous motoneurons are axotomized (Herrera and Grinnell, 1981). Within a few days or weeks, and lasting for at least a year or more, the mean quantal output and release per unit length increase sharply (Table 3). It seems likely that this synaptic enhancement phenomenon is related to the contralateral sprouting response studied in the c.p. and sartorius by Rotshenker and his colleagues (Rotshenker, 1979, 1982; Rotshenker and Reichert, 1980). This sprouting response occurs earlier when the contralateral nerve section is closer to the spinal cord, and it is also seen when the denervated muscle is removed, so it apparently involves some interaction between motoneurons

TABLE 3. Effects of Contralateral Denervation on Transmitter Release in Ringer Containing 0.3 mM Ca^{2+} plus 1 mM Mg^{2+} [a]

	m	IQR	m/100 μm	IQR
Contra denervated	4.8	2.05–9.22 (n = 28)	0.88	0.41–1.92 (n = 26)
Normal	1.65	0.97–2.87 (n = 45)	0.28	0.16–0.54 (n = 31)
		$p < .001$		$p < .005$

[a] (m = median quantal content, IQR = interquartile range)

rather than a systemic response to denervated tissue. The enhancement of terminal output per unit length that we observe is also accompanied by an increase in polyneuronal innervation, but not by the increase in terminal sprouting reported by Rotshenker. Hence it seems possible, in our experiments and perhaps to some degree in Rotshenker's as well, that the increase in apparent polyneuronal innervation is due principally to the enhancement of efficacy of preexistent weak inputs to polyneuronally innervated junctions, causing many of them to become detectable. In support of this hypothesis is the finding that if formamide is used to uncouple contraction from excitation in the normal frog sartorius (del Castillo and Escalona de Motta, 1978), a much higher proportion of the unblocked junctions show polyneuronal innervation than have previously been recognized (Herrera, 1984).

The ability of sartorius junctions to increase their mean output by two times or more, presumably owing to an interaction between axotomized and normal motoneurons in the spinal cord, without having been directly manipulated in any way and without any change in terminal size or appearance (at the light microscopic level), shows that the "intrinsic" strength of a motoneuron is alterable. This parameter is probably determined by a number of factors, including developmental history, number of terminals maintained, amount of feedback from the innervated muscle fibers, location within the motoneuron pool, CNS connectivity and activity pattern, and doubtless many other influences. It can be postulated that the combination of all of these factors determines the level of synapse maintenance and transmitter release capability for any given motoneuron, and this may be relatively slow to change. The slowly reversing change in efficacy brought about by a reduction in terminal arborization suggests that mature motoneurons programmed to maintain a certain number of synapses can increase the effectiveness of each when the number of synapses is reduced. Eventually, however, a new equilibrium is reached with more normal synaptic strength, adjusted presumably on the basis of a reduced level of muscle fiber feedback.

In addition to this important feedback from the periphery, the potent effect of axotomy on contralateral homologous motoneurons suggests that a significant fraction of the regulation of motoneuron efficacy occurs in the spinal cord. In view of this, and if one accepts that synaptic enhancement and sprouting are related phenomena, it is attractive to consider the possibility that either response might be greater still if motoneurons within the same motor pool are axotomized. Thus one might predict that intact motoneurons in partially denervated muscle would show marked enhancement of synaptic efficacy prior to, or associated with, their well-known sprouting response. [This is not to ignore the importance of peripheral sprouting stimuli, thought to be released by denervated muscle or endplates (Brown et al., 1981), which might be the same or different from the growth-regulating stimulus described previously.]

MORPHOLOGICAL CORRELATES OF DIFFERING TRANSMITTER RELEASE EFFICACY

Without knowing what it is about a motoneuron cell body that determines its overall transmitter-releasing capabilities, it is still interesting to ask whether there are detectable morphological correlates for the differences in release per unit length ("release efficacy") of different terminals. As mentioned before, differences of tenfold or more in release efficacy in different terminals cannot be correlated with morphological differences at the light microscopic level (Grinnell and Herrera, 1980; Herrera, Grinnell, and Wolowske, 1985a). It is true that some terminals are conspicuously "fatter" than others and tend to show high levels of release per unit length. Others are unusually thin, with a varicose appearance; these tend to be very weak. However, for the vast majority of terminals, large differences in release do not have an obvious morphological correlate. An initial ultrastructural comparison also showed no apparent differences (Grinnell and Herrera, 1980).

In a more thorough analysis at the ultrastructural level, however, there do appear to be consistent differences between strong and weak terminals (Herrera, Grinnell, and Wolowske, 1985a,b). This has been pursued in two ways: (1) by comparing single random cross sections through approximately 100 different c.p. terminal branches with sections from a similar number of sartorius terminals [the latter, on the average, release less than half as much transmitter per unit length (Grinnell and Herrera, 1980)]; and (2) by making semiserial reconstructions on nine different identified terminals of known release efficacy from two sartorius muscles whose contralateral homologues had been denervated two months earlier. The c.p. versus sartorius comparison showed that differences in mean synaptic efficacy were in fact not correlated with terminal diameter, cross-sectional area, terminal shape, or vesicle density at active zones (located by the presence of a postsynaptic fold opening, and often by increased presynaptic membrane density.) However, there was a clear-cut positive correlation between synaptic efficacy and (1) numbers of mitochondria per terminal cross section; (2) the amount of terminal membrane per cross section that is within 0.2 μm of the muscle fiber surface and unobstructed by Schwann cell processes ("contact width"); and (3) the amount of perimeter membrane of the terminal within 0.2 μm of the postsynaptic cell, including portions with glial processes interposed ("potential contact width"). Figure 9 summarizes these significant differences diagrammatically.

The more thorough analysis of ultrastructure in the nine identified sartorius terminals, which differed in release per length over an eighteenfold range, confirmed these findings. Within this group of terminals, those with greater release efficacy did tend to have larger diameters, but even better correlated with strength were the three parameters just mentioned. Of these, the critical factor is likely to be the width of the terminal in close

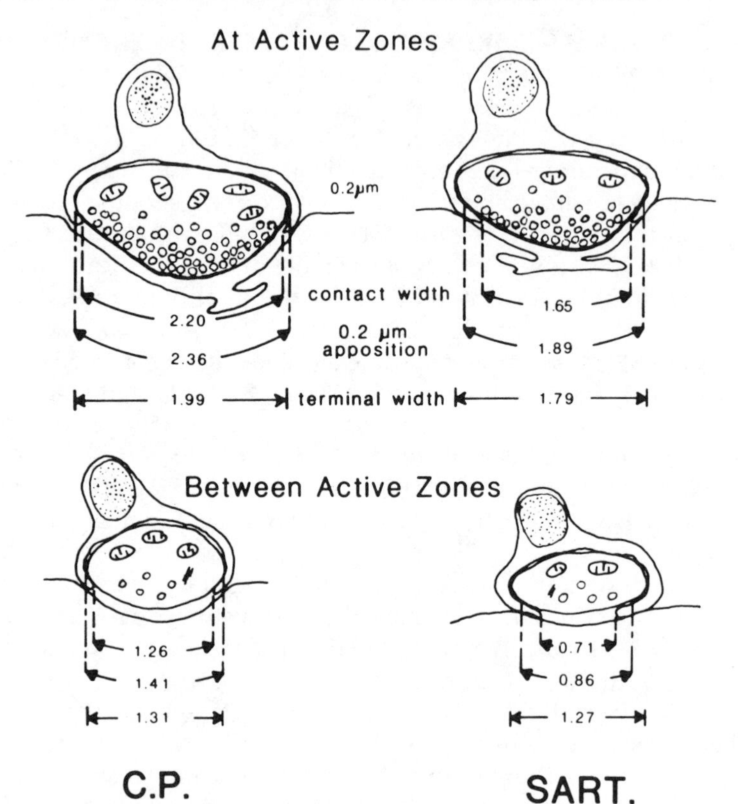

FIGURE 9. Diagrammatic summary of differences in ultrastructure of c.p. and sartorius terminals at active zones and in the regions between active zones. The major significant differences, possibly correlated with differences in transmitter release, were in contact width, extent of wrapping by Schwann cell processes, and numbers of mitochondria.

contact with the postsynaptic membrane at active zones. As Figure 9 indicates, this increase in contact width is accomplished in two ways: the stronger terminals appear to be more deeply invaginated into the muscle fiber, and there is less interposition of Schwann cell processes between the pre- and postsynaptic membranes. Whether these morphological differences can fully explain the differences in release is not clear. The range of release/unit length values is much greater than the range in contact widths. In fact, if one assumes that active zones span the entire width of close contact, then release per unit length varies approximately in proportion to (active zone)$^{2.5}$. The consistent correlation of morphology with release efficacy nevertheless suggests that it is likely to be responsible for at least part of the differences in synaptic strength. Moreover, since the strongest terminals in this series were absolutely stronger than the strongest normal junctions found to date, it seems probable that their extremely high release levels were the result of the earlier axotomy of contralateral homologous motoneurons. This suggests that a plastic increase in transmitter release can be achieved somehow by an increase in size of individual active zones without, necessarily, an increase in the number of active zones. The sequence and rate at which these changes occur is not known, but it is attractive to think that a metabolic change in a motoneuron, leading to synthesis and transport to the terminal of increased amounts of transmitter releasing machinery, results in expansion of the active zones, retraction of glial processes, and deeper invagination into the synaptic gutter. These observations are consistent with findings in other preparations. For example, in several invertebrate preparations, the size and numbers of presynaptic active zones have been correlated with synaptic strength (Govind and Chiang, 1979; Govind and Meiss, 1979; Bailey and Chen, 1983); whereas in human patients with Lambert-Eaton myasthenic syndrome, Fukunaga et al. (1982) have shown that reduced transmitter release is associated with disorganization and reduction in the number of the aggregates of presynaptic particles thought to represent Ca^{2+} channels (Heuser and Reeser, 1973; Pumplin, Reese, and Llinas, 1981). It is also of interest that differing levels of neurohormonal release from pituitary cells are correlated with the amount of glial wrapping of the neurosecretory terminals (Tweedle and Hatton, 1982; Tweedle, 1983).

PHYSIOLOGICAL CORRELATES OF PLASTICITY

It is equally valid to ask whether, in addition to morphological differences, there are distinguishable differences in physiology between strong and weak terminals. For example, are strong terminals more fully invaded by action potentials than weak ones? Are invading action potentials longer in duration or of greater peak amplitude? Is the Ca^{2+} influx/action potential greater in stronger terminals? Do different terminals have different surface charge affecting Ca^{2+} entry? Is Ca^{2+} more effective in causing release in some cases than others? Is there a difference in Ca^{2+} buffering and metabolism? Is the difference entirely presynaptic, or are there differences in postsynaptic sensitivity to transmitter? The answers to most of these questions are not known, and many will be difficult to approach experimentally. However, we have investigated several parameters, and can assess some of these possibilities.

All of the differences in efficacy we have studied appear to be presynaptic in origin. Single quantum or miniature endplate potential (mEPP) amplitudes vary linearly with input resistance, suggesting that all quanta are the same size and postsynaptic membranes are equally sensitive.

It is noteworthy that terminals with higher quantal output also exhibit higher mEPP frequencies. In fact there is a good linear correlation between these two variables, which is the same whether studied in terminals from the c.p. or the sartorius, which differ by 2.4-fold in their mean quantal output (Grinnell and Herrera, 1980). This immediately implies that the difference in evoked release is not simply a difference in action potential amplitude or duration. Moreover, if one assumes that the probability of spontaneous quantal release is equal throughout a terminal's branches, the good correlation between mEPP frequency and EPP quantal content suggests that both strong and weak terminals are invaded to an equivalent degree (Grinnell and Herrera, 1980), for if weak terminals were being only partially invaded, their mEPP frequency should be disproportionately high compared with the evoked output. This conclusion is supported by the extracellular recording of action potential currents all the way to the ends of terminal branches (Katz and Miledi, 1965). [Parenthetically, it is interesting to note that a different frog muscle, the extensor 1. dig. IV of the toe, also shows a good correlation between $\bar{m}$ and mEPP frequency, but with a different slope; for a given $\bar{m}$, the mEPP frequency is much lower (Weakly and Yao, 1983). The explanation for this difference is not clear.]

Although stronger terminals release more quanta per action potential at any given external [Ca^{2+}], both strong and weak terminals show an approximately fourth power dependence of quantal content on external [Ca^{2+}]. Moreover, the increase in release of mEPPs with depolarization is dependent on external [Ca^{2+}] to the same degree (Fig. 10). This indicates that active Ca^{2+} inside the terminal acts in the same way to evoke release in strong and weak terminals (Grinnell and Herrera, 1980).

Since both EPPs and mEPPs appear to be subserved by the same release mechanisms (Katz, 1969), the difference in mEPP frequency per unit length between strong and weak terminals suggests that the intraterminal Ca^{2+} concentration may be regulated at a higher level in strong terminals. Since the intraterminal Ca^{2+} concentration is tightly regulated in the micromolar range, a small change in resting Ca^{2+} concentration would provide a powerful mechanism for biochemical regulation (Carafoli and Crompton, 1978). The apparent difference in intraterminal Ca^{2+} might be the result of a difference in chronic leak of Ca^{2+} into the terminal and/or to a difference in buffering of Ca^{2+} or removal of Ca^{2+} from the terminal.

We approached this question by investigating the kinetics of Ca^{2+} buffering in strong versus weak neuromuscular junctions (Pawson and Grinnell, 1984; and unpublished observations). At the normal frog neuromuscular junction, four distinct phases of enhanced release have been documented following a train of stimuli: two phases of facilitation, augmentation, and posttetanic potentiation (Rahamimoff, 1968; Magleby, 1973; Lev-Tov and Rahamimoff, 1980; Zengel and Magleby, 1982). Each of these four synaptic plasticities have separate and characteristic decay constants, suggesting that there is a multiple compartment sequestration of Ca^{2+} in presynaptic terminals (Magleby and Zengel, 1982).

We have adopted the methodology of Lev-Tov and Rahamimoff (1980) to study the posttetanic potentiation (PTP) of mEPP frequency. Following the assessment of synaptic strength in a low Ca^{2+} Ringer, identified junc-

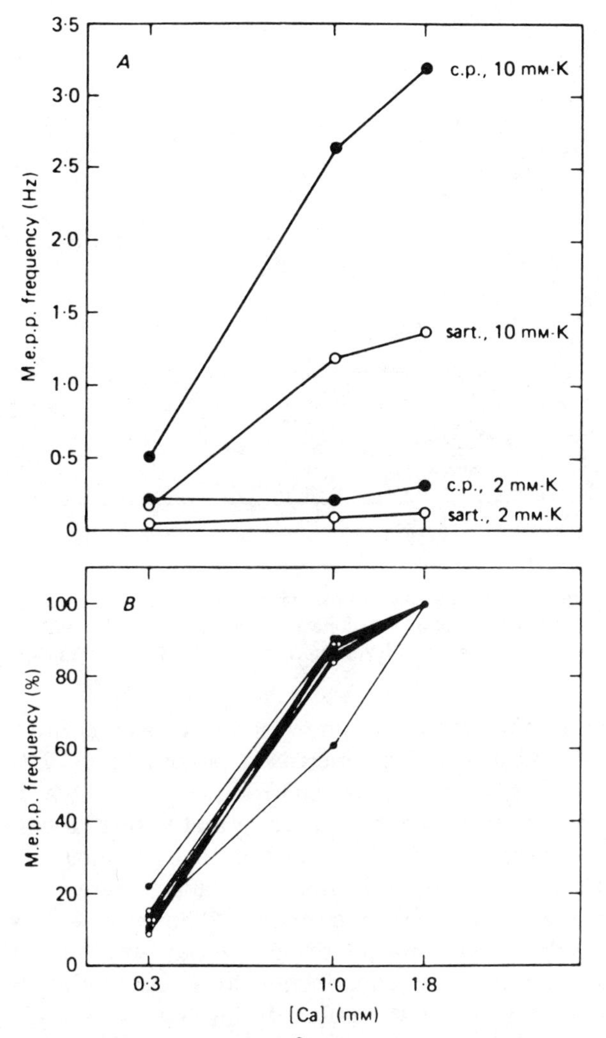

FIGURE 10. (*A*) The effect of changing [Ca^{2+}] on mEPP frequency at c.p. and sartorius (sart.) junctions in 2 m*M* and 10 m*M* K^+. Each c.p. point is an average of data from seven junctions; each sartorius point is an average of data from four junctions. (*B*) Same data, but with mEPP frequency plotted as percent of the frequency in 1.8 m*M* Ca^{2+} Ringer. (From Grinnell and Herrera, 1980.)

tions were studied in a Ringer solution (0.1 m*M* Ca^{2+}, 5.0 m*M* Mg^+) in which little if any evoked release occurs, even during a tetanus (50 Hz, 40 sec). All junctions show a progressive increase in mEPP frequency during a tetanus. Thereafter mEPP frequency typically decays back to the baseline level in a multiexponential fashion. At present, we are studying the slowest exponent, which represents the time constant of PTP. Our results (Fig. 11) demonstrate that there is a strong positive correlation between increas-

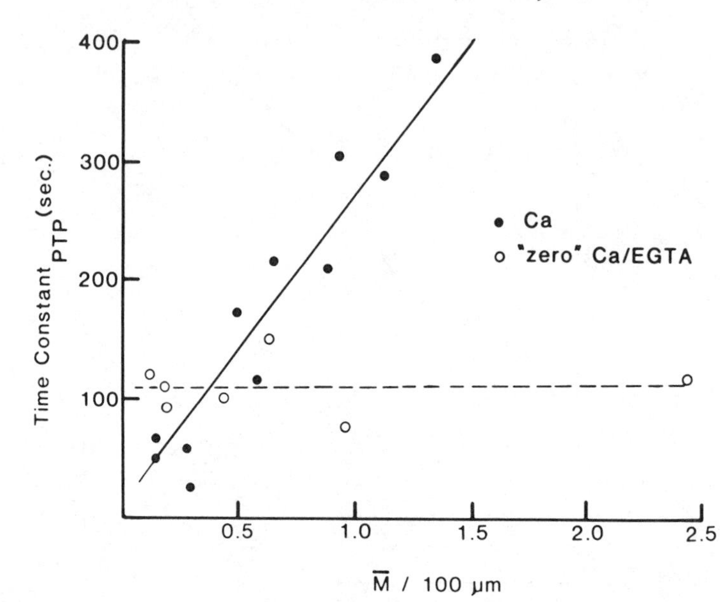

FIGURE 11. Time course of decay of posttetanic potentiation (PTP) as a function of mean release per unit length of sartorius junctions in Ringer containing 0.1 mM Ca^{2+}, 5 mM Mg^{2+} (n = 11, r = 0.96), and in Ringer with zero calcium, 1 mM EGTA, and 2 mM Mg^{2+} (n = 7, r = 0.07).

ing synaptic strength and increasing PTP time constant. The consistency of the findings indicates that an increased potential for PTP represents a heretofore unrecognized attribute of strong neuromuscular junctions.

We interpret these results as indicating that strong junctions have (1) a reduced Ca^{2+}-buffering capacity, and/or that (2) they have a proportionately greater Ca^{2+} influx during the tetanus, thereby leading to prolonged movement of Ca^{2+} through the Ca^{2+}-buffering/extrusion systems. To determine the relative importance of the Ca^{2+}-buffering system to our findings, we repeated the experiments in a "zero" Ca^{2+}/1 mM EGTA Ringer where there is no Ca^{2+} influx during the tetanus (100 Hz, 40 sec). Under these experimental conditions, there is still a tetanus-associated increase in mEPP frequency, presumably caused by a Na-mediated translocation of Ca^{2+} within the terminal (Lev-Tov and Rahamimoff, 1980). In a reversed Ca^{2+} gradient, we find that all junctions show a similar time constant for the decay of PTP (Fig. 11). Therefore, strong and weak junctions do not differ in their buffering capacity to a *moderate* increase in Ca^{2+} levels. Nor does Na^{+} influx appear to be a determinant of synaptic strength. More important, the elaboration of strength-dependent differences in Ca^{2+}-buffering is dependent on a Ca^{2+} influx during the tetanus, suggesting that a differential Ca^{2+} influx/impulse may underlie the differences in synaptic efficacy.

Support for this conclusion comes from another observation. In going from a low Ca^{2+} to a zero Ca^{2+}/EGTA Ringer, stronger junctions show a substantial decrease in the resting mEPP frequency, whereas weak junctions are relatively unaffected by the withdrawal of external Ca^{2+}. Apparently, strong junctions have a greater chronic Ca^{2+} leak. At this point, we do not know the pathway of this resting leak. However, if the amount of chronic leak is reflective of the number of voltage-dependent Ca^{2+} channels, then we have further evidence that strong terminals may have a greater Ca^{2+} influx/action potential. There are no doubt many other possible physiological differences between strong and weak terminals, but this demonstration of the importance of the Ca^{2+}-influx and the differences in intraterminal Ca^{2+} metabolism provide us with an important foundation for further study.

THE UNIFORMITY OF TRANSMITTER RELEASE IN DIFFERENT PORTIONS OF A TERMINAL

Implicit in our description of synaptic efficacy and its regulation is the assumption that release from any given portion of the terminal is approximately the same as release from any other equivalent length of the same terminal, that is, that the probability of release from any active zone within the same terminal is roughly equivalent. This assumption underlies most models of transmitter release (Heuser and Reese, 1973; Bennett and Florin, 1974; Wernig, 1975). However, using an extracellular recording technique, Bennett and Lavidis (1979, 1982) have reported that in iliofibularis junctions of the toad, *Bufo marinus,* release can be highly nonuniform, with up to 30–50% of the terminal's release coming from within recording distance of one electrode, conceivably from one site (Bennett and Lavidis, 1979). They report a consistent pattern of high release from portions of branches within 20 μm of the last myelin segment, with a sharp decline in their release probability distal to that in each branch (Bennett and Lavidis, 1982). If most release can be from just one or a few sites along a terminal, this would greatly alter the interpretation of plastic changes in release efficacy. Hence we have sought to ascertain in our own preparation whether this is the case.

In these experiments (D'Alonzo and Grinnell, 1982, 1985, and unpublished), we have chosen to record with two intracellular electrodes, one at either end of a given terminal, in fibers sufficiently small that single quantum EPPs are 1.0 mV or more in amplitude. With quiet microelectrodes and an on-line computer to determine the amplitude of the response at each electrode by fitting a curve to 10 voltage readings taken just before, during, and after the peak of each response, it is possible to obtain ratios of the amplitudes of single quantum EPPs at the two electrodes that allow localization of the site of release to within about ± 10–20 μm. By recording a few thousand single quantum EPPs, and comparing the distribution of their release locations with the morphology of the terminal, it is possible to determine the relative probability of release from different portions of the terminal. In the several terminals studied to date, we have found no evidence that release deviates by more than 5–6% from uniformity. That is, if the fraction of the terminal's total branch length located within any 10–20 μm bin between the two electrodes is compared with the fraction of release coming from that same region, there is excellent correspondence between the two (D'Alonzo and Grinnell, 1982, 1985). Figure 12 shows an example of this. Although we cannot yet exclude the possibility that release is in certain cases very nonuniform, in the terminals we have studied this does not seem to be the case, suggesting that it is valid to consider the release from a given terminal as being distributed relatively uniformly throughout its branches. On the other hand, our results may not be inconsistent with those of Bennett and Lavidis (1979, 1982). To the extent that we see any deviation from uniformity, it suggests that the ends of branches often do release less than the proximal portions. If the extracellular electrodes of Bennett and Lavidis were recording from an area with a radius of 20–25 μm, in a region with several branches, it would probably be possible for much of a terminal's release to be detectable at that spot.

CONCLUSIONS

Frog neuromuscular junctions differ greatly in size, strength, and even number per fiber. Although we have explored only a few of the factors influencing these properties, it is clear that certain rules govern at least part of this variability. There are several clear forms of regulatory interaction between motoneurons and the fibers they innervate, dependent on the release efficacy of the terminal and feedback from the muscle fiber. Given a knowledge of the intrinsic properties of different motoneurons and the size of the motor units they innervate, it is now possible to make realistic predictions regarding muscle fiber size, terminal length and strength, the location of single or double endplates on a fiber, and the properties of different portions of their terminals. Moreover, the strength of nerve terminals can be altered for long periods of time in ways revealing both peripheral and central regulation of transmitter release properties. Morphological as well as physiological correlates of differing efficacy have been found. Thus some feeling for the dynamic equilibrium matching motoneurons and muscles is beginning to emerge.

However, to the extent we understand these regulatory relationships, it is in the form of empirical rules. We know only a few of these, and in none do we understand the mechanisms involved. It is important to emphasize a few of the many ways in which we do not understand the system. In the first place, we do not know the basis for intrinsic differences between motoneurons, or for differences between different terminals of the same axon. We need to know how these differences arise during development,

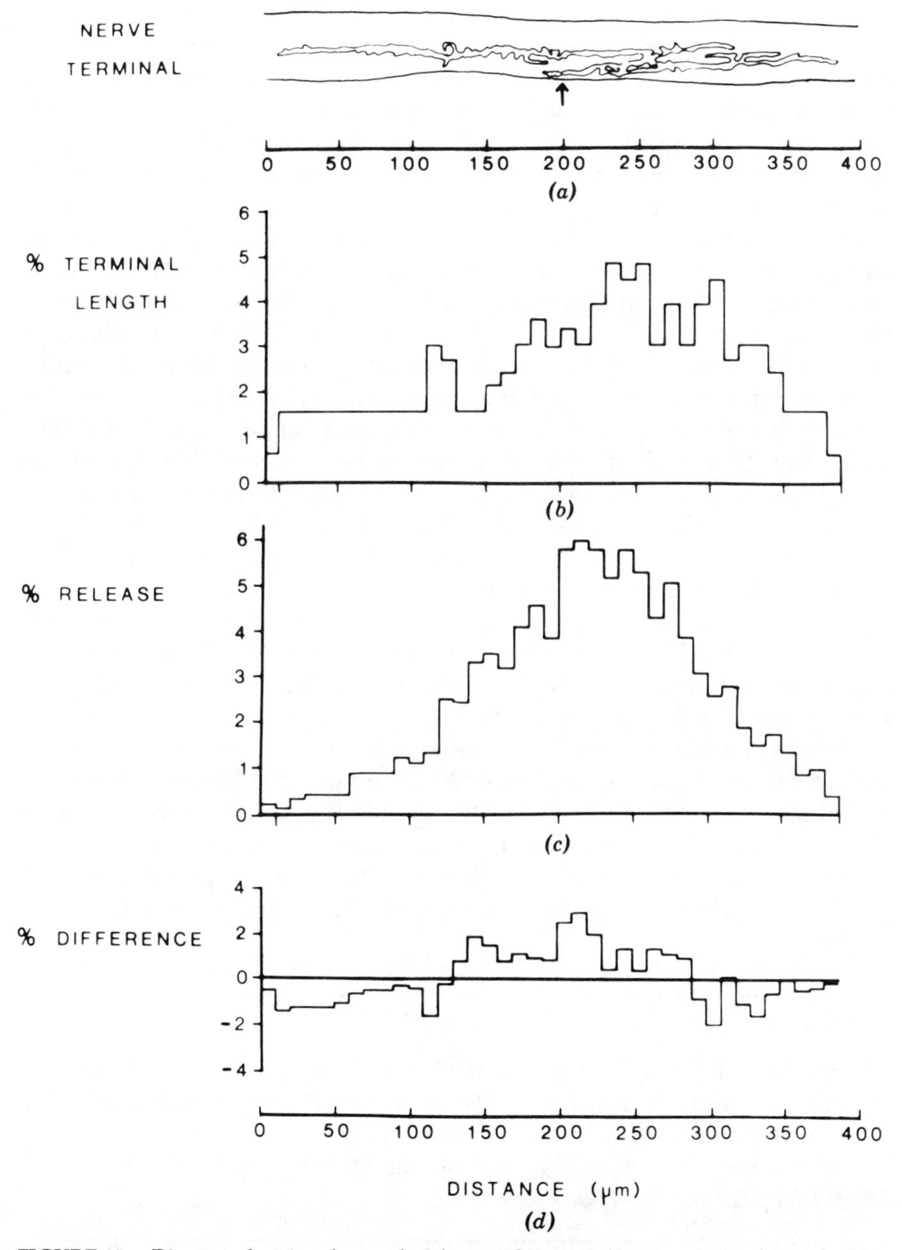

FIGURE 12. Diagram showing the results of a sample experiment measuring the uniformity of transmitter release from different portions of a single nerve terminal. In Ringer containing 0.3 m*M* Ca^{2+}, 2 m*M* Mg^{2+}, and 2 μ*M* neostigmine to block acetylcholinesterase, the amplitudes of single quantum EPPs were measured with intracellular electrodes (E_1 and E_2) located just beyond either end of the terminal. The ratio of EPP amplitudes at E_1 and E_2 is dependent on the distance of the postsynaptic event from the two electrodes. The distribution of these ratios is plotted in part C. Following the experiment, the terminal was stained with

and the mechanisms involved. We find larger active zones and higher chronic calcium influx in stronger terminals, but we do not know whether they are related, their relative importance in governing release, or what other morphological and physiological differences we might have missed. Terminal length appears to follow clear-cut rules independent of the motoneuron of origin or its intrinsic transmitter release capabilities. As a result, terminals of the same length may release very different amounts of transmitter. Why should terminal lengths be relatively closely adjusted for muscle fiber size, when release is not? If release efficacy can be sharply altered by enlarging active zones or changing calcium fluxes, what can be gained by having larger terminals that release less transmitter per unit length? Perhaps the answer lies in the efficacy of different terminals under conditions of normal activity, about which we know very little, but which almost certainly is not the regular 0.5–1/sec impulse frequency that we have used in most of these studies. Finally, in view of the apparent effectiveness of strong synaptic inputs in preventing or eliminating polyneuronal innervation in half sartorius muscles (Herrera and Grinnell, 1981) and the ability of strong single synapses in *Xenopus* pectoralis fibers to prevent dual innervation (Nudell and Grinnell, 1983), why is it that such a high percentage of c.p. fibers in mature frogs is polyneuronally innervated and that such junctions tend to be relatively strong and on large muscle fibers? It is clear that we do not understand fully the reasons for polyneuronal innervation or for synapse elimination in frogs.

Despite these and many more unknowns, we are encouraged to feel progress is being made in understanding the rules that govern the properties of motor nerve terminals and the muscle fibers they innervate. With the knowledge of these rules, mechanisms can be sought. And long experience has taught us that much of what is found at the frog neuromuscular junction will also prove to be true in most other synapses in most other animals.

Acknowledgments

We are grateful to Frances Knight, Bibbi Wolowske, Don Simpson, and Brad Smith for technical assistance. This research has been supported by research grants from NIH (NS06232) and the Muscular Dystrophy Association, and by postdoctoral fellowships from NIH (A. A. Herrera) and the Muscular Dystrophy Association (A. J. D'Alonzo and P. A. Pawson).

nitro-blue tetrazolium (Letinsky and DeCino, 1980) and its morphology carefully drawn (*a*). The arrow indicates the point where the preterminal axon first contacted the muscle. The amount of terminal present at different distances between the two electrodes is shown in (*b*). The relative frequency of quantal release in any given 10-μm-wide bin can then be compared with the percentage of the terminal located in the same 10-μm length of the fiber. If release is equally probable from all portions of the terminal, plots (*b*) and (*c*) should be identical. The extent of their deviation is shown in (*d*).

REFERENCES

Antony, M. T. and D. A. Tonge (1980) Effects of denervation and botulinum toxin on muscle sensitivity to acetylcholine and acceptance of foreign innervation in the frog. *J. Physiol.* **303**:23–31.

Bailey, C. H. and M. Chen (1983) Morphological basis of long-term habituation and sensitization in *Aplysia. Science* **220**:91–93.

Bennett, M. R. (1983) Development of neuromuscular synapses. *Physiol. Rev.* **63**:915–1048.

Bennett, M. R. and T. Florin (1974) A statistical analysis of the release of acetylcholine at newly formed synapses in skeletal muscle. *J. Physiol.* **238**:93–107.

Bennett, M. R. and N. A. Lavidis (1979) The effect of calcium ions on the secretion of quanta evoked by an impulse at nerve terminal release sites. *J. Gen. Physiol.* **74**:429–456.

Bennett, M. R. and N. A. Lavidis (1982) Variation in quantal secretion at different release sites along developing mature motor terminal branches. *Dev. Brain Res.* **5**:1–9.

Bennett, M. R. and A. G. Pettigrew (1975) The formation of synapses in amphibian striated muscle during development. *J. Physiol. (Lond.)* **252**:203–239.

Berg, D. K. (1982) Cell death in neuronal development: regulation by trophic factors. In *Neuronal Development*, N. C. Spitzer, ed. Plenum, New York, pp. 297–331.

Brown, M. C., R. L. Holland, and W. G. Hopkins (1981) Motor nerve sprouting. *Annu. Rev. Neurosci.* **4**:17–42.

Brown, M. C., J. K. S. Jansen, and D. Van Essen (1976) Polyneuronal innervation of skeletal muscle in new-born rats and its elimination during maturation. *J. Physiol. (Lond.)* **261**:387–422.

Burke, R. E. (1981) Motor units: anatomy, physiology, and functional organization. In *Handbook of Physiology. The Nervous System.* Sect. 1, Vol. 2, Part 1, Ch. 10, pp. 345–422, Am. Physiol. Soc., Bethesda, Md.

Carafoli, E. and M. Crompton (1978) The regulation of intracellular calcium by mitochondria. *Ann. N.Y. Acad. Sci.* **307**:269–284.

D'Alonzo, A. J. and A. D. Grinnell (1982) Uniformity of transmitter release along the length of frog motor nerve terminals. *Soc. Neurosci. Abstr.* **8**:493.

D'Alonzo, A. J. and A. D. Grinnell (1985) Profiles of evoked release along the length of frog motor nerve terminals. *J. Physiol.* **359**:235–258.

del Castillo, J. and G. Escalona de Motta (1978) A new method for excitation-contraction uncoupling in frog skeletal muscle. *J. Cell Biol.* **78**:782–784.

Dribin, L. B. and J. N. Barrett (1980) Conditioned medium enhances neuritic outgrowth from spinal cord explants. *Dev. Biol.* **74**:184–195.

Fukunaga, H., A. G. Engel, M. Osame, and E. H. Lambert (1982) Paucity and disorganization of presynaptic membrane active zones in the Lambert-Eaton myasthenic syndrome. *Muscle Nerve* **5**:686–697.

Govind, C. K. and R. G. Chiang (1979) Correlation between presynaptic dense bodies and transmitter output at lobster neuromuscular terminals by serial section electon microscopy. *Brain Res.* **161**:377–388.

Govind, C. K. and D. E. Meiss (1979) Quantitative comparison of low and high-output neuromuscular synapses from a motoneuron of the lobster (*Homarus americanus*). *Cell Tiss. Res.* **198**:455–463.

Grinnell, A. D. and A. A. Herrera (1980) Physiological regulation of synaptic effectiveness at frog neuromuscular junctions. *J. Physiol. (Lond.)* **307**:301–317.

Grinnell, A. D. and A. A. Herrera (1981) Specificity and plasticity of neuromuscular connections: Long-term regulation of motoneuron function. *Prog. Neurobiol.* **17**:203–282.

Grinnell, A. D., M. S. Letinsky, and M. B. Rheuben (1979) Competitive interaction between foreign nerves innervating frog skeletal muscle. *J. Physiol. (Lond.)* **289:**241–262.

Grinnell, A. D. and L. O. Trussell (1983) Synaptic strength as a function of motor unit size in the normal frog sartorius. *J. Physiol.* **338:**221–241.

Haimann, C., A. Mallart, J. Tomas i Ferre, and N. F. Zilber-Gachelin (1981) Patterns of motor innervation in the pectoral muscle of adult *Xenopus laevis:* Evidence for possible synaptic remodelling. *J. Physiol. (Lond.)* **310:**241–256.

Harris, J. B. and R. R. Ribchester (1979) The relationship between endplate size and transmitter release in normal and dystrophic muscles of the mouse. *J. Physiol. (Lond.)* **296:**245–265.

Herrera, A. A. (1984) Polyneuronal innervation and quantal transmitter release in formamide-treated frog sartorius muscles. *J. Physiol.* **355:**267–280.

Herrera, A. A. and A. D. Grinnell (1980) Transmitter release from frog motor terminals depends on motor unit size. *Nature* **287:**649–651.

Herrera, A. A. and A. D. Grinnell (1981) Contralateral denervation causes enhanced transmitter release from frog motor nerve terminals. *Nature* **291:**495–497.

Herrera, A. A. and A. D. Grinnell (1985) Effects of changes in motor unit size on transmitter release at the frog neuromuscular junction. *J. Neurosci.* (in press).

Herrera, A. A., A. D. Grinnell, and B. Wolowske (1985a) Ultrastructural correlates of naturally occurring differences in transmitter release efficacy in frog motor nerve terminals. *J. Neurocytol.* **14:**193–202.

Herrera, A. A., A. D. Grinnell, and B. Wolowske (1985b) Ultrastructural correlates of experimentally altered transmitter release efficacy in frog motor nerve terminals. *Neuroscience* (in press).

Heuser, J. E. and T. S. Reese (1973) Evidence for recycling of synaptic vesicle membrane during transmitter release at the frog neuromuscular junction. *J. Cell Biol.* **57:**315–344.

Heuser, J. E., T. S. Reese, M. J. Dennis, Y. Jan, L. Jan, and L. Evans (1979) Synaptic vesicle exocytosis captured by quick freezing and correlated with quantal transmitter release. *J. Cell Biol.* **81:**275–300.

Hoffman, H. (1951) A study of the factors influencing innervation of muscle by implanted nerves. *Aust. J. Exp. Biol. Med. Sci.* **29:**280–307.

Jansen, J. K. S., W. Thompson, and D. P. Kuffler (1978) The formation and maintenance of synaptic connection as illustrated by studies of the neuromuscular junction. *Prog. Brain Res.* **48:**3–18.

Katz, B. (1948) The electrical properties of the muscle fiber membrane. *Proc. R. Soc. (Lond.) B* **135:**506–534.

Katz, B. (1966) *Nerve, Muscle, and Synapse.* McGraw-Hill, New York.

Katz, B. (1969) *The Release of Neural Transmitter Substances.* Charles C Thomas, Springfield, Ill.

Katz, B. and S. W. Kuffler (1941) Multiple motor innervation of the frog's sartorius muscle. *J. Neurophysiol.* **4:**209–223.

Katz, B. and R. Miledi (1965) Propagation of electric activity in motor nerve terminals. *Proc. R. Soc. (Lond.) B* **161:**453–482.

Kuffler, D. P., W. Thompson, and J. K. S. Jansen (1980) The fate of foreign endplates in cross-innervated rat soleus muscle. *Proc. R. Soc. (Lond.) B* **208:**189–222.

Kuno, M., S. A. Turkanis, and J. N. Weakly (1971) Correlation between nerve terminal size and transmitter release at the neuromuscular junction of the frog. *J. Physiol. (Lond.)* **213:**545–556.

Letinsky, M. S. and P. DeCino (1980) Histological staining of pre- and postsynaptic components of amphibian neuromuscular junctions. *J. Neurocytol.* **9:**305–320.

Lev-Tov, A. and R. Rahamimoff (1980) A study of tetanic and post-tetanic potentiation of miniature end-plate potentials at the frog neuromuscular junction. *J. Physiol.* **309**:247–273.

Lømo, T. (1976) The role of activity in the control of membrane and contractile properties of skeletal muscle. In *Motor Innervation of Muscle,* S. Thesleff, ed. Academic Press, New York, pp. 289–321.

Luff, A. R. and U. Proske (1976) Properties of motor units of the frog sartorius muscle. *J. Physiol. (Lond.)* **258**:673–685.

Magleby, K. L. (1973) The effect of repetitive stimulation on facilitation of transmitter release at the frog neuromuscular junction. *J. Physiol.* **234**:327–352.

Magleby, K. L. and J. E. Zengel (1982) A quantitative description of stimulation-induced changes in transmitter release at the frog neuromuscular junction. *J. Gen. Physiol.* **80**:613–638.

Mark, R. F. (1980) Synaptic repression at neuromuscular junctions. *Physiol. Rev.* **60**:355–395.

Morrison-Graham, K. (1983a) An anatomical and electrophysiological study of synapse elimination at the developing frog neuromuscular junction. *Dev. Biol.* **99**:298–311.

Morrison-Graham, K. (1983b) Sprouting and regeneration of frog motoneurons during synapse elimination. *Dev. Biol.* **99**:312–317.

Nudell, B. M. and A. D. Grinnell (1982) Inverse relationship between transmitter release and terminal length in synapses on frog muscle fibers of uniform input resistance. *J. Neurosci.* **2**:216–224.

Nudell, B. M. and A. D. Grinnell (1983) Regulation of synaptic position, size, and strength in anuran skeletal muscle. *J. Neurosci.* **3**:161–176.

Pawson, P. and A. D. Grinnell (1984) Posttetanic potentiation in strong and weak neuromuscular junctions: physiological differences caused by a differential Ca^{2+}-influx. *Brain Research* **323**:311–315.

Pumplin, D. W., T. S. Reese, and R. Llinas (1981) Are the presynaptic particles the calcium channels? *Proc. Natl. Acad. Sci. U.S.A.* **78**:7210–7213.

Purves, D. (1976) The formation and maintenance of synaptic connections. In *Function and Formation of Neural Systems,* G. S. Stent, ed. Dahlem Konferenzen, Berlin, pp. 21–49.

Purves, D. and J. W. Lichtman (1980) Elimination of synapses in the developing nervous system. *Science* **210**:153–157.

Rahamimoff, R. (1968) A dual effect of calcium ions on neuromuscular facilitation. *J. Physiol.* **195**:471–480.

Rotshenker, S. (1979) Synapse formation in intact innervated cutaneous-pectoris muscles of the frog following denervation of the opposite muscle. *J. Physiol.* **292**:535–547.

Rotshenker, S. (1982) Transneuronal and peripheral mechanisms for the induction of motor neuron sprouting. *J. Neurosci.* **2**:1359–1368.

Rotshenker, S. and U. J. McMahan (1976) Altered patterns of innervation in frog muscles after denervation. *J. Neurocytol.* **5**:719–730.

Rotshenker, S. and F. Reichert (1980) Motor axon sprouting and site of synapse formation in intact innervated skeletal muscle of the frog. *J. Comp. Neurol.* **193**:413–422.

Sayers, H. and D. A. Tonge (1983) Persistence of extra-junctional sensitivity to acetylcholine after reinnervation by a foreign nerve in frog skeletal muscle. *J. Physiol.* **335**:569–575.

Trussell, L. O. (1983a) The regulation of synaptic strength in motor units of the frog. Ph.D. thesis, University of California, Los Angeles.

Trussell, L. O. (1983b) Synaptic properties of motor units in the frog cutaneous pectoralis muscle. *Abstr. Soc. Neurosci.* **9**:858.

Trussell, L. O. and A. D. Grinnell (1985) The regulation of synaptic strength within motor units of the frog. *J. Neurosci.* **5**:243–254.

Tweedle, C. D. (1983) Ultrastructural manifestations of increased hormone release in the neurohypophysis. *Prog. Brain Res.* **60**:259–272.

Tweedle, C. D. and G. I. Hatton (1982) Magnocellular neuropeptidergic terminals in neurohypophysis: Rapid glial release of enclosed axons during parturition. *Brain Res. Bull.* **8**:205–209.

Van Essen, D. C. (1982) Neuromuscular synapse elimination. In *Neuronal Development*, N. C. Spitzer, ed. Plenum, New York, pp. 333–376.

Weakly, J. N. and M. Yao (1983) Synaptic efficacy at singly and dually innervated neuromuscular junctions in the frog, *Rana pipiens. Brain Res.* **273**:319–323.

Wernig, A. (1975) Estimates of statistical release parameters from crayfish and frog neuromuscular junctions. *J. Physiol.* **244**:207–221.

Wernig, A., M. Peacot-Dechavassine, and H. Stover (1980) Sprouting and regression of the nerve at the frog neuromuscular junction in normal conditions and after prolonged paralysis with curare. *J. Neurocytol.* **9**:277–303.

Zengel, J. E. and K. L. Magleby (1982) Augmentation and facilitation of transmitter release. *J. Gen. Physiol.* **80**:583–611.

Chapter Four

CHANGES OF FORM AND FUNCTION IN REGENERATING NEURONS

MELVIN J. COHEN

Department of Biology
Yale University
New Haven, Connecticut

Differentiation of a mature neuron into the three general structual compartments of soma, dendrites, and axon is paralleled by a functional specialization of the associated membrane. This is reflected by the ability of different neuronal compartments to support either regenerative spikes or slow graded potentials. In my laboratory we have been concerned with factors that control the emergence of specified form in nerve cells, and whether changes in neuronal form may be correlated with alteration in functional properties as well. We are also interested in the stability of this structural and functional relationship in the various neuronal compartments during the lifetime of the animal.

We have chosen to investigate these questions of structural and functional differentiation in nerve cells by utilizing the changes evoked as a result of selective injury to specific neuronal regions. We have chosen preparations that enable us to look at these questions in identified giant central neurons. Here, the form of the neuron and its membrane properties can be monitored in single cells by intracellular recording and dye injection procedures. The data presented in this chapter derive from studies carried out on the giant interneurons in the terminal abdominal ganglion of the cricket *Achaeta domesticus* and from the giant reticulo-spinal neurons (Muller cells) in the brain of the larval lamprey, *Petromyzon marinus.*

When a neuron is injured, most commonly by axotomy, there is first a partial retrograde degeneration in the proximal axonal stump and then an initial sprouting from that region (Cajal, 1928). The induced neuritic outgrowth emerges primarily from the remnant of the proximal axon and not from the other structural compartments of the neuron. Why does the induced regenerative growth occur in the proximal axonal stump adjacent to the site of injury? Can other regions of the cell, such as soma or dendrites, be induced to sprout as well in these differentiated nerve cells?

THE SITE OF INJURY-INDUCED NEURITIC OUTGROWTH IN CRICKET INTERNEURONS

Growth from Axon and Soma

If the axon of a medial giant interneuron (MGI) in the cricket is cut in the ventral nerve cord at a distance greater than 1 mm from its cell body in the terminal abdominal ganglion, then within a matter of hours sprouts will emerge from the proximal axonal stump (Roederer and Cohen, 1983a). The sprouts remain confined to the lesion area and form a neuroma (Fig. 1*a–c*). If on the other hand, the giant axon is crushed in the connective rather than cut, it will sprout from the axon, grow across the lesion and continue regenerating through several ganglia of the ventral cord (Fig. 1*d*).

At approximately one month after cutting the axon, sprouting from the cut end in the neuroma stops. This is followed by neurite-like outgrowth emerging from the soma of these giant cells as seen in Figure 2. Many invertebrate central neurons, and the MGI in particular, are known to be monopolar with only a single initial neurite emerging from the cell body (see Fig. 3). This sprouting from the soma was therefore a somewhat startling finding. Under these conditions, the normally monopolar neuron soma of the insect assumed a form reminiscent of the soma-dendritic complex of the vertebrate central neuron. This was our first indication that the growth induced by injury to a neuron is not necessarily localized to the site of injury. If, for some reason, growth from the injured compartment stops, the new membrane may be inserted into other regions of the neuron, to produce sprouts there. In the case mentioned previously, there is a shift in growth from the injured axon to neurites that emerge from the normally smooth cell body. Both the form of the usually stable dendritic tree (Roederer and Cohen, 1983a), and the electrical properties of the soma membrane (Roederer and Cohen, 1983b), remain unchanged during these responses to distant axotomy.

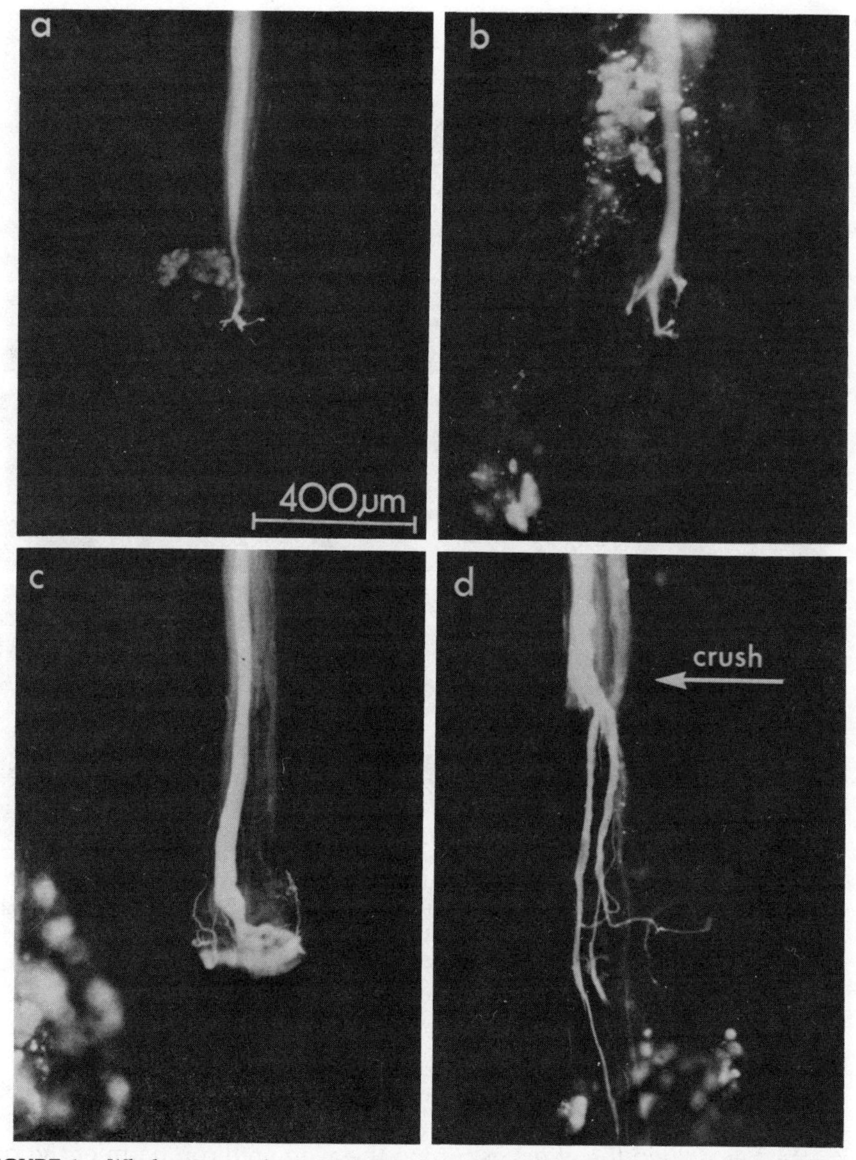

FIGURE 1. Wholemounts of proximal axonal segments from cricket giant interneurons in the connective between the last two abdominal ganglia stained by intracellular injection of Lucifer Yellow. (*a*) Sprouts emerging from the cut end of a MGI six days after transection. (*b*) MGI axon 10 days after cut. (*c*) End of cut MGI axon 15 days after section. (*d*) Fills of MGI axon and another giant interneuron 47 days after crushing the connective where indicated by the arrow. Note that after a crush, the regenerating neurites have crossed the lesion to enter the next rostral abdominal ganglion where they form branches. Calibration is the same for the entire figure. (From Roederer and Cohen, 1983.)

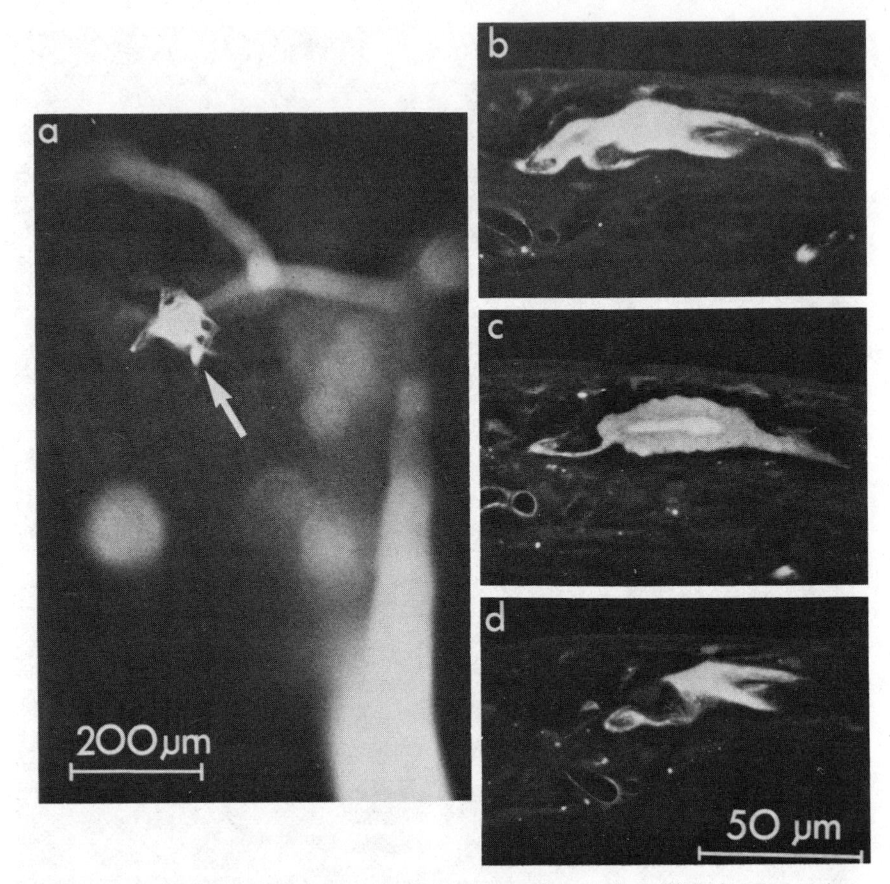

FIGURE 2. (*a*) Wholemount of the last abdominal ganglion in the cricket showing an interneuron filled with Lucifer Yellow 47 days after distant axotomy. The arrow points to the cell body which shows several neuritic sprouts. (*b–d*) Serial sections through the soma in (*a*) viewed with epifluorescence optics showing the outgrowth of neurites from the soma in greater detail. (From Roederer and Cohen, 1983a.)

Induced Dendritic Growth in the Cricket

We next examined the effect of cutting the axon within 200 μm of the cell body. The proximal axonal remnant dies back to the cell body and frequently disappears entirely. Within two days after close axotomy, sprouts begin to emerge from the normally stable dendritic tree as shown in Figure 3 (Roederer and Cohen, 1983a). The new sprouts originating from the old dendritic tree are readily seen because the branching pattern of the MGI is normally stable and well characterized after development (Mendenhall and Murphey, 1974).

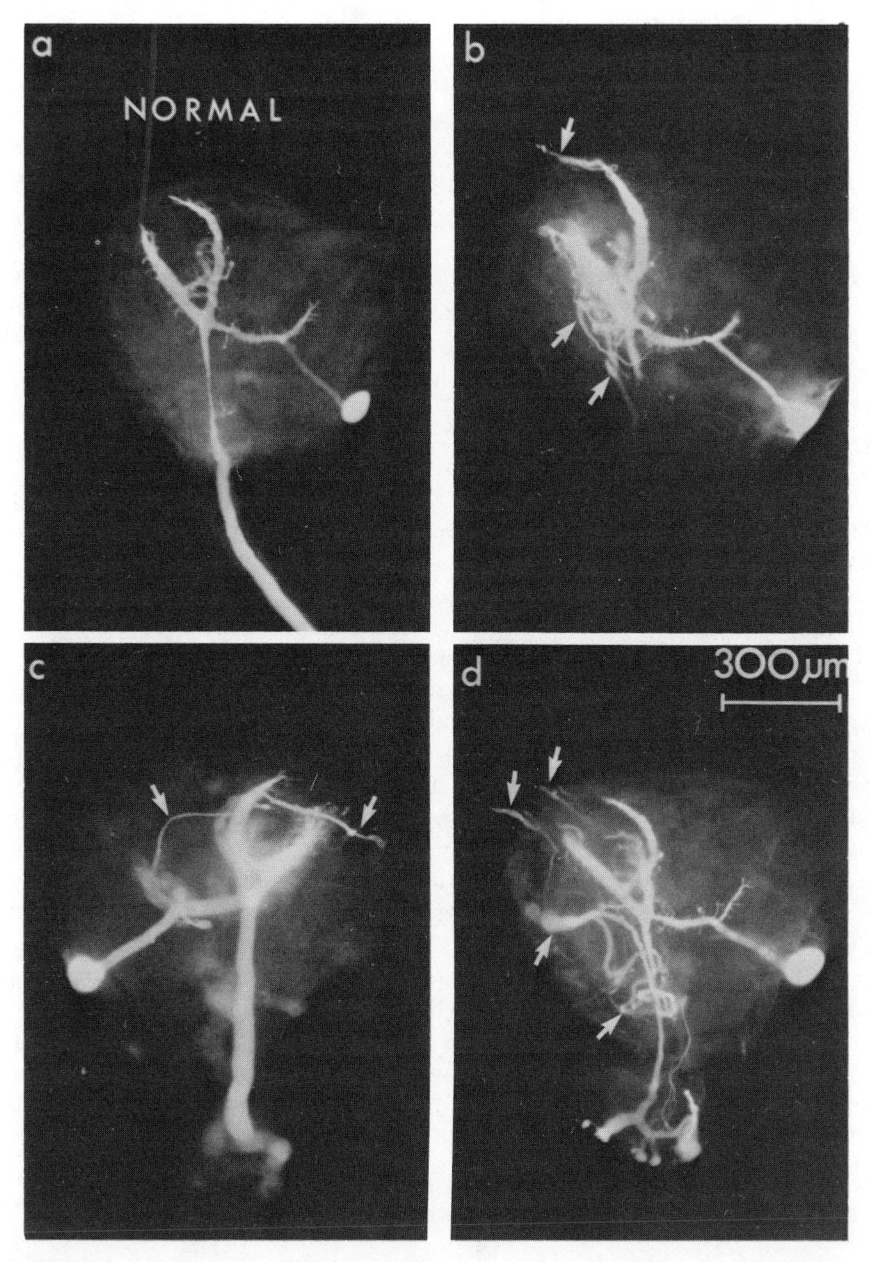

FIGURE 3. Wholemounts of the cricket terminal abdominal ganglion showing MGIs injected with Lucifer Yellow. (*a*) An intact normal neuron with the dendritic arborization characteristic of this cell. (*b–d*) Examples of sprouting from the dendritic tree induced by axotomy close to the ganglion at 6, 10, and 14 days. (From Roederer and Cohen, 1983a.)

Soma Excitability Following Close Axotomy

The soma membrane of the MGI in the normal adult cricket is electrically inexcitable (Murphey, 1973). This is also true during regenerative growth caused by distant axotomy, as mentioned before. However, within six hours after close axotomy, the soma is capable of generating action potentials (Roederer and Cohen, 1983b). Ion substitution studies indicate that these spikes are sodium dependent. By 48 hours after close axotomy, the cell bodies of the close axotomized interneurons have lost this excitability and revert once again to the inexcitable state characteristic of the normal neuron. This transient appearance of electrical excitability appears within a few hours after close axotomy and is gone by two days, just as new sprouts begin to emerge from the old dendritic tree.

Neuritic Outgrowth Induced by Injury in the Lamprey

The giant reticulo-spinal neurons (Muller cells) in the brain of the lamprey have long been known to regenerate at least part of their axons when they are severed in the spinal cord (Maron, 1959; Rovainen, 1976; Wood and Cohen, 1981). Here, as in the giant interneurons of the cricket just discussed, injury to the axons in the spinal cord several millimeters from the cell bodies in the brain results in regenerative sprouting which is localized primarily to the proximal axonal stump. There is little effect on the dendritic tree with this form of distant axotomy (Wood and Cohen, 1981). However, if the axons of these cells are cut in the brain within 500 μm of their cell bodies, there is profuse sprouting from the old dendritic tree by two weeks after injury, as seen in Figure 4 (Hall and Cohen, 1983). As with close axotomy in the cricket, the proximal axonal stump dies back completely in most instances, and there is no definitive regenerative growth at the site where the axon normally originates from the soma. The neuritic outgrowth from the dendrites proceeds in a linear manner at a mean rate of approximately 100 μm per day up to 80 days after injury. Axonal lesions at an intermediate distance from the soma (1–1.4 mm) result in sprouting from both the proximal axonal stump and the dendritic tree. The total sprout length per cell is the same whether it all originates from the dendrites, or whether it arises from both dendritic and axonal compartments (Table 1).

Effects of Dendrotomy

The anterior group of bulbar Muller cells have large cell bodies that can be seen in the living animal when the cranium is opened to expose the walls of the IV ventricle. The major dendritic trunk of these cells projects laterally (see Fig. 4). A substantial part of the dendritic tree may be separated from its cell body by an incision into the wall of the IV ventricle just lateral to the

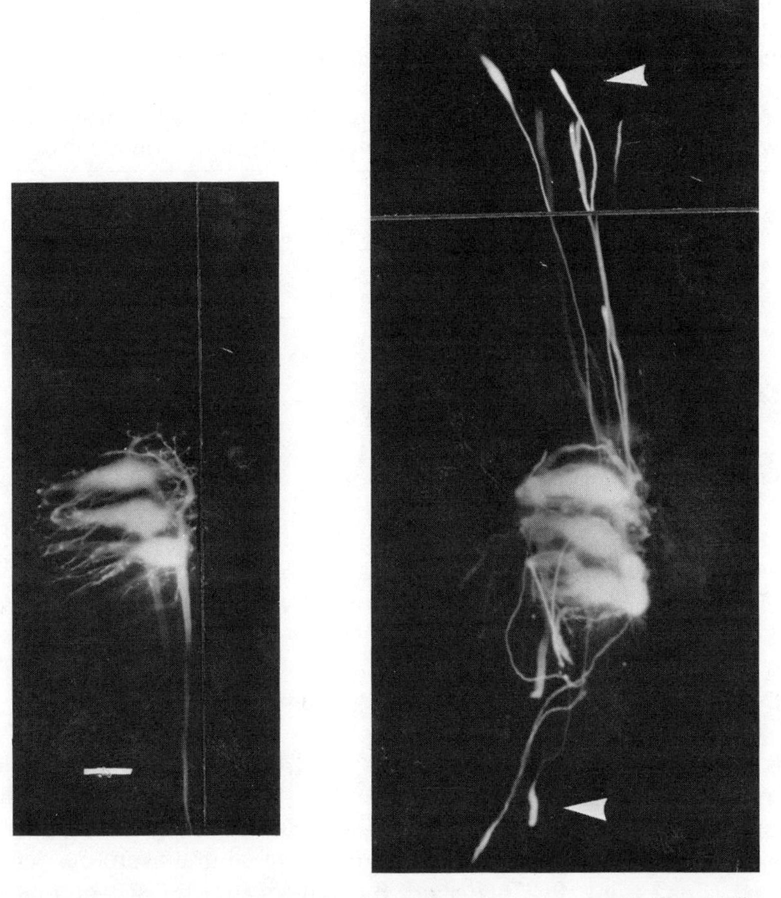

FIGURE 4. Wholemounts of the lamprey brain showing three normal bulbar Muller cells injected with Lucifer Yellow (left). Note that normal cells have a dendritic domain that is restricted to a region close to the cell bodies. On the right is a wholemount from another animal showing three similar neurons whose axons were cut close to the cell bodies 49 days previously. Note the lengthy sprouts extending from the old dendritic tree. Arrows point to swollen tips of some of the sprouts. Scale bar, 100 μm. (From Hall and Cohen, 1983.)

cell bodies. The cranium is then closed and the effects of dendrotomy examined from days to months later by reopening the skull and staining the cells with intracellular injection of the fluorescent dye, Lucifer Yellow. Much to our surprise, cells dendrotomized in this manner showed no neuritic outgrowth and no obvious degenerative changes when examined at various times after dendrotomy. We had assumed that whatever signals were induced by axotomy close to the soma would be further enhanced by an injury even closer to the cell body in the main dendritic trunk, but this was not so. The dendrotomized neurons simply remain morphologically

TABLE 1. A comparison of the amount of neuritic sprouting from the dendritic tree of giant bulbar neurons in the lamprey in cells subjected to axotomy close to the soma (200–500 μm) versus cells that underwent intermediate axotomy (1000–1400 μm from the soma). The total sprout length per cell was essentially the same despite the different apportionment of new growth between axonal and dendritic compartments in the two groups.

	AXONAL SPROUTS	DENDRITIC SPROUTS	TOTAL SPROUTS	N
	(MEAN SUMMED SPROUTS PER CELL IN μm)			
AXOTOMY 200-500μm FROM SOMA	463* (12%)	3401 (88%)	3864 (100%)	21 (13 animals)
AXOTOMY 1000-1400 μm FROM SOMA	2798 (72%)	1108 (28%)	3906 (100%)	27 (9 animals)
SIGNIFICANCE	p<01	p<01	NOT SIGNIFICANT	

*All sprouts arising from the periaxonal region were counted as axonal sprouts. (Modified from Hall and Cohen, 1983.)

stable and do not show new neuritic outgrowth anywhere in the cell (Hall and Cohen, 1984).

Dendrotomy Preceded by Distant Axotomy

Bulbar interneurons were subjected to dendrotomy after first having their axons sectioned distally in the spinal cord 15–20 days previously. As stated before, distant axotomy alone in these neurons results in regenerative sprouting only from the proximal stump of the injured axon; the dendritic tree remains essentially unaffected. By 15 days after the second operation (dendrotomy), there was clear evidence of neuritic outgrowth from the remainder of the old dendritic tree (Hall and Cohen, 1984). Therefore, for dendrotomy to induce new growth from the remainder of the dendritic tree, it must be preceded by interruption of the axon.

FACTORS GOVERNING NEURITIC SPROUTING

Two Signals Evoked by Neuron Injury

There appear to be at least two different factors accounting for the response of a neuron to injury. The first involves the integrity of the axon. With the variety of surgical lesions used in the studies previously described, the integrity of the axon must be breached before a net new outgrowth of neurites occurs anywhere in the cell. This axotomy signal turns on metabolic pathways in the soma related to chromatolytic changes shown, in many neurons, to involve increases in RNA and protein synthesis (Grafstein, 1975; Grafstein and McQuarrie, 1978). A second factor may act locally immediately adjacent to the lesion and determine where the newly synthesized membrane is inserted. The two signals could be quite distinct: The axon factor reaches the soma and acts on either cytoplasmic or nuclear components to increase membrane synthesis, that is, it turns on growth. The signal local to the lesion acts near the injury site on elements of the cytoskeleton or plasma membrane and makes that region receptive to the insertion of new membrane. This second factor then determines where in the neuron remnant the regenerative growth will occur.

Evidence for the Two-Signal Hypothesis

A key finding supporting the two-signal hypothesis is the observation that, in the lamprey, dendrotomy alone does not evoke growth anywhere in the remainder of the neuron. The cell simply remains morphologically stable following this operation. However, if the axon is first cut distally in the cord and the dendrotomy done several days later, then neuritic outgrowth occurs from the remnant of the dendritic tree as well as from the proximal axon stump (Hall and Cohen, 1984). Apparently, dendrotomy alone induces the local changes in the injured dendrites that makes them receptive to the insertion of newly synthesized membrane. However, in the absence of axotomy, growth has not been turned on in the cell and there is no new membrane available to be inserted into the receptive areas of the injured dendrites. An initial distant axotomy in the cord turns on membrane synthesis in the soma, so that when the dendrites are subsequently injured and become receptive to the insertion of new membrane, that membrane is available, and dendritic sprouting occurs.

Axotomy close to the soma in both the cricket and the lamprey causes prolific sprouting from the dendritic tree. The interpretation of these findings is that both signals are evoked in the close as well as the distant lesions. But with close axotomy, the more localized factor is near enough to the dendritic tree to exert its effect, and the dendrites now become receptive to the newly synthesized membrane induced by breaching the integrity of the axon.

Shunting of Membrane to Available Growth Sites

Another element of close axotomy favoring outgrowth from the dendrites is the dying back of the proximal axonal stump such that it frequently disappears completely. When this occurs there is no clear regenerative axonal growth from the site where the original axon emerged from the soma. This locus is apparently no longer receptive to new membrane, and the potential for regenerating the axon as it originally formed in the differentiating neuron is lost. The newly synthesized membrane is therefore shunted into the remaining receptive region of the neuron, the dendritic remnant, where it now gives rise to new neurites. With either close or intermediate axotomy, if a short segment of the axonal stump remains, then there is an apportionment of new growth between the dendritic tree and the axon stump (Roederer and Cohen, 1983a; Hall and Cohen, 1983). This implies that an intermediate lesion is still close enough to the dendritic tree that the local signal for membrane acceptance can exert its effect on the dendrites and make them receptive to the newly synthesized membrane. At the same time, the remnant of the axonal stump still maintains the axon as a viable target for new membrane, and the local factor exerts its influence to cause sprouting from there as well.

The evidence that a fixed amount of newly produced membrane is shunted into whichever receptive sites are available comes from both the cricket and lamprey studies. With close or intermediate lesions, where some remnant of the proximal axonal stump remains, it is apparent in the cricket that there is a reciprocal relationship in the amount of new sprouting occurring in the axonal and dendritic compartments (Roederer and Cohen, 1983a). The appearance of outgrowth from the soma in the cricket when the neuromalike growth at the end of the proximal axonal stump has stopped also fits in with the idea that membrane is shunted into alternate sites if the primary growth loci are no longer receptive. It remains a puzzle in this instance why growth emerges from the soma rather than the dendrites. The soma sprouting after distant axotomy also seems not to fit the "local signal hypothesis," because the lesion is at a long distance from the soma. However, as seen in Figure 2, the processes emerging from the cell body appear different from regenerative growth induced either from the axon or the dendrites. The soma processes are larger in diameter and have the appearance of extensions of the cell cytoplasm more in the form of pseudopodia than of neurites. There may be yet another mechanism operating here.

The strongest case for an apportionment of a constant amount of newly produced membrane to whatever sites in the cell will incorporate it into neuritic sprouts comes from the lamprey. The mean length of new sprouts per cell was the same in animals induced to sprout by intermediate axotomy as in those caused to grow by close axotomy. This was so, despite the fact that most sprouting occurred from the dendritic tree with the close lesions, while a significant amount of sprouting occurred from both the axon and dendrites following intermediate axotomy as shown in Table 1 (Hall and Cohen, 1983).

The transient appearance of sodium spikes in the soma of the cricket interneuron within six hours after close axotomy might also be explained by a shunting of newly synthesized membrane to a target not normally receptive to it. Close axotomy is frequently associated with a complete dying back of the proximal axonal stump, thereby removing it as the normal recipient for the newly synthesized membrane. This membrane, presumably containing sodium channels in numbers sufficient to support regenerative electrical activity, then could be inserted into the soma membrane and be responsible for that membrane transiently being capable of spiking. By 48 hours after close axotomy, when sprouting begins from the dendritic tree, the ability of the soma to support spikes has disappeared. This could be correlated with a shunting of the continually synthesized membrane (with its sodium channels) into the newly receptive target, the dendritic tree.

INJURY CURRENT AS A LOCAL SIGNAL FOR INITIAL DIE-BACK AND SUBSEQUENT GROWTH

The Injury Current

In considering possible candidates for the local signal that causes existing plasma membrane to incorporate the newly synthesized membrane into sprouts, the ionic fluxes associated with the injury current seem a likely possibility. We have shown that an electric current approaching 1 mA/cm^2 enters the cut face of the lamprey spinal cord immediately after injury (Borgens, Jaffe, and Cohen, 1980). Over the next two days, this current drops to approximately 4 $\mu A/cm^2$ and then remains relatively constant for the balance of the six-day measuring period. The current tends to peak opposite the visible cut ends of the giant reticulospinal axons in the cord (Fig. 5). Ion substitution experiments showed that approximately one-half of the current is carried by sodium ions, and much of the remainder is due to an influx of calcium ions.

Retrograde Degeneration

The initial massive influx of cations, especially calcium, into the cut ends of axons in the lamprey spinal cord has been proposed as a causal agent in the retrograde axonal degeneration (die-back) seen in this preparation following spinal transection (Roederer, Goldberg, and Cohen, 1983). The role of intracellular calcium in the assembly and disassembly of fibrous cytoskeletal elements is well documented. High levels of intracellular calcium break down microtubules in axons (Kirschner and Williams, 1974). An elevated intracellular calcium level is also related to the breakdown of neurofilaments by activation of a calcium-dependent protease (Pant, Terakawa, and Gainer, 1979). The breakdown of cytoskeletal elements underlying Wallerian degeneration in isolated axonal segments has also been related to elevated intracellular calcium levels (Schlaepfer, 1974; 1977).

We have shown that the retrograde die-back in severed giant reticulospinal axons in the lamprey cord may reach 2 mm by five days after injury.

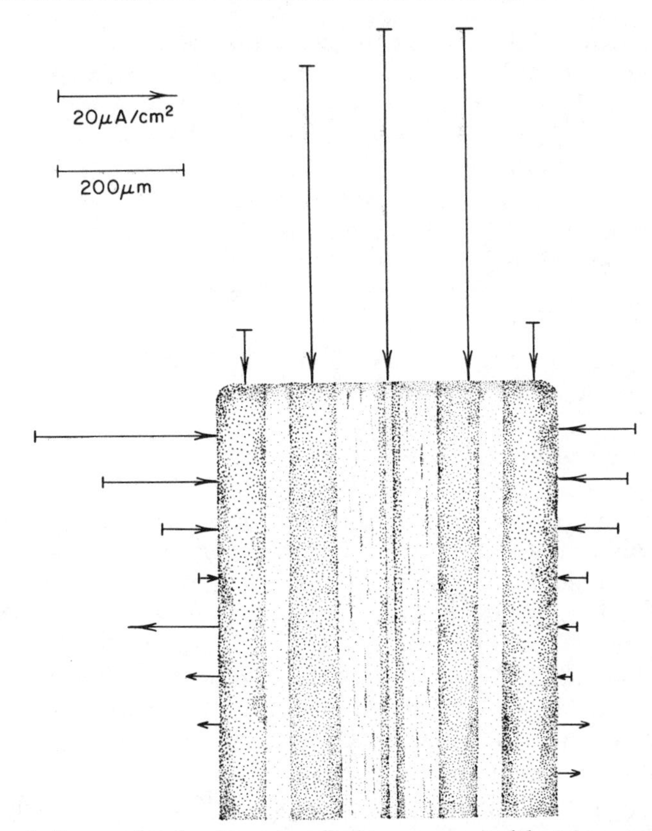

FIGURE 5. A diagram showing the perpendicular component of the injury current entering or leaving the cut end of the lamprey spinal cord four hours after transection. The lengths of the arrows are proportional to the density of current. The giant axons are indicated as light longitudinal bands. (From Borgens, Jaffe, and Cohen, 1981.)

The extent of this die-back readily accounts for the total breakdown of the proximal axonal stump in the lamprey following axotomy within 0.5 mm of the soma (close axotomy). We can substantially reduce this die-back by applying an external electric field that counteracts the endogenous inward-flowing injury current (Roederer, Goldberg, and Cohen, 1983). Such an applied current would reduce the inward flux of cations and thus reduce the cytoskeletal destruction giving rise to the degeneration, as shown in the proposed model seen in Figure 6. Subsequent work has indeed confirmed that the die-back of severed giant axons in the lamprey is associated with major disruption of the cytoskeleton (McHale and Cohen, 1983). The reduction of die-back following application of an imposed electric current may also be partly responsible for the increased rate of regeneration seen when electric fields are applied to the lamprey spinal cord.

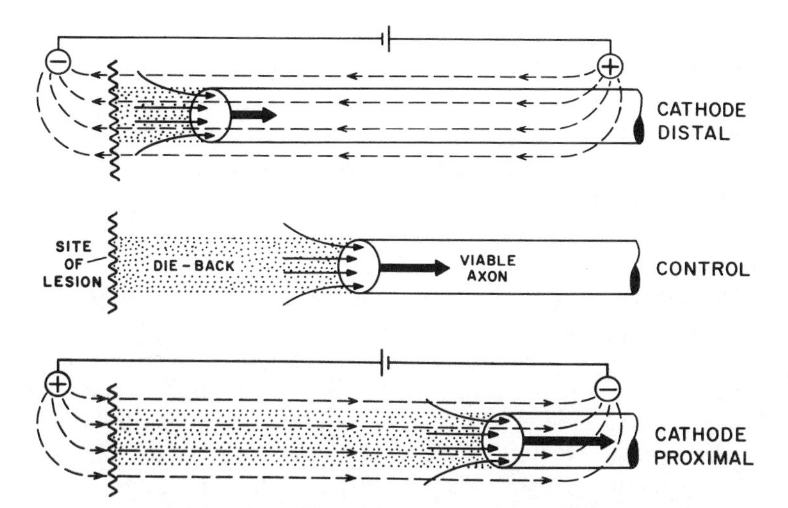

FIGURE 6. A proposed model indicating the possible relationship between axonal die-back and current entering the cut end of a single giant axon. The net inward current (single broad solid arrow) results in the die-back seen in each of the three treatment groups. In the control, only the endogenous DC current induced by the injury (solid thin arrows) enters the axon to cause the intermediate extent of die-back seen in this group. With the cathode located proximal to the lesion, the applied DC current (dashed lines) sums with the injury current to produce a net increase in the current entering the cut axon. This results in an increased length of die-back as compared to the controls. With the cathode distal to the lesion, the applied current opposes the endogenous injury current to yield a decrease in the net inward current and a resulting reduction in axonal die-back. (From Roederer, Goldberg, and Cohen, 1983.)

Ionic Influx and Growth

The dying back of severed giant axons in the lamprey cord takes place over a five-day period (Roederer, Goldberg, and Cohen, 1983). The reversal of this process by seven days after the lesion is signaled by the appearance of regenerative sprouts which indicate the onset of axonal regeneration (McHale and Cohen, 1983). This occurs at a time when the injury current has dropped to a relatively low level, such that the continual leakage of cations into the cut axon end may now facilitate membrane insertion and result in regenerative neuritic outgrowth.

There is ample evidence that steady ionic currents are associated with growth in the nervous system (see Borgens, 1982, for review). This is particularly true for an inward flux of calcium ions. Meiri et al. (1982) have reported that regenerating giant axons in the cockroach central nervous system show an increased conductance to calcium. This is accompanied by a shift from sodium-dependent action potentials to action potentials based on a calcium influx in the region of the axon injury. The sprouting of neuroblastoma cells in culture has also been correlated with an inward flux of calcium ions (Anglister et al., 1982). Llinas and Sugamori (1979) have reported an increased calcium conductance when dendrites begin to emerge from developing Purkinje cells in the cerebellum. Based on this observation, they have suggested that calcium is involved in adding new membrane to the ends of growing neurites. They point out the similarity between membrane incorporation in neuron growth cones and presynaptic terminals, drawing attention to the involvement of calcium in both these instances.

There is an intriguing but elusive area of inquiry involving the possible longer-ranging roles played by the ions active in the transient electrophysiological phenomena of neurons. In other tissue, some of these ions, calcium in particular, seem implicated in a wide variety of phenomena involving fertilization, cell elongation, movement, and cell division (Jaffe, 1981). From the foregoing evidence, it seems likely that the calcium influx associated with the injury current may be involved in the degenerative and regenerative responses evoked by traumatic breaching of the neuron plasma membrane. It is tempting to speculate that similar mechanisms may be involved in the development of neuronal form. This would be reflected in the interplay of membrane retrieval and insertion that underlies the development of the embryonic nervous system, as well as the dynamic stability of the mature neural state.

Acknowledgments

Supported by NIH Spinal Trauma Grant 2P50 NS 10174-12 and NIH Physiology Training Grant GM 27527.

REFERENCES

Anglister, L., I. C. Farber, A. Shakar, and A. Grinvald (1982) Localization of voltage-sensitive calcium channels along developing neurites: Their possible role in regulating neurite elongation. *Dev. Biol.* **94**:351–365.

Borgens, R. B. (1982) What is the role of naturally produced electric current in vertebrate regeneration and healing? *Int. Rev. Cytol.* **76**:245–298.

Borgens, R. B., L. F. Jaffe, and M. J. Cohen (1980) Large and persistent electrical currents enter the transected lamprey spinal cord. *Proc. Natl. Acad. Sci. U.S.A.* **213**:611–617.

Cajal, S. Ramon y (1928) *Degeneration and Regeneration of the Nervous system*, Vol. 2. Oxford University Press, London, pp. 487–516.

Grafstein, B. (1975) The nerve cell body response to axotomy. *Exp. Neurol.* **48**:32–51.

Grafstein, B. and I. G. McQuarrie (1978) Role of the nerve cell body in axonal regeneration. In *Neuronal Plasticity*, C. W. Cotman, ed. Raven Press, New York, pp. 155–195.

Hall, G. F. and M. J. Cohen (1983) Extensive dendritic sprouting induced by close axotomy of central neurons in the lamprey. *Science* **222**:518–521.

Hall, G. F. and M. J. Cohen (1984) Dendritic injury evokes dendritic sprouting only in axotomized lamprey central neurons. *Abstr. Soc. Neurosci.* **10**:1018.

Jaffe, L. F. (1981) The role of ion currents in establishing developmental gradients. In *International Cell Biology*, H. G. Schweiger, ed. Springer-Verlag, New York, pp. 507–511.

Kirshner, M. W. and R. C. Williams (1974) The mechanism of microtubule assembly *in vitro. J. Supra. Mol. Struct.* **2:**412–428.

Llinas, R. and M. Sugamori (1979) Calcium conductances in Purkinje cell dendrites: their role in development and integration. In *Developmemnt and Chemical Specificity of Neurons, Progress in Brain Research,* Vol. 51, M. Cuenod, G. W. Kreutzberg, and F. E. Bloom, eds. Elsevier/North-Holland Biomedical Press, Amsterdam, pp. 323–334.

Maron, K. (1959) Regeneration capacity of the spinal cord in *Lamprey fluviatilis* larvae. *Folia biol. (Buenos Aires)* **7:**179–189.

McHale, M. K., and M. J. Cohen (1983) Early degenerative and regenerative responses in transected spinal axons of the lamprey. *Abstr. Soc. Neurosci.* **9:**982.

Meiri, H., I. Parnas, and M. Spira (1981) Membrane conductances and action potential of a regenerating axon tip. *Science* **211:**709–711.

Mendenhall, B. and R. K. Murphey (1974) The morphology of cricket giant interneurons. *J. Neurobiol.* **5:**565–580.

Murphey, R. K. (1973) Characterization of an insect neuron which cannot be visualized *in situ.* In *Intracellular Staining in Neurobiology,* S. B. Kater and C. Nicholson, eds. Springer-Verlag, New York, pp. 135–150.

Pant, H. C., S. Terakawa, and H. Gainer (1979) A calcium activated protease in squid axoplasm. *J. Neurochem.* **32:**99–102.

Roederer, E. and M. J. Cohen (1983a) Regeneration of an identified central neuron in the cricket: I. Control of sprouting from soma, dendrites, and axon. *J. Neurosci.* **3:**1835–1847.

Roederer, E. and M. J. Cohen (1983b) Regeneration of an identified central neuron in the cricket: II. Electrical and morphological responses of the soma. *J. Neurosci.* **3:**1848–1859.

Roederer, E., N. H. Goldberg, and M. J. Cohen (1983) Modification of retrograde degeneration in transected spinal axons of the lamprey by applied DC current. *J. Neurosci.* **3:**153–160.

Rovainen, C. M. (1976) Regeneration of Muller and Mauthner axons after spinal transection in larval lampreys. *J. Comp. Neurol.* **168:**545–554.

Schlaepfer, W. W. (1974) Calcium-induced degeneration of axoplasm in isolated segments of rat peripheral nerve. *Brain Res.* **69:**203–215.

Schlaepfer, W. W. (1977) Structural alterations of peripheral nerve induced by the calcium ionophore A23187. *Brain Res.* **136:**1–9.

Wood, M. R. and M. J. Cohen (1981) Synaptic regeneration and glial reactions in the transected spinal cord of the lamprey. *J Neurocytol.* **10:**57–79.

Chapter Five

AUDITORY EXPERIENCE INFLUENCES THE DEVELOPMENT OF SOUND LOCALIZATION AND SPACE CODING IN THE AUDITORY SYSTEM

ERIC I. KNUDSEN

Department of Neurobiology
Stanford University School of Medicine
Stanford, California

Sound localization is based on the association of sets of auditory cues with locations in space. A number of cues contribute to the percept of sound location, including binaural differences in timing and intensity and monaural spectrum cues. How does the auditory system come to associate these various cues with appropriate locations in space?

The problem faced by the auditory system is difficult because the relationship between auditory cues and sound locations is complex. All localization cues are frequency dependent. For example, a given interaural intensity difference will correspond to very different locations in space depending on the frequency at which the difference is measured. Moreover, different cues change at different rates and along different spatial axes: some cues change with the azimuth (left-right location) of the sound source while others change with both the azimuth and elevation (up-down location) of the source. The problem of sound localization is complicated further by the fact that the correspondence between auditory cues and locations in space varies during the lifetime of an animal. The size and shape of the head and ears are the major factors in determining sound localization cues, and as an animal grows it therefore experiences new and changing localization cues.

Because of the complexity of sound localization cues and their variation over time (and between individuals), it seems unlikely that the neural circuitry that underlies sound localization is established in a highly predetermined fashion. A more plausible hypothesis is that sensory experience plays an important role in directing the formation of appropriate connections. The experiments discussed in this chapter demonstrate that this is the case. Specifically, they show that both sound localization behavior and a neural representation of auditory space are shaped by early experience.

EXPERIMENTAL PARADIGM

The animal used in these experiments was the barn owl. The barn owl is a good model for studying sound localization because its range of audible frequencies approximates that of humans, its external ears are immobile, and, most important, its sound localization ability is unsurpassed in the animal kingdom (Payne, 1971; Konishi, 1973; Knudsen et al., 1979). The technique used to manipulate auditory experience was monaural occlusion. An ear plug attenuates and alters the timing of sounds reaching the ear drum, thereby changing the correlations between auditory cues and locations in space (Knudsen et al., 1984a). As a consequence, the perceived location of a sound source shifts toward the side of the unplugged ear (i.e., a sound localization error is induced). In barn owls, because of a vertical asymmetry in the directional sensitivity of their ears (Payne, 1971; Knudsen, 1980), a plug in the left ear causes a localization error to the right and above the sound source, while a plug in the right ear induces an error to the left and below the source (Knudsen and Konishi, 1979).

Sound localization accuracy was assessed by comparing the way in which owls oriented their heads toward auditory and visual stimuli (Knudsen et al., 1982; 1984a). Normally, an owl orients identically to both types of stimuli. However, an owl that has recently had one ear plugged orients differently to sounds as opposed to lights, and this difference was assumed to represent an error in sound localization. Based on the performance of control owls, localization errors greater than 3° in azimuth or elevation were considered abnormal.

In the experiments that follow, barn owls were raised with one ear chronically occluded. Foam rubber ear plugs were sutured into the external meatus of birds ranging in age from 27 to 186 days (posthatching). A 27-day-old bird is still quite immature: the skull and ear canals have not reached full size, and the feathers that will form the sound-collecting sur-

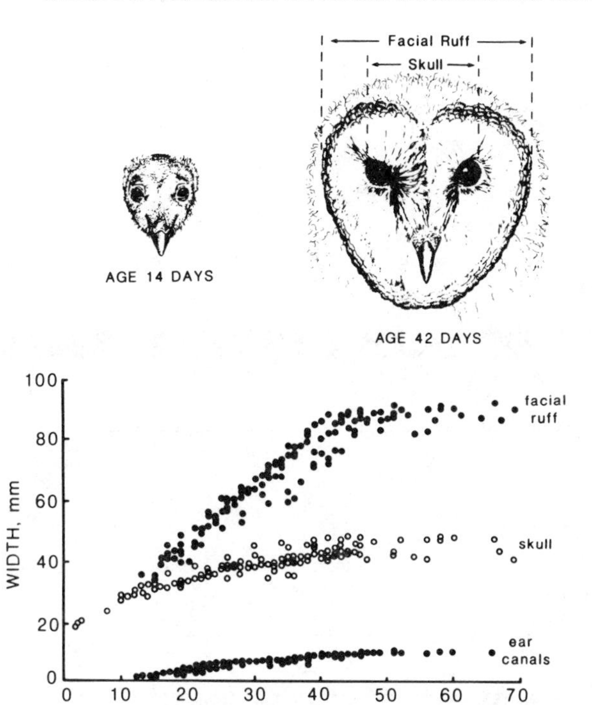

FIGURE 1. Growth of the head and ears of the barn owl. The width of the skull and facial ruff and the inner diameter of the external ear canals are plotted as a function of age posthatching for eight barn owls. These features determine the magnitude of monaural and binaural localization cues. The sketches are scaled drawings of the owl's face as it appears at 14 and 42 days of age, respectively. (From Knudsen et al., 1984a.)

faces of the external ears (the facial ruff) are just beginning to appear (Fig. 1). These structures continue to grow rapidly until the age of 45–50 days. Because the dimensions of the head and ears determine the magnitudes and characteristics of sound localization cues, owls younger than 45 days of age experience changing cues, whereas owls over 50 days of age experience stable adultlike cues. This consideration becomes important in the interpretation of the differential effects of monaural occlusion on sound localization by owls of various ages.

EFFECTS OF EXPERIENCE ON SOUND LOCALIZATION BEHAVIOR

The effect of long-term monaural occlusion on sound localization depends critically on the age at which the ear is occluded. Figure 2*A* shows results from monaurally occluding a 186 day old owl, that is, one that was fully grown. Before its ear was plugged the owl localized sounds with normal

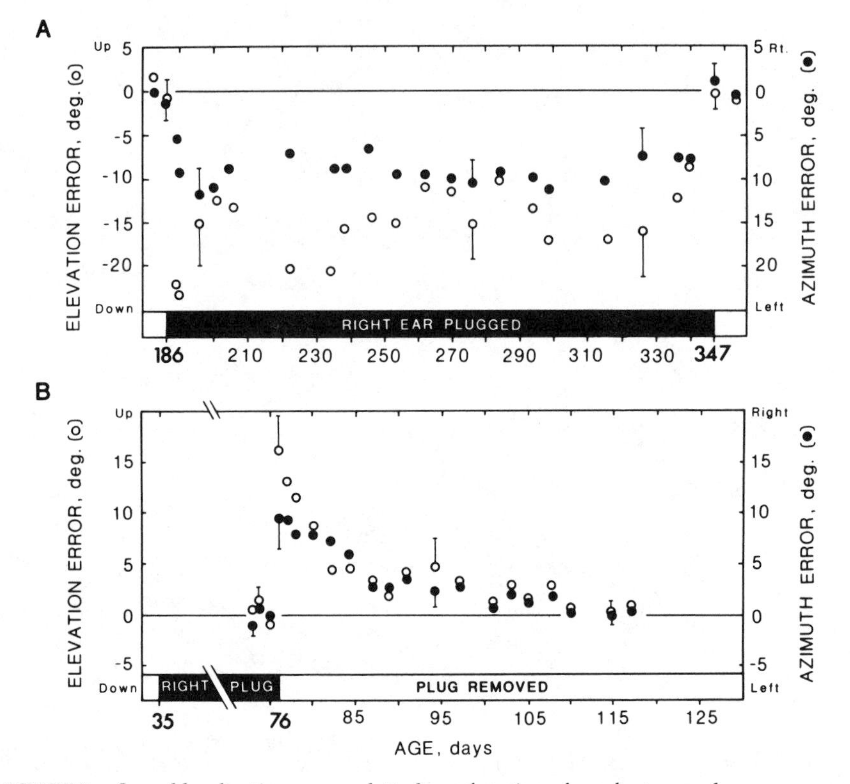

FIGURE 2. Sound localization errors plotted as a function of age for two owls, one monaurally occluded when fully grown (*A*), the other monaurally occluded while still immature (*B*). The azimuth (closed circles) and elevation (open circles) components of the localization errors are plotted separately in each graph. Representative standard error bars are shown periodically. The owl in (*A*) shows no evidence of adjusting its sound localization in response to the abnormal cues induced by the ear plug. The owl in (*B*) learned to localize accurately with the ear plug in place. When the ear plug was removed, it exhibited a large localization error which it corrected over a period of weeks.

accuracy. As expected, when the plug was placed in the owl's right ear, the owl exhibited a large localization error to the left of and below the sound source. Although the magnitude of the error fluctuated substantially during the period of monaural occlusion, there is no indication that the owl was correcting its error. On the day the ear plug was removed, after more than five months of monaural occlusion, the owl immediately localized sounds with normal accuracy. Thus prolonged exposure to abnormal auditory cues seemed to have no effect on the auditory circuits underlying sound localization in a bird of this age.

When performed on younger owls, the same experiment yields entirely different results (Knudsen et al., 1982). The data in Figure 2*B* are from an owl that had its right ear plugged at 35 days of age. As soon as the owl was old enough to be trained and tested (72 days), it was localizing sounds with

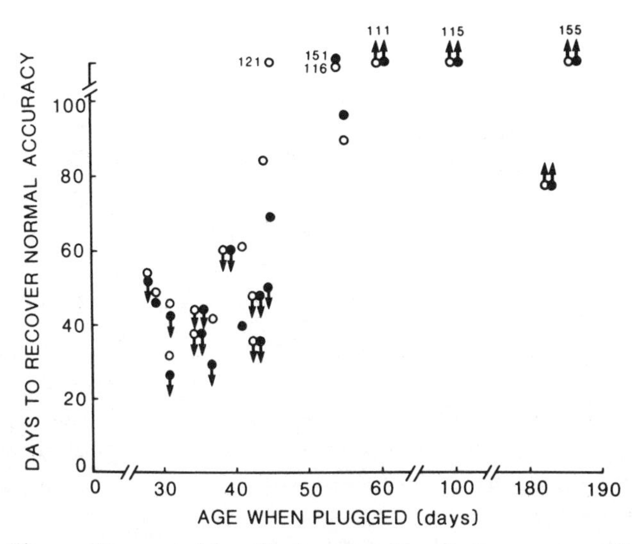

FIGURE 3. The sensitive period for adjusting sound localization accuracy. The number of days required for an owl to regain normal localization accuracy (error less than 3°) is plotted against the age of the owl at the time one ear was plugged. Closed symbols represent the day that normal accuracy in azimuth was attained; open symbols signify the day that normal accuracy in elevation was attained. A downward pointing arrow indicates that the owl exhibited normal accuracy in that dimension on the first day it was tested. An upward pointing arrow signifies that normal accuracy was never achieved through the last day of testing, indicated by the number above (in fact, these birds made no adjustment of their localization accuracy at all). *All* birds plugged before 60 days of age adjusted their sound localization, whereas *no* bird plugged after 60 days of age adjusted its sound localization. (From Knudsen et al., 1984a.)

normal accuracy. This implies that the owl had altered its associations between auditory cues and locations in space and had learned to localize sounds accurately using the abnormal cues imposed by the ear plug. This interpretation was confirmed when the ear plug was removed and normal hearing was restored: the owl exhibited a large sound localization error opposite in direction to the one originally induced by the insertion of the plug.

These results, when considered together, indicate that the auditory system can adjust its interpretation of localization cues through experience, but only during a restricted period in early life. To determine more precisely the extent of this sensitive period, owls were monaurally occluded at various ages and ability to recover accurate localization was evaluated (Knudsen et al., 1984a). As summarized in Figure 3, owls occluded at up to 45 days of age recovered accurate sound localization quickly (in under 5 weeks in some cases); those occluded between 45 and 55 days of age recovered normal accuracy much more slowly; and owls occluded at 60 days of age or older never showed signs of adjusting their sound localization. Thus there appears to be a discrete period in development during which abnormal auditory experience can lead to changes in sound localization.

The termination of this sensitive period at about 55 days of age might be triggered by an age-dependent mechanism. On the other hand, the time course of growth of the owl's head and ears is consistent with the alternative hypothesis that sensory experience brings the sensitive period to a close. Since the head and ears reach adult size at about 45 days of age (Fig. 1), owls at this age begin to experience adult sound localization cues. The data in Figure 3 show that when an owl is occluded just beyond this age, it adjusts its accuracy slowly, and owls two weeks older do not adjust at all. This coincidence between the maturation of the head and ears and the termination of the sensitive period could mean that exposure of the auditory system to stable adult cues, rather than age *per se*, brings the sensitive period to a close. Consistent with this hypothesis is the finding that birds that are monaurally occluded toward the end of the sensitive period continue to adjust their sound localization for weeks or even months past the age of 60 days (Fig. 3). If the capacity for adjustment were terminated by a purely age-dependent mechanism, these birds would have stopped adjusting their sound localization at approximately 60 days of age.

Preventing the animal from experiencing normal adult cues seems to keep the associations between localization cues and locations in space in a modifiable state. This is evidenced both by the continued adjustment by monaurally occluded birds past the normal end of the sensitive period (described previously) and by the readjustment by much older birds when the ear plug is removed. The owl represented in Figure 2*B* had its ear plug removed at 76 days of age, well beyond the age at which plug insertion has any effect. Yet, this owl rapidly corrected the error that resulted from ear plug removal. We studied the readjustment of sound localization accuracy following ear plug removal at various ages and found that owls readjusted their accuracy until about 270 days of age, after which readjustment essentially ceased even if the bird still had a large sound localization error (Knudsen et al., 1984b). Thus, although the period of plasticity might be extended by preventing exposure to normal adult cues, at approximately 270 days of age the neural connections responsible for sound localization appear to crystallize, whether they are correct or not. This age, which coincides with sexual maturation, is interpreted as the end of a critical period for the consolidation of associations between auditory cues and locations in space.

EFFECTS OF EXPERIENCE ON A NEURAL CODE OF AUDITORY SPACE

Behavioral experiments demonstrate that the percept of a sound's location in space can be altered by early auditory experience. The neurophysiological experiments discussed in this section show that early experience can also modify a neural representation of auditory space. The experiments were conducted in the optic tectum, the avian homologue of the mammalian superior colliculus. The optic tectum was studied because the neural code for auditory space is fairly well understood in this structure (Knudsen, 1982; Palmer and King, 1982; Middlebrooks and Knudsen, 1984). The code is based upon spatially selective neurons that respond best to sound sources located in a restricted region of space, or "best area" (see Konishi, Chapter 19). Such units are organized within the optic tectum according to the locations of their best areas so that they form a physiological map of auditory space (Knudsen, 1982). An important additional property of these units is that they are bimodal: they respond both to auditory and visual stimuli from a particular region of space.

The auditory and visual receptive fields of a typical unit in the owl's optic tectum are shown in Figure 4. Although the auditory and visual receptive fields differ greatly in size, the center of the auditory best area is closely aligned with the visual receptive field. The locations of auditory best areas and visual receptive fields vary systematically with the recording site in the tectum. Units at the rostral end of the tectum have fields located directly in front of the animal; units at the caudal end have fields located far to the contralateral side. Dorsally located units have high fields and ventrally located units have low fields. As a result, the azimuth of a stimulus is mapped along the rostrocaudal axis of the tectum, and the elevation of a stimulus is mapped along the dorsoventral axis. A close mutual alignment of auditory best areas and visual receptive fields occurs throughout most of the tectum, demonstrating that the auditory and visual maps of space are in register.

Because these tectal units are bimodal, the activity of a unit does not distinguish the modality of a stimulus, but only its location in space. In other words, tectal units code for stimulus location independent of stimulus modality. The set of auditory cues to which a tectal unit is tuned are those cues that have been associated with the location coded by that unit. Thus the alignment of auditory best areas with visual receptive fields provides a neurophysiological measure of the accuracy with which auditory cues are associated with appropriate locations in space. In the experiments that follow, we asked the question: What happens to this alignment in animals that have been raised under the same abnormal auditory conditions that are known to cause alterations in sound localization behavior?

The responses of a tectal unit recorded in an owl raised with one ear occluded are illustrated in Figure 5. An ear plug was inserted in the left ear of this bird at 41 days of age and was removed 301 days later. The unit was recorded on the day the ear plug was removed, and its receptive field properties were tested before and after plug removal. Notice that with the ear plug still in place, the auditory best area and receptive field were well aligned with the visual receptive field. This indicates that the spatial tuning

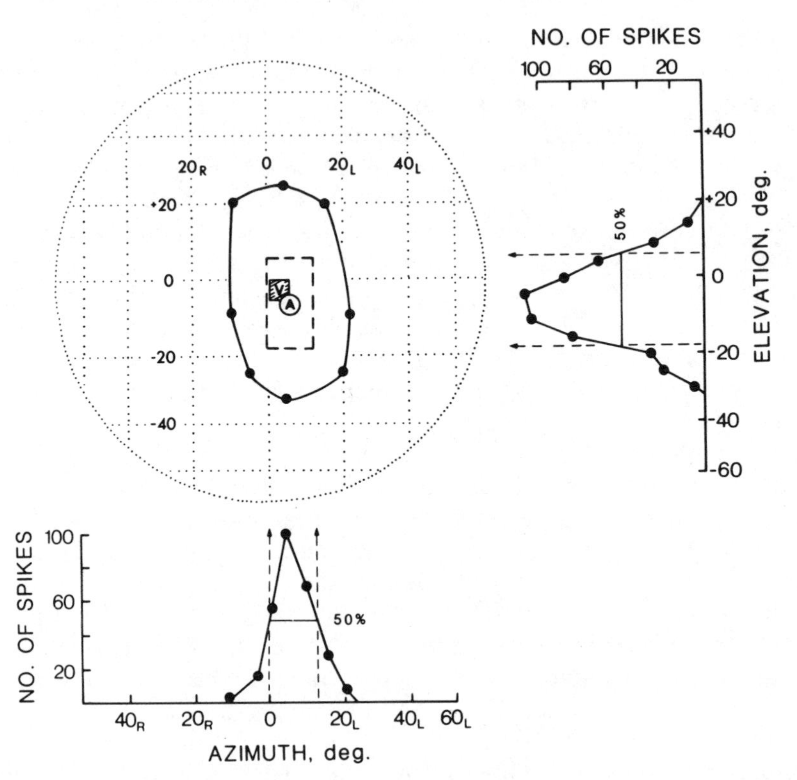

FIGURE 4. The auditory and visual receptive fields of a bimodal unit in the optic tectum of the barn owl. In the upper left, the auditory field (indicated by solid circles) and the visual field (hatched and marked with a V) are plotted in double pole coordinates of space. The unit's auditory best area is defined as the region of space in which a noise burst elicited more than 50% of the maximum number of spikes and is marked by the dashed lines on the globe. The center of the best area is indicated by the encircled A. The response of the unit (number of spikes) to 10 repetitions of a standard noise burst is plotted as a function of speaker location in azimuth (below) and elevation (right). The noise was 20 dB above the unit's threshold. (From Knudsen, 1982.)

of the unit had been adjusted to select for the altered auditory cues that corresponded to a sound source at the location of its visual receptive field. Consistent with this, removal of the ear plug resulted in a dramatic shift in the locations of the auditory best area and receptive field so that they were no longer aligned with the visual receptive field. The magnitude and direction of the misalignment was that expected from sound localization errors measured behaviorally in birds treated in a similar way. Based on the data presented by Konishi (Chapter 19), the misalignment in azimuth between auditory best areas and visual receptive fields implies a change in the tuning of units to interaural time delays, and the misalignment in elevation implies a change in tuning to interaural intensity differences.

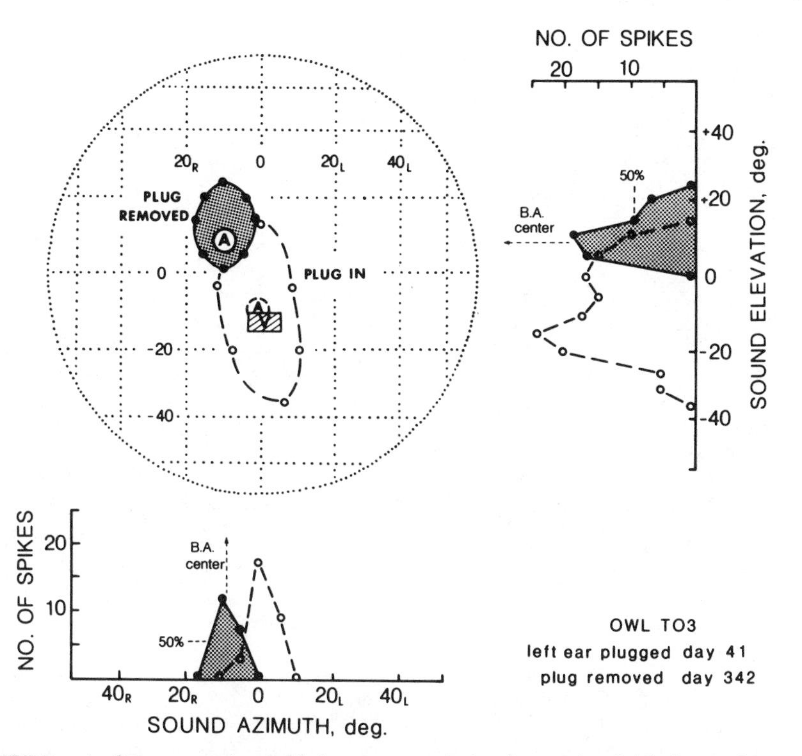

FIGURE 5. Auditory receptive field, best area, and visual receptive field of a single unit in the optic tectum measured before and after a chronic plug in the left ear was removed. The unit was recorded in the deep layers of the right tectum (contralateral to the plugged ear) on the day the ear plug was removed. The grid represents space in double pole coordinates. The auditory receptive field and best area center measured with the plug still in the ear are plotted with dashed lines and a circled A, respectively. The visual receptive field is represented by the hatched rectangle marked with a V. The auditory receptive field and best area center of this same unit were measured immediately after the ear plug was removed and are plotted with solid lines and circled A (stippled). The spike counts shown below and to the right are from eight presentations of a noise burst at 20 dB above the unit's threshold. These counts were used to determine the best area of this unit before and after the removal of the ear plug. (From Knudsen, 1983.)

The misalignment between auditory and visual fields following ear plug removal was stable over a period of at least six months in birds raised with one ear occluded past the end of the critical period. This stability of field misalignment permitted a detailed assessment of changes that had occurred in the auditory map of space (Knudsen, 1983). Data from mapping experiments conducted on one of these owls are shown in Figure 6. In both tecta of this bird, auditory best areas were shifted uniformly about 10° to the left of visual receptive fields (Fig. 6*A*). This corresponds to a rostralward shift of the auditory space map in the right tectum and a caudalward

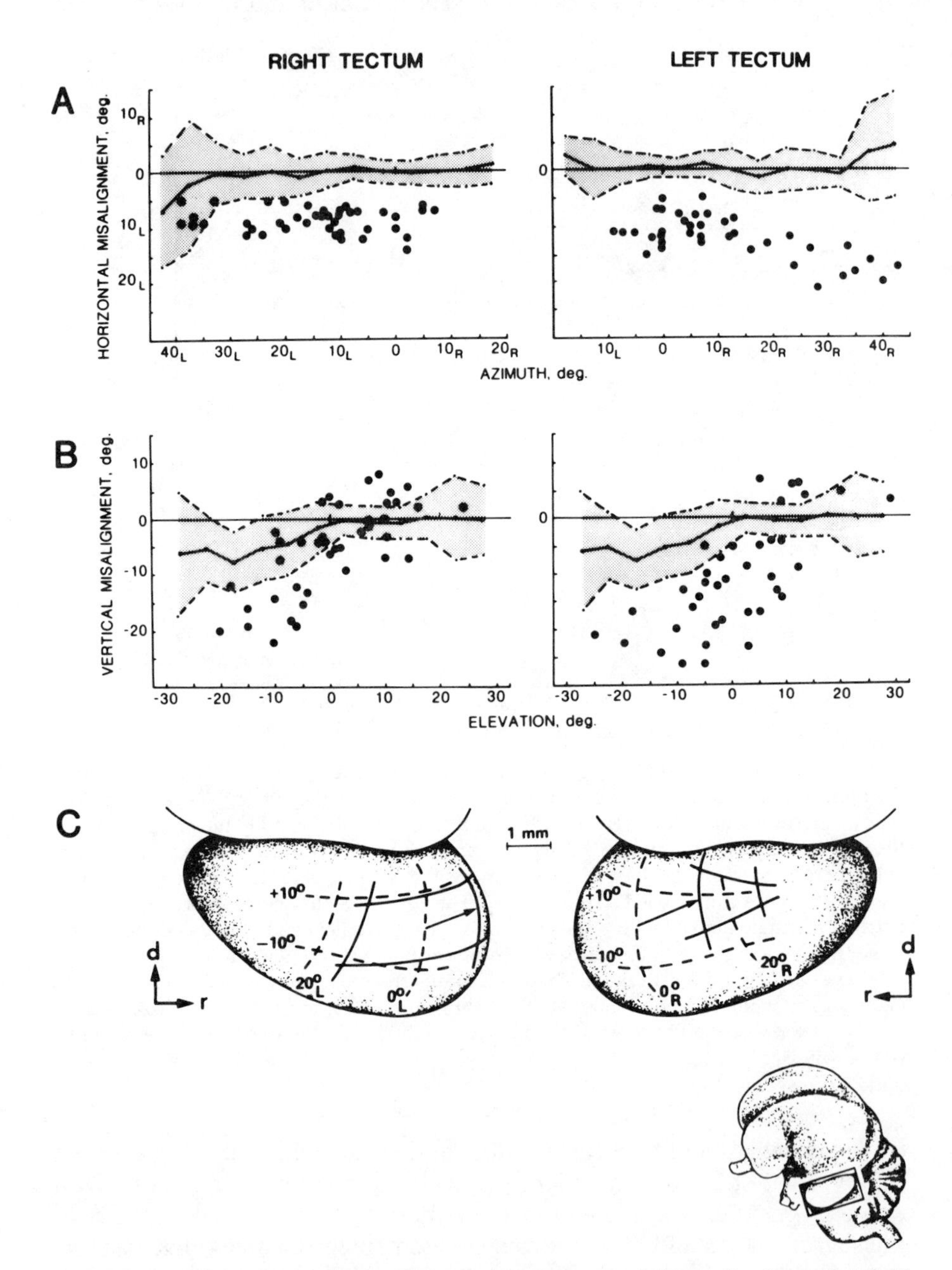

FIGURE 6. The misalignment of auditory best area centers relative to visual receptive field centers of single units, and the shifts in the auditory maps of space in the optic tecta of an owl subjected to a right-ear occlusion throughout the critical period. Normal field misalignment (mean and standard deviation) measured in control owls is indicated by the stippled area). The data from the experimental owl were collected over a period of 4 weeks begining 124 days after the ear plug had been removed. (*A*) The horizontal misalignment of auditory best area centers is plotted (open circles) as a function of the azimuth of visual receptive field centers

shift in the left tectum (Fig. 6C). In both tecta the representation of 0° azimuth (the midsagittal plane) had shifted about 1.5 mm. In addition to shifting the map, monaural occlusion resulted in a compression of the representation of lower auditory elevations in the ventral portion of the tectum as seen by the increasing vertical misalignments between auditory best areas and visual receptive fields at lower elevations (Fig. 6*B*). Thus both the position and the relative magnification properties of the auditory space map were modified by abnormal auditory experience. These data do not imply that the underlying changes in neural connectivity took place in the optic tectum itself. In fact a number of anatomical and physiological results suggest that the changes occur earlier in the auditory pathway, but these data are still preliminary.

In the tecta of this particular owl, visual receptive fields were centered an average of 9.7° ± 2.5° to the right and 6.1° ± 9.2° above auditory best areas. When tested behaviorally, this owl exhibited a mean sound localization error of right 9.5° and up 3.5°. The good agreement between these values suggests that the alignment of the auditory map in the tectum offers a reliable neural correlate of the experience-dependent plasticity that has been documented behaviorally.

CONCLUSION

The percept of a sound's location is created by associating sets of auditory cues with locations in space. Certainly, the basic neuronal circuitry necessary for extracting and interpreting localization cues develops according to genetic determinants. However, within the constraints imposed by these determinants, there remains a substantial amount of flexibility subject to the shaping forces of experience. Early in life this plasticity is sufficiently great to allow the association of highly abnormal auditory cues with locations in space. In the course of normal development this capacity for ad-

---

for the right and left tecta. (*B*) The vertical misalignment of auditory best area centers is plotted (open circles) as a function of the elevation of visual receptive field centers for the right and left tecta. (*C*) The shift and distortion of the auditory maps that correspond to the misalignment shown in (*A*) and (*B*) are depicted on lateral views of the left and right tecta. The dashed coordinate lines represent the auditory map in normal owls. The solid coordinate lines represent the shifted auditory map. These coordinate lines were derived by calculating the shift in the tectum of the representations of the corner locations, that is, $0°_{Az}$, $+10°_{El}$; $0°_{Az}$, $-10°_{El}$; $20°_{Az}$, $+10°_{El}$; $20°_{Az}$, $-10°_{El}$. This was done by computing the mean field misalignment for units with visual receptive fields located within 10° of each location. The mean field misalignment for equivalent locations in control birds was subtracted, yielding the actual induced shift. These shifts were plotted onto the tectal surface using the visual map. The vectors indicate the shift in the representation of auditory $0°_{AZ}$, $0°_{El}$ in the tecta. The positions and lengths of these vectors were confirmed by electrolytic lesions that were placed at visual $0°_{Az}$, $0°_{El}$ and auditory $0°_{Az}$, $0°_{El}$ on each side. Abbreviations: L, left; R, right; d, dorsal; r, rostral. (From Knudsen, 1983.)

justment enables the auditory system to fine-tune its localization circuitry and to respond to changes in auditory cues brought about by growth of the head and ears. After the owl reaches adult size, the plasticity of this system decreases substantially, and the calibration of the sound localization circuitry becomes extremely resistant, if not immune, to further change.

REFERENCES

Knudsen, E. I. (1980) Sound localization in birds. In *Comparative Studies of Hearing in Vertebrates,* A. N. Popper and R. R. Fay, eds. Springer, Berlin-Heidelberg-New York, pp. 287–322.

Knudsen, E. I. (1982) Auditory and visual maps of space in the optic tectum of the owl. *J. Neurosci.* **2**:1177–1194.

Knudsen, E. I. (1983) Early auditory experience aligns the auditory map of space in the optic tectum of the barn owl. *Science* **222**:939–942.

Knudsen, E. I., G. G. Blasdel, and M. Konishi (1979) Sound localization by the barn owl measured with the search coil technique. *J. Comp. Physiol.* **133**:1–11.

Knudsen, E. I., S. D. Esterly, and P. F. Knudsen (1984a) Monaural occlusion alters sound localization during a sensitive period in the barn owl. *J. Neurosci.* **4**:1001–1011.

Knudsen, E. I., P. F. Knudsen, and S. D. Esterly (1984b) A critical period for the recovery of sound localization accuracy following monaural occlusion in the barn owl. *J. Neurosci.* **4**:1012–1020.

Knudsen, E. I., P. F. Knudsen, and S. D. Esterly (1982) Early auditory experience modifies sound localization in barn owls. *Nature* **295**:238–240.

Knudsen, E. I. and M. Konishi (1979) Mechanisms of sound localization by the barn owl (*Tyto alba*). *J. Comp. Physiol.* **133**:13–21.

Konishi, M. (1973) How the owl tracks its prey. *Am. Sci.* **61**:414–424.

Middlebrooks, J. C. and E. I. Knudsen (1984) A neural code for auditory space in the cat's superior colliculus. *J. Neurosci.* **4**:2621–2634.

Palmer, A. R. and A. J. King (1982) The representation of auditory space in the mammalian superior colliculus. *Nature* **299**:248–249.

Payne, R. S. (1971) Acoustic location of prey by barn owls (*Tyto alba*). *J. Exp. Biol.* **54**:535–573.

Chapter Six

SOURCES OF INTRASPECIES AND INTERSPECIES CORTICAL MAP VARIABILITY IN MAMMALS

Conclusions and Hypotheses

MICHAEL M. MERZENICH

Coleman Laboratory
and
Departments of Otolaryngology and Physiology
University of California at San Francisco
San Francisco, California

We have mapped the representation of skin surfaces in cerebral cortical fields 3b and 1 in a series of normal adult owl and squirrel monkeys and have found those maps to be idiosyncratic in topographic detail (Merzenich et al., 1985a; Merzenich, 1985). This chapter summarizes these data briefly and then considers the possible origins of the substantial intraspecies and interspecies variability recorded. There are a number of potential sources of representational variability in individuals of the same or of different species, including (1) genetic differences; (2) inherent imprecision in the early development of neuronal networks and internetwork connections; (3) differences in experience through a critical period of early shaping of neuronatomical connections; and (4) alteration of functional cortical representations by experience or by other peripheral or central factors in post-critical-period life.

Studies that we have conducted over the past several years strongly suggest that map alteration by experience effected throughout life constitutes a principal source of differences of functional map detail in adult individuals of any one species (Merzenich et al., 1983a, b; 1984a, b; 1985b; Merzenich, 1985). Given these findings, it can be argued that the details of cortical maps in different related mammals would be largely determined by niche-related dominant-use considerations.

In this review, we briefly summarize evidence for cortical map variability by example. Results of a series of studies demonstrating that cortical maps in adult monkeys are locally functionally plastic are summarized. Finally, implications of these results for individual and species differentiation are summarized, with a brief attempt to reconcile this new view of cortical organization and function with the currently theoretically dominant static view of central nervous system organization.

CORTICAL MAP VARIATION IN NORMAL ADULT MONKEYS

Typical maps of the cortical representation of the surfaces of the hand in parietal areas of 3b and 1 are illustrated, by example, for two adult owl monkeys and for two adult squirrel monkeys in Figure 1. These maps were derived by defining receptive fields for neurons in each of 135–343 parallel microelectrode penetrations introduced over a fine (50–300 μm) grid completely covering the cortical zone of representation of the hand in the indicated field. In these maps each outlined area is the cortical zone in which receptive fields were centered on the skin surface indicated in the hand drawing at the right.

It is important to note that the boundaries in these maps are resolved in great detail. The boundaries of the territories representing the glabrous surface of a digit, for example, were commonly defined at about 10–30 perimeter sites (see Fig. 2). Although there is a significant (plus or minus about 25–150 μm) error in the definition of the boundaries of these zones at any point around their perimeter, the boundaries are crossed many times. Hence the average error is very small, the overall error in estimated areas of representation of given skin surfaces is small, and the resolution of the internal topography of different skin surfaces is relatively detailed.

A glance at these maps illustrates that there are basic commonalities in different maps. Thus digits and palmar pads are represented in an orderly medial-to-lateral ulnar-to-radial sequence, and the hand has the same basic orientation in every area 3b or area 1 map, with digit tips represented toward their rostral and caudal borders, respectively. On the other hand, all topographic map *details* are variable. Thus (1) there is substantial variability in the overall territory of representation of hand surfaces in different monkeys. (2) There are substantial differences in the absolute or proportional areas of magnification of different hand surfaces. Compare, for ex-

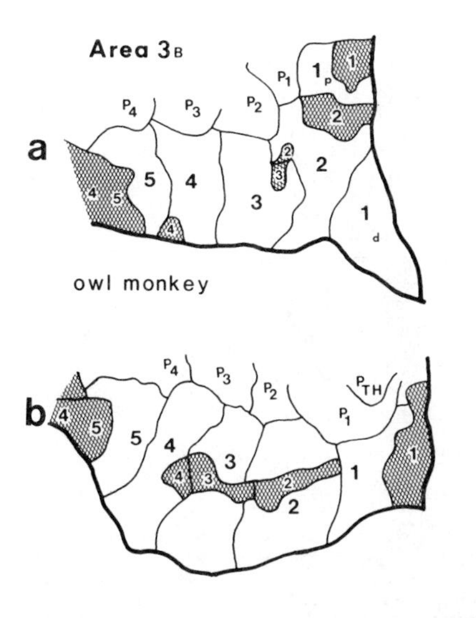

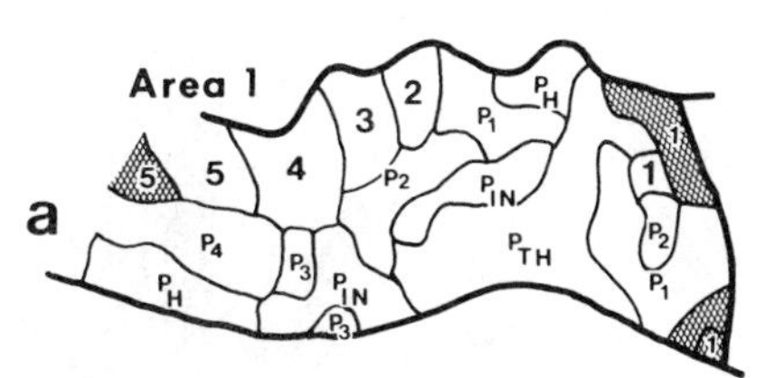

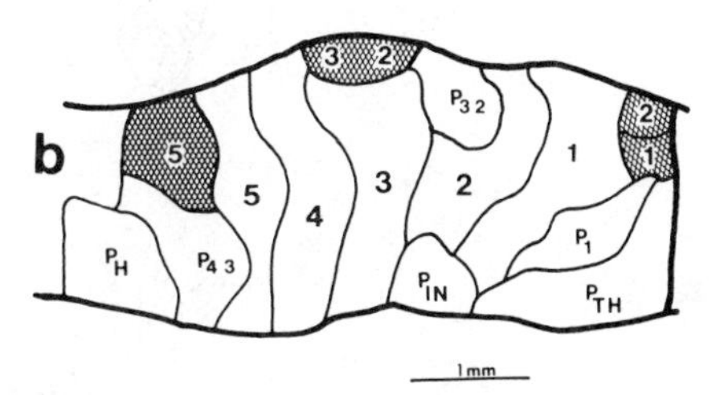

CORTICAL REPRESENTATIONAL AREAS

		GLABROUS DIGITS				
		1	2	3	4	5
OWL MONKEYS Area 3b (n=9)	MEAN (n=9)	.78	1.04	1.10	1.05	.66
	RANGE	.53 1.41	.77 1.35	.73 1.44	.77 1.44	.47 .88
SQUIRREL MONKEYS Area 3b (n=5)	MEAN	1.54	1.80	1.83	1.18	.77
	RANGE	.90 2.30	1.40 2.43	1.45 2.80	.82 1.52	.69 .84
SQUIRREL MONKEYS Area 3b (n=5)	MEAN	.43	.44	.67	.80	.79
	RANGE	.10 1.24	.17 .94	.27 1.33	.48 1.30	.37 .98

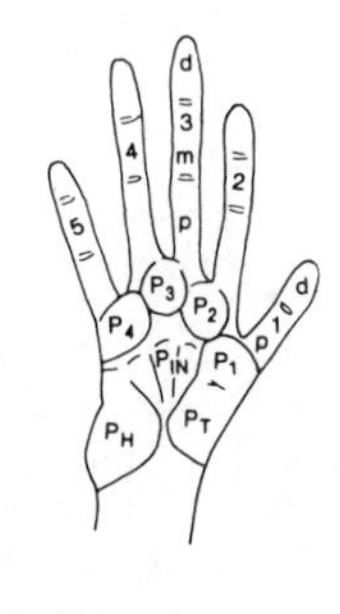

FIGURE 1. Cortical maps of the hand surfaces in area 3b of two adult owl monkeys (upper left) and in area 1 of two adult squirrel monkeys (upper right). Outlined zones are cortical sectors in which receptive fields were centered on skin surfaces indicated in the hand drawing at the lower right. Zones representing dorsal (hairy) hand surfaces are crosshatched. Rostral is toward the bottom of the drawing; lateral is toward the right. At the bottom, the mean areas of representation of the glabrous digits and the ranges of representational areas recorded in a series of normal adult owl and squirrel monkeys are recorded. Glabrous digital representational areas actually vary less than most other map features. For example, note the striking variations in the territories of representation of dorsal hand surface, or of palmar surfaces in the maps shown above. (Redrawn from Merzenich, 1985.)

ample, the areas of representation of digit 1 or of the dorsal surfaces of different digits in the illustrated cases (Fig. 1). (3) There are many differences in internal topographic sequences and representational neighborhood relationships. Note, for example, the differences in the relationships of the representations of digits one and two with other skin surfaces and with each other. (4) The skin surfaces represented along all of the boundaries of the hand surface area in these fields are idiosyncratic in detail.

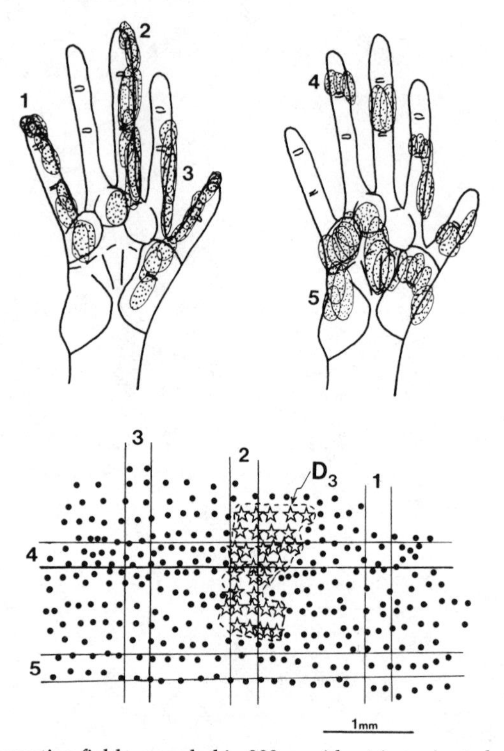

FIGURE 2. Top: Receptive fields recorded in 300-μ-wide strips oriented rostrocaudally (1, 2, 3) and mediolaterally (4, 5) across a typical map (below). Note that some receptive fields defined for penetrations in these strips were on the dorsal hand; they are not illustrated in these drawings. Bottom: The recording grid for a representative cortical map in this series. Dots and stars denote the 320 penetration sites. Stars mark penetration at which receptive fields were on the glabrous surface of the third digit. Note that with this grain of recording (1) the border of such representational zones (e.g., of digit 3) is defined at many perimeter locations. The average error in defining the border position is therefore small. (2) Internal representational detail is well defined.

The dimensions of some of these differences for a population of five squirrel monkeys and nine owl monkeys are summarized in the tables at the bottom of these maps (Fig. 1). Note that all of the previously summarized variable map dimensions apply to abolute or proportional areas or to magnification factors. As a general rule, representational areas by any of these measures commonly vary severalfold in area 3b and manyfold in area 1.

Despite the fact that maps are substantially variable, they all have a basic internal topographic order in which neighborhood relationships are preserved. That is illustrated by example by showing receptive field sequences for typical rows of penetrations in a representative map (Fig. 2). There are some discontinuities in such penetration sequences, both between digits and with receptive fields that shift from predominantly dorsal to predominantly glabrous locations. However, if all surrounding sites were to be examined, overlapping continuities could be seen for any cortical site.

Sur and colleagues (1980a) quantified this receptive field overlap relationship for the representation of an owl monkey, demonstrating that away from lines of discontinuity there is a roughly linear decline in the *percentage* of overlap of receptive fields as a function of distance across the cortex, with no overlap seen at distances greater than about 500–600 μm. These functions are independent of the body surface represented or of the sizes of receptive fields recorded.

Thus, whatever the process by which the large individual differences in cortical maps are generated, they operate with preservation of topographic relationships and, specifically, with at least a rough maintenance of this receptive field percentage overlap "rule."

POSSIBLE SOURCES OF INDIVIDUAL MAP VARIABILITY

There are at least six arguable sources of cortical map variability. First, there are inherent errors in deriving these maps. What of the differences in maps recorded could be due to these experimental errors? Second, there are possible genetically determined differences. Third, the genetic control of map formation (e.g., of the establishment of neural connections to the representation) may not be highly specified. A significant amount of variability might be attributable to inherent imprecision in the development of anatomical connections. Fourth, differences in experience early in life (prior to a "critical period" of development) have been shown to influence the formation of the adult form of thalamocortical connections (Harris, 1981; Hubel et al., 1977; Movshon and Van Sluyters, 1981; Simons et al., 1983; Sur et al., 1980a). Thus map variability might be to some extent attributable to differences in very early experience. Fifth, features of cortical maps have long been attributed to features in the sensory periphery. Thus, for example, central map proportionalities have long been arguably related to the peripheral distributions of receptors (Bard, 1983; Hines, 1929; Woolsey et al., 1942). If the periphery is a determinant of cortical map structure, then peripheral differences (also changing with age or skin use) might underlie cortical map differences. Sixth, studies conducted principally within the somatosensory cortex indicate that the details of cortical maps are alterable by use in adult primates. If this use-dependent alteration is of an appropriate magnitude, it might be the dominant determinant of differences in map detail in different individual adult monkeys.

It is probable that several of these factors contribute to some extent to variability in cortical map structure. However, we argue that there is apparently a surprising capacity for map alteration by experience in adult monkeys. It appears likely that this use-dependent alterability dominates the functional map forms recorded in different individual adults (Merzenich et al., 1985a). At the same time, it appears almost certain that this use-dependent development, maintenance, and adjustment of the details of functional maps by sensory (and possibly other) inputs is limited by a genetically constructed and precritical-period shaping of neuroanatomical maps.

If results in the somatosensory cortex are general to other cortical areas, as seems likely, then map alterability by use probably largely accounts both for individual differences in the forms of maps in adults—and presumably for what individuals can derive from them idiosyncratically. They must also largely account for at least many of the differences recorded between individuals of different but related species.

Map Alteration by Use in Adult Monkeys

A number of different studies have now shown that somatosensory cortical maps can be altered in detail in adult monkeys. Some of the features of this use-dependent alteration are illustrated for one such experiment in Figure 3, which shows a map of the hand surface in area 3b of an adult owl monkey derived 76 days after digits 2 and 3 were amputated. In such cases there is an expansion of the representation of adjacent digital and palmar surfaces, which "occupy" the cortical territory formerly representing the now-missing digits (Merzenich et al., 1984a). When reorganization is completed, the margins of the skin surface representation bordering those of amputated surfaces have moved hundreds of microns across the cortex, and over a region of cortex more than 2 mm across neurons at all sites develop "new" peripheral receptive fields. With expansion of representation, there is a corresponding increase in representational magnification (cortical area/skin surface area) with rough maintenance of the percentage overlap rule of Sur and colleagues (Sur et al., 1980a). That is another way of stating that as territories of representation expand, there is a corresponding decrease of receptive field sizes.

Similar changes have been recorded following peripheral nerve transection and ligation (Merzenich, 1983a). In that model, the time course of cortical map reorganization was studied (Merzenich, 1983b). Neurons throughout the large cortical zone normally representing the median nerve were driven by newly effective inputs within about three weeks of the nerve transection. However, substantial internal map reorganization occurred subsequent to complete initial "reoccupation" of the deprived zone. Cortical maps also reorganize following induction of restricted cortical lesions in somatosensory cortical fields (Jenkins et al., 1982). After lesions completely destroy the cortical representation of given skin surfaces (e.g., a digit or digits), the skin surface representations reappear in the cortical zone bordering the lesion. In establishing a new cortical territory of representation, that of skin surfaces formerly represented in the perilesion zone is correspondingly reduced in area and degraded in topographic detail. Cortical map changes have also been tracked in animals following nerve

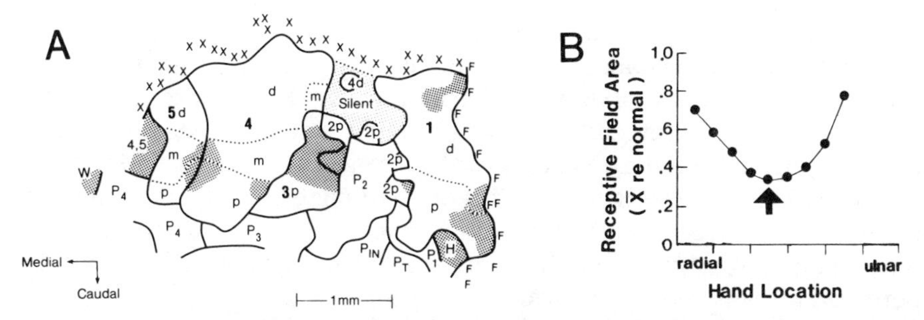

FIGURE 3. One class of experiment revealing that cortical maps of the hand are dynamically alterable. (*A*) A map of the hand for a monkey studied 72 days after digits 2 and 3 were amputated following digital nerve transection and ligation (preventing nerve regeneration into the stump). Note the great expansion of representation of digits 4 and 1, both of which are nearly twice their normal representational size. Palmar pad 2, subjacent to the missing digits, has also expanded greatly. The digit 5 representation remains of nearly normal size. Contrast with normal maps in Figure 1. Abbreviations as in that Figure. X = penetration sites at which "deep" but not cutaneous inputs drive cortical neurons; F = penetrations in which receptive fields were on the face; H = hand; W = wrist. (*B*) The mean receptive field area as a function of hard locations. The arrow marks the approximate center of the reorganized zone. See text for further description. (Redrawn from Merzenich et al., 1984.)

transection and repair (Wall et al., 1985). In those preparations the central nervous system appears to partially sort and reorganize peripherally scrambled inputs, and central reorganization processes appear to underlie recovery of function and its acceleration by use following such lesions (Dellan, 1981; Wall et al., 1985; Wynn-Parry, 1973). Finally, we have growing evidence that hand use in normal monkeys can substantially alter cortical maps (Jenkins, Merzenich, Ochs, unpublished studies). Thus, for example, a substantial enlargement of hand surfaces with inverse changes in receptive field sizes have been recorded following heavy daily tactile stimulation over a several month period.

All of these results have been derived in *adult* monkeys. They unequivocally demonstrate that cortical maps of the skin surface are not static in adult monkeys. *There are many forms of these maps possible,* and the form recorded in any individual at any given time must at least largely reflect the monkey's dominant hand uses up to that point in its life.

The theoretical implications of this finding are great. However, they are largely beyond the scope of this brief review (see Merzenich et al., 1983a, b; 1984a, b; 1985a, b; Merzenich, 1985 for further discussion).

Influence of Peripheral Factors on Map Formation

When cortical maps were first recorded, it was found that the most densely innervated and functionally most heavily used sectors of sensory epithelia were differentially enlarged in the maps. It was hypothesized that the magnifications of representation in cortical maps likely directly reflect peripheral innervation densities (Woolsey et al., 1942). In fact, since the magnifications of representations can be altered by use, there can be no static isomorphic relationship between sensory epithelia and the cortical map (Merzenich et al., 1983a; 1985b; Merzenich, 1985). Recent evidence also reveals that the magnifications of different cortical fields are actually field specific, in both the visual and somatosensory system (see Merzenich and Kaas, 1980 for review). Given this often dramatic variation in representational magnification between fields, it appears unlikely that the details of organization of any one of them are related or determined directly by peripheral innervation densities.

On the other hand, there *are* clear global effects. Obviously, if a monkey has a large, densely innervated, and functionally heavily used tail, a substantial tail representation will be found (see Welker, 1974); if not, not.

Experience in Early Life: Effects on Formation of "Adult" Connections of Cortical Maps

There is growing evidence that initially very crude neuroanatomical projections to the cortex are pruned by activity in a precritical period in early life (Harris, 1981; Hubel et al., 1977; Movshon and Van Sluyters, 1981; Simons et al., 1983). However, it has been argued that not very specific activity is required for normal neuroanatomical development, with a predictable result occurring with eyes open or whiskers intact across the period. Such observations indicate that precritical period experience might not normally be a very important source of functional representational variation.

Imprecision in the Developmental Process

Recent studies indicate that the formation of connections in the brain are not strictly regulated line by line and synapse by synapse, but to the contrary, are only approximately controlled by cell adhesion molecules (see Edelman, 1983, 1984 for review). At least in early development, the system appears to develop with an in-built orderly disorder. Again, there is some apparent pruning of connections in at least some cortical zones (see preceding section) through the late developmental period. However, the net result is probably still not highly specified, as *connections must support the cortical functional dynamism described earlier.* Unless there is a capacity for the growth of terminals and/or cell processes throughout life on a scale hitherto only dreamed of, then a substantial divergence and convergence of connections far beyond that indicated by functional maps must actually be present. The combination of an initially inherently imprecise formation of connections, coupled with a pruning of connections designed to establish a substrate for use-dependent dynamism of representational detail might constitute a substantial underlying source of cortical map variability. At the least, it presumably establishes the potential *limits*, field by field, of variability of cortical map detail by use (Merzenich et al., 1983b, 1984a, 1985b; Merzenich, 1985).

Might There Be Real Genetic Differences between Individuals? Are There Genetic Differences Controlling Differences in the Details of Cortical Maps between Individuals of the Same or Related Species?

Historically, maps have been derived with the presumption that there is a strict species-specific form of the map in each cortical area. An underlying assumption has been that cortical maps are genetically determined, as different in different species as the external physiogomy of different mammals. Thus, for example, for primate S1 fields, a "spider monkey map," a "galago map," a "macaque map," and so on, have been described (Benjamin and Welker, 1957; Carlson and Fitzpatrick, 1982; Hirsch and Coxe, 1958; Krishnamurti et al., 1976; Merzenich et al., 1978; Nelson et al., 1980a; Penfield and Rasmussen, 1950; Pubols and Pubols, 1971; Sur et al., 1980b; Woolsey et al., 1942; Woolsey, 1952, 1954). Taking the hand representation alone, the variability recorded in areas 3b and 1 maps in owl or squirrel monkeys is greater than that described for different studied primate species. In most of these studies, one or two or three relatively coarse-grained maps were derived in different individual monkeys. Common features of representation were sought and formed the basis of a composite or representational general map for the species. The hypothesis that an extreme genetic determinism underlies map formation was fostered principally by Sperry, who viewed the cortex as a precisely hard-wired machine in which all details must necessarily be prescribed point by point (Sperry, 1963). For reasons already cited (i.e., manifest functional map dynamism and moderately nonspecific formation of neural connections in development), this position is now unsupportable.

There is actually no clear proof that *any* differences in cortical map detail in different primate species are strictly genetically determined. It seems likely that there are some instances in which structural differences are recorded (e.g., the appearance of ocular dominance columns or "barrels" in some but not in other species) where that is the case. However, despite the fact that this seems to be a commonly held notion, the genetic origins of such differences have not been proven. Consider the case of the monkey with or without a tail, cited before. It should not be assumed that the different form of a topographic map in the tailed or tailless species is genetically determined in the cortex. More likely the genetic difference is in the tail.

How Many Observed Differences Are Simply Attributable to Errors in Defining Map Detail?

There are some unavoidable errors in deriving maps of representations using available neurophysiological methods. It is important to note that these errors cannot contribute very significantly to the differences in map detail in different adult monkeys in these studies. With the grain of definition with which these maps are defined, the boundaries of representation of given skin surfaces (outlined in the maps) are highly resolved. There can be small errors (of the order of tens of microns) in siting of microelectrode penetrations on the cortical surface. At any one site on a map boundary perimeter, there can be errors of the order of about ± 25–150 μm. There can be errors in definition of the actual boundaries of receptive fields. However, for any given skin representational area, map perimeter boundaries are usually defined at numerous sites. Thus the overall boundary estimate is relatively accurate, and because the *average* error in perimeter definition is practically nil, errors in estimations of area are small. There has been good consistency in the definition of receptive fields when they have been derived by different experimenters in our group; boundary definitions of receptive fields sometimes differ substantially, but the definition of the sites of the centers of the receptive fields (upon which map definitions are based) rarely differ (Stryker et al., 1985). Thus we believe mapping techniques generate errors that are only a small fraction of the magnitude of differences recorded in different individual monkeys.

SOME CONCLUSIONS

1. Although there are a number of factors that might influence cortical map variability, two likely dominate. First, there is inherent imprecision in the development of cortical afferent connections, and the subsequent precritical period pruning of connections must still result in a somewhat crude neuroanatomical map. This initial rough neuroanatomical map may be somewhat variable. Second, and more important, detailed functional maps are created from these relatively crude neuroanatomical maps, *and these functional maps are alterable by experience throughout life.*
2. For closely related species, then, dominant-use patterns must predominantly control the forms of maps most prevalent for each species. This constitutes an inherent basis for species and individual adaptation.
3. Neuroscience has long struggled to understand the sources of individual variation in given (especially the human) species. These kinds of studies indicate that genetic and precritical factors control the establishment of neural substrates which are functionally modifiable by use. The basic process accounting for map change by use might constitute *the* basic adaptive neural process.

Acknowledgements

This work is dedicated to Ted Bullock. It was conducted in collaboration with many colleagues in the Departments of Otolaryngology and Physiology at UCSF and at the Laboratory of Dr. Jon Kaas in the Departments of Psychology and Anatomy at Vanderbilt University. The author thanks Joseph Molinari for assistance in prepa-

ration of this manuscript. The described work was conducted with the support of NIH Grant NS-10414, the Coleman Fund, and Hearing Research, Inc.

REFERENCES

Bard, P. (1983) Studies on the cortical representation of somatic sensibility. *Bull. N.Y. Acad. Med.* **14**:585–607.

Benjamin, R. M. and W. I. Welker (1957) Somatic receiving areas of cerebral cortex of squirrel monkey (*Saimiri sciureus*). *J. Neurophysiol.* **20**:286–299.

Carlson, M. and K. A. Fitzpatrick (1982) Organization of the hand area in the primary somatic sensory cortex (SmI) of the prosimian primate, *Nycticebus coucang. J. Comp. Neurol.* **204**:280–295.

Dellan, A. L. (1981) *Evaluation of Sensibility of Re-education of Sensation in the Hand.* Williams and Wilkins, London.

Edelman, G. M. (1983) Cell adhesion molecules. *Science* **9**:450–457.

Edelman, G. M. (1984) Modulation of cell adhesion during induction, histogenesis and perinatal development of the nervous system. *Ann. Rev. Neurosci.* **7**:339–378.

Harris, A. (1981) Neural activity and development. *Ann. Rev. Physiol.* **43**:689–710.

Hines, M. (1929) On cerebral localization. *Physiol. Rev.* **9**:462–574.

Hirsch, J. F. and W. S. Coxe (1958) Representation of cutaneous tactile sensibility in cerebral cortex of *Cebus. J. Neurophysiol.* **21**:481–498.

Hubel, D. H., T. N. Wiesel, and S. LeVay (1977) Plasticity of ocular dominance columns in monkey striate cortex. *Proc. R. Soc. Lond.* **278**:377–409.

Jenkins, W. M., M. M. Merzenich, J. M. Zook, B. C. Fowler, and M. P. Stryker (1982) The are 3b representation of the hand in owl monkeys reorganizes after induction of restricted cortical lesions. *Soc. Neurosci. Abstr.* **8**:141.

Krishnamurti, A., F. Sanides, and W. I. Welker (1976) Micro-electrode mapping of modality-specific somatic sensory cerebral neocortex in slow loris. *Brain Behav. Evol.* **13**:367–383.

Merzenich, M. M. (1985) Development and maintenance of cortical somatosensory representations: Functional "maps" and neuroanatomical repertoires. In *Touch,* K. Barnard and T. B. Brazelton, eds. International University Press, New York (in press).

Merzenich, M. M., J. H. Kaas, M. Sur, and C.-S. Lin (1978) Double representation of the body surface within cytoarchitectonic areas 3b and 1 in "S1" in the owl monkey (*Aotus trivirgatus*). *J. Comp. Neurol.* **181**:41–74.

Merzenich, M. M. and J. H. Kaas (1980) Principles of organization of sensory-perceptual systems in mammals. In *Progress in Psychobiology and Physiological Psychology,* J. M. Sprague and A. N. Epstein, eds. Vol. 9. Academic Press, New York, pp. 1–42.

Merzenich, M. M., J. H. Kaas, J. T. Wall, R. J. Nelson, M. Sur, and D. J. Felleman (1983a) Topographic reorganization of somatosensory cortical areas 3b and 1 in adult monkeys following restricted deafferentation. *Neuroscience* **8**:33–55.

Merzenich, M. M., J. H. Kaas, J. T. Wall, M. Sur, R. J. Nelson, and D. J. Felleman (1983b) Progression of change following median nerve section in the cortical representation of the hand in areas 3b and 1 in adult owl and squirrel monkeys. *Neuroscience* **10**:639–665.

Merzenich, M. M., R. J. Nelson, M. P. Stryker, M. S. Cynader, A. Schoppmann, and J. M. Zook (1984a) Somatosensory cortical map changes following digit amputation in adult monkeys. *J. Comp. Neurol.* **224**:591–605.

Merzenich, M. M., W. M. Jenkins, and J. C. Middlebrooks (1984b) Observations and hypotheses on special organizational features of the central auditory nervous system. In *Dynamic*

Aspects of Neocortical Function, G. Edelman, M. Cowan, and E. Gall, eds. Wiley, New York, pp. 397–424.

Merzenich, M. M., R. J. Nelson, J. H. Kaas, M. P. Stryker, J. M. Zook, M. S. Cynader, and A. Schoppmann (1985a) Variability in hand surface representations in areas 3b and 1 in adult owl and squirrel monkeys. *J. Comp. Neurol.* (submitted).

Merzenich, M. M., J. H. Kaas, M. Sur, and G. M. Edelman (1985b) The selection and dynamics of cerebral cortical maps. *Science* (to be submitted).

Movshon, J. A. and R. C. Van Sluyters (1981) Visual neural development. *Ann. Rev. Psychol.* **32**:477–522.

Nelson, R. J., M. Sur, D. J. Felleman, and J. H. Kaas (1980a) Representations of the body surface in postcentral parietal cortex of *Macaca fascicularis*. *J. Comp. Neurol.* **192**:611–643.

Penfield, W. and T. Rasmussen (1950) *The Cerebral Cortex of Man*. Macmillan, New York.

Pubols, B. H. and L. M. Pubols (1971) Somatotopic organization of spider monkey somatic sensory cerebral cortex. *J. Comp. Neurol.* **141**:63–76.

Simons, D. J., D. Durham, and T. A. Woolsey (1984) Functional organization of mouse and rat SmI barrel cortex following vibrissal damage on different postnatal days. *Somatosen. Res.* **1**:207–245.

Sperry, R. W. (1963) Chemoaffinity in the orderly growth of nerve fiber patterns and connections. *PNAS* **50**:703–710.

Stryker, M. P., W. M. Jenkins, and M. M. Merzenich (1985) Anesthetic state does not affect the map of the hand representation within area 3b somatosensory cortex in owl monkey. *J. Comp. Neurol.* (submitted).

Sur, M., M. M. Merzenich, and J. H. Kaas (1980a) Magnification, receptive-field area, and hypercolumn size in areas 3b and 1 of somatosensory cortex in owl monkeys. *J. Neurophysiol.* **44**:295–311.

Sur, M., R. J. Nelson, and J. H. Kaas (1980b) Representation of the body surface in somatic koniocortex in the prosimian galago. *J. Comp. Neurol.* **189**:381–402.

Wall, J. T., M. M. Merzenich, M. Sur, R. J. Nelson, D. J. Felleman, and J. H. Kaas (1985) Somatosensory cortex organization following transection and regeneration of the median nerve in adult owl monkeys. *J. Neurosci.* (submitted).

Welker, W. I. (1974) Principles of organization of the ventrobasal complex in mammals. *Brain Behav. Evol.* **7**:253–336.

Woolsey, C. N., W. H. Marshall, and P. Bard (1942) Representation of cutaneous tactile sensibility in the cerebral cortex of the monkey as indicated by evoked potentials. *Bull. Johns Hopkins Hosp.* **70**:399–441.

Woolsey, C. N. (1952) Patterns of localization in sensory and motor areas of the cerebral cortex. In *The Biology of Mental Health and Disease*. Milbank Memorial Fund, Hoeber, New York, pp. 193–206.

Woolsey, C. N. (1954) Localization patterns in a lissencephalic primate (*Hapale jacchus*). *Am. J. Physiol.* **178**:686.

Wynn-Parry, C. B. (1973) *Rehabilitation of the Hand*. Butterworths, London.

Part Two

MODES OF COMMUNICATION

Commentary

C. LADD PROSSER

University of Illinois
Urbana, Illinois

The subjects presented in this section on modes of communication are extremely diverse. Such breadth of coverage is appropriate because of Ted Bullock's wide-ranging interests.

Chapter 7 presents results obtained by the relatively new technique of patch clamping. It is an analysis of calcium channels in lymphocytes. Such a biophysical analysis is a step toward identifying the molecular properties of membranes in all excitable systems. Ted was instrumental in introducing the giant barnacle muscle fibers to membrane neurophysiologists. The barnacle fibers were used to establish calcium inward currents. Here Hagiwara and Fukushima demonstrate that cells in the immune system also possess calcium channels and examine their relationship to antibody secretion.

Chapters 8 and 9 exemplify the evolutionary view that has permeated many of Ted's publications. Electrotonic coupling harks back to the electrical theory of synaptic transmission. It was during his career that both electrical and chemical transmission came to be recognized. One of Ted's early interests was in earthworm giant fibers, in which transmission across septa is by way of gap junctions. Loewenstein reviews current knowledge of gap junctions in his chapter. Conduction in epithelia, as demonstrated here by Josephson, probably preceded that in discrete neurons in the course of evolution.

Chapter 10 relates to Ted's career in its emphasis on humoral functions in nervous systems. The connection between neurons as secretory cells and as producers of transmitters was emphasized in his work with Horridge. Strumwasser reviews the neurohormonal commands that mediate a reproductive behavior in *Aplysia*. Chapter 11 is concerned with plasticity in nervous systems. In his studies on electrocommunication, Ted has emphasized the variations in coding according to behavioral interactions between fish. Walters and colleagues, in their chapter, emphasize that even simple reflexes are modifiable by experience.

Communication is what neurobiology is about. The modes of communication include membrane conductances, patterns of neuronal spikes and graded potentials, electrical coupling between cells, electrical and chemical transmission at synapses, secretion, and modification of neural function.

Chapter Seven

ION CHANNELS OF LYMPHOCYTES

S. HAGIWARA and Y. FUKUSHIMA

Department of Physiology
Ahmanson Laboratory of Neurobiology of the Brain Research Institute
and
Jerry Lewis Neuromuscular Research Center
University of California
Los Angeles, California

One of the authors (S. Hagiwara) has had a quarter-century-long association with Ted Bullock. In fact, Ted is the very first American scientist he became acquainted with in his life. When he joined the laboratory of Ted Bullock in 1955 the author was working mostly on synaptic transmission and sensory transduction. Later, his interests shifted to membrane biophysics of ion channels, particularly Ca^{2+} channels.

In 1963 Hoyle and Symth described the giant muscle fiber system in the barnacle *Balanus nubilis* Darwin. Muscle fibers wider than 1 mm and longer than 5 cm could be obtained from this animal. Until that time the squid giant axon was the only preparation in which one could change the intracellular as well as the extracellular medium. One could perform the same procedures with the barnacle muscle fiber. These barnacles are distributed along the Pacific coast of the American continent and can be found in relatively shallow waters north of the Puget Sound in the United States or south of Valparaiso in Chile. At the latitude of Los Angeles one must dive 50–100 feet to collect the specimens, and this requires special skill and luck. Ted was consulted and offered an excellent idea: collecting the specimens by diving from the offshore oil drilling structures in Santa Barbara. This was actually performed by Dr. R. Fay, who is now the President of Pacific Biomarine Company. The method was very successful, and enough specimens were supplied during the following six months until Dr. Fay found locations where he could collect the specimens more easily.

Ca channels were found and analyzed extensively in the barnacle muscle fiber. At that time most membrane physiologists believed that Ca channels were a rather exceptional membrane channel found only in crustacean muscle fibers. During the following decade Ca channels were found in a wide variety of nerve and muscle preparations. In fact, at this moment it is difficult to mention nerve and muscle tissues that do not have Ca channels. Ca channels are also found in a number of non-nerve and muscle tissues (for a review see Hagiwara and Byerly, 1981, 1983; Hagiwara, 1983). As a matter of fact, one of the major topics for this chapter concerns the Ca channels that are found in non-nerve-muscle tissue, lymphocytes.

Immunology and neurobiology are very different branches of science. However, we can find a number of phenomenological similarities in nervous systems and immune systems. An immunologist, Golub (1982), emphasizes these similarities by stating that both systems recognize and react to the outside world, as well as learn, remember, and even forget. The cell membrane has been one of the major concerns for immunology as well as for research in neurobiology. However, the two disciplines use very different experimental approaches and differ in their interpretations of phenomena occurring at the cell membrane. Electrophysiological techniques, particularly intracellular electrode techniques, have been employed very successfully in neurobiology. A few attempts (Taki, 1970) have been made to apply these techniques to the study of membrane properties of lymphocytes, which play the major role in immune reactions. Because of the small size (about 10 μm in diameter) of lymphocytes, it was extremely difficult to record their membrane potentials reliably. Recent introduction of the whole cell variation of the patch electrode voltage clamp technique has overcome this difficulty (Hamill et al., 1981).

Immunologists classify lymphocytes into two groups, B- and T-lymphocytes. B-lymphocytes originate from the bone marrow, which is the equivalent of the bursa of Fabricius in birds. They secrete antibodies when they are stimulated by appropriate antigens. T-lymphocytes originate from the thymus, and they act as helpers and suppressors to B-cells when forming antibodies. Some of the T-cells participate in the cell-mediated cytotoxic reactions (cytotoxic T-lymphocytes).

Ca CHANNELS OF MYELOMA CELLS

To study membrane properties of lymphocytes we used clonal cell lines. This facilitated studies because cells are obtainable by tissue culture techniques. The first preparation we studied was mouse myeloma cell line S194 (Fukushima and Hagiwara, 1983). Multiple myeloma is a neoplastic disease of B-lymphocytes. We found inward currents in this myeloma cell during voltage clamp. Traces in Figure 1 (left) were obtained when the external solution consisted of 116 m*M* NaCl, 5 m*M* KCl, 25 m*M* CaC_2, 1 m*M* $MgCl_2$, 17 m*M* glucose, and 10 m*M* Hepes (pH = 7.4). The electrode was filled with a solution containing 150 m*M* KCl, 1 m*M* $MgCl_2$, 5 m*M* EGTA, 2.5 m*M* $CaCl_2$ (pCa = 7), and 10 m*M* Hepes (pH = 7.4). Replacing the 116 m*M* NaCl in the external saline with an equimolar tetraethylammonium chloride did not alter the inward currents. In contrast, the currents disappeared reversibly when 25 m*M* $CaCl_2$ was replaced with $MnCl_2$ [Fig. 1 (right)]. The results suggest strongly that the current is carried by Ca^{2+} rather than Na^+. This was further confirmed by observing increases of the inward current with an increase of the external Ca^{2+} concentration.

The properties of this Ca channel differ from those of Ca channels often found in neuromuscular systems in the following two aspects. First, as in other Ca channels Ba^{2+} and Sr^{2+} can carry currents (Fig. 2). In many other preparations Ba^{2+} carries significantly greater current as compared with Ca^{2+} or Sr^{2+}. Thus Ba^{2+} is often called *supercalcium*. In this preparation, however, Sr^{2+} always carries a greater current at a given membrane potential than do Ba^{2+} and Ca^{2+}. This property resembles that of the Ca channel in mouse eggs (Okamoto et al., 1977; Yamashita, 1982). Second, the inward current decayed during the maintained voltage pulse. Time courses of decays at a given membrane potential were similar in Ca^{2+}, Sr^{2+}, and Ba^{2+} solutions (Fig. 2). The decay of the Ca current in nerve and muscle systems is often due to the influx of Ca^{2+} ions during the inward current. The increase of the internal Ca^{2+} induces the counteracting outward K current (Meech, 1974) or results in Ca-induced blockade of the Ca channel (Tillotson, 1979). These effects are practically absent in Ba^{2+}. The fact that the decay is similar in Ca, Sr, and Ba media argues against this possibility. When the current was detected by altering the holding potential, the current quickly diminished at the membrane potential at which the inward Ca current is hardly activated (Fig. 3). In other words, the Ca channel of the myeloma cell has voltage-dependent inactivation as in the case of the voltage-gated Na channel of the nervous system.

Ca CHANNELS AND IMMUNOGLOBULIN SECRETION

When we discovered the Ca channel in the myeloma cell line S194 we worried that only a small percentage of the cells showed significant Ca currents while only just recognizable current was found in the remaining cells. Thus we had some doubt about the biological significance of the Ca channel in immune function. Then we realized that the clone S194 synthesizes immunoglobulins but does not secrete them. We then obtained hy-

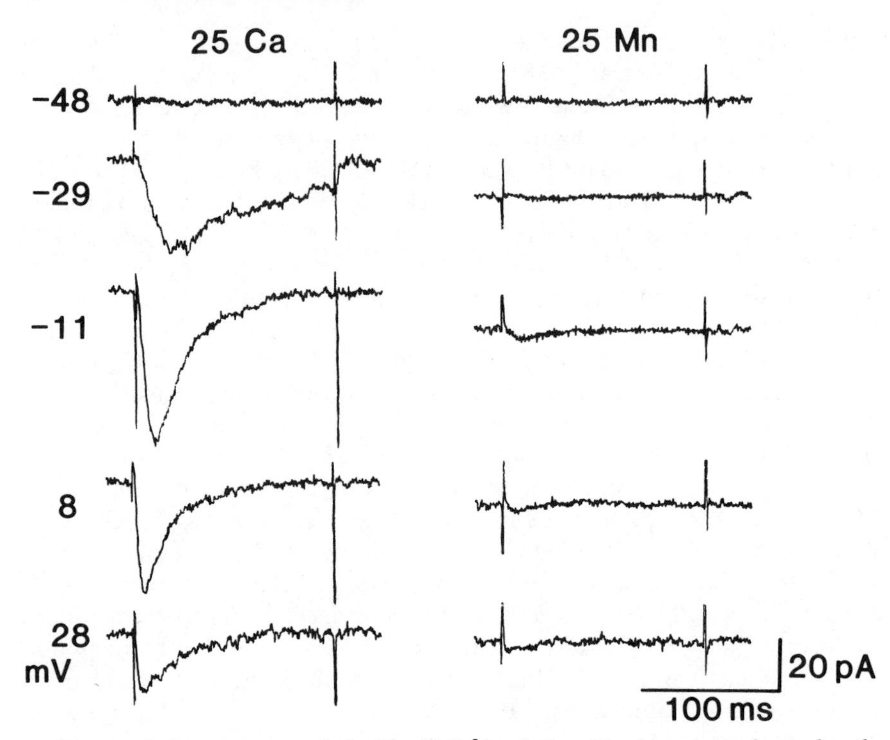

FIGURE 1. Left, current records in 25 mM Ca^{2+} solution. Membrane was clamped to the potential (mV) shown on the left of each trace; the holding potential was −96 mV. Right, current records in 25 mM Mn^{2+} solution. All Ca^{2+} in the external solution was replaced by Mn^{2+}. In both sets of current records, the transients due to the capacitance of the electrode were compensated by an analogue subtraction.

bridomas constructed by fusion of S194 and splenic B-lymphocytes. One line (MAb2-1) secretes immunoglobulin G, and the other line (7B) secretes immunoglobulin M. In these hybridomas we invariably found significant Ca currents. This encouraged us to make further investigations into the relationship between the Ca channel and immunoglobulin secretion. Although the current was significant, it varied among different cells. We first thought that the difference might be due to the size of the cell. Experiments were performed in 25 mM Ca media, and the current-voltage relations at the peak of the inward current were constructed as in the method of Fig. 2*B*. The maximum peak inward current was then plotted against the diameter of the cell (Fig. 4). The result indicates that the variation is independent of the cell size.

Thereafter we noticed that there was a correlation between the Ca current and the time after the cells were transferred into fresh culture medium. This led us to perform the following experiments. Hybridomas were cultured in suspension in RPMI 1640 medium supplemented with 10% fetal calf serum. Culture medium was renewed every four days. Each

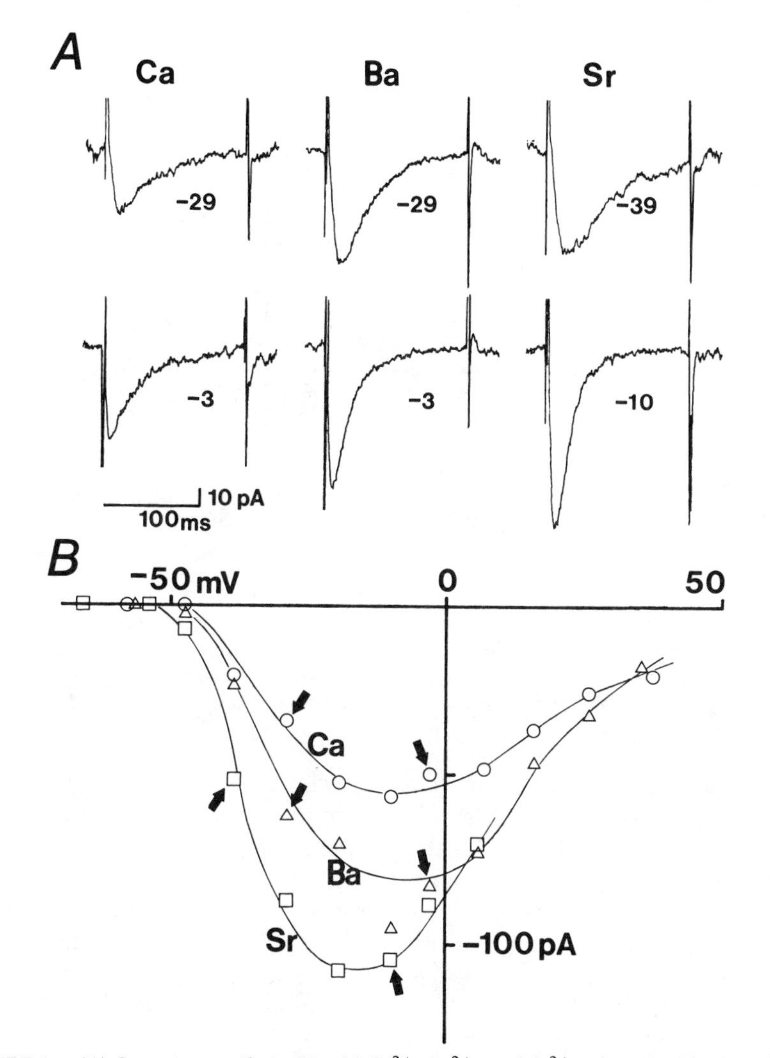

FIGURE 2. *(A)* Current records in 25 m*M* Ca^{2+}, Ba^{2+}, and Sr^{2+} solutions. The membrane potential level (mV) is indicated adjacent to each record. The data from these current records are indicated by arrows in *(B)*. *(B)* Current-voltage relationships at the peak of the inward current in 25 m*M* Ca^{2+}, Ba^{2+}, or Sr^{2+} solutions. (From Fukushima and Hagiwara, 1983.)

time the cell density was adjusted to start at about 20×10^4 cells per milliliter. The cell density increased sigmoidally and reached saturation the third day. Immunoglobulin content in the supernatant was measured each day by a sandwich radioimmunoassay. The increment of immunoglobulin in two successive days was obtained as the amount secreted per day. This was then divided by an appropriate cell number to obtain the immunoglobulin secretion per cell per day as illustrated in Figure 5 (open circles). The average value of the maximum peak calcium current obtained each day

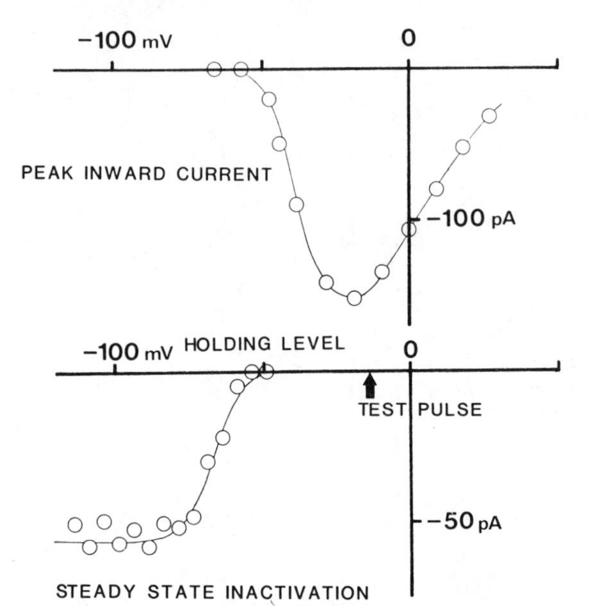

FIGURE 3. Upper, current-voltage relation at the peak of the imward current in 10 m*M* Ca^{2+} solution. Lower, steady-state inactivation in 10m*M* Ca^{2+} solution. The pilot shows the amplitude of peak inward current obtained from a constant test command pulse to −10 mV (arrow) from various holding potentials. (From Fukushima and Hagiwara, 1983.)

was plotted on the same figure (closed circles). We find a good parallelism between the two events. A similar experiment was performed with the initial cell density of 46 × 10^4 cells per milliliter. Then both curves shifted to the left, but the parallelism was maintained. At this stage we do not know whether the evidence demonstrates the role of the Ca channel in immunoglobulin secretion or if the parallelism is merely coincidental. One further experimental result suggests that the former could be true. The Ca channel blocker D600 also blocks the Ca channel of hybridomas at concen-

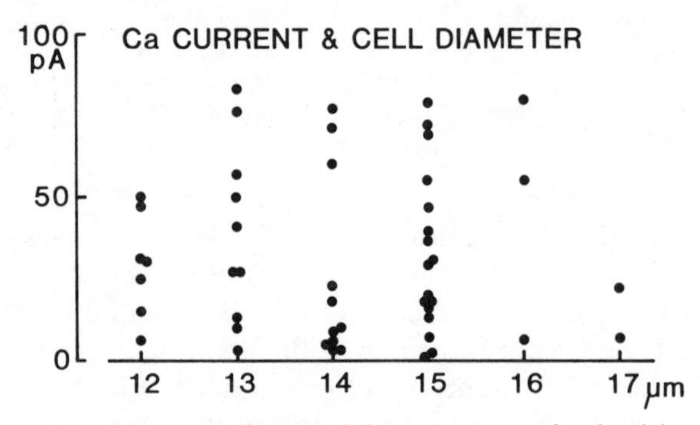

FIGURE 4. Correlation between cell size and the mximum amplitude of the peak inward Ca current in 25 m*M* Ca solution. Results show there was no correlation.

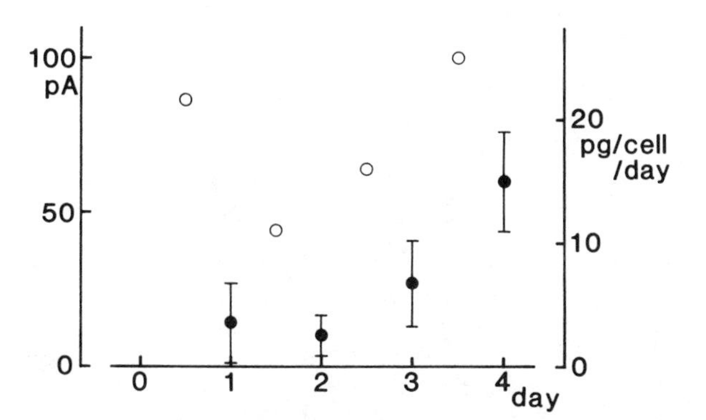

FIGURE 5. Ca current and immunoglobulin secretion in hybridoma secreting IgG measured simultaneously after renewal of the culture medium. Closed circles indicate the maximum amplitude of the peak inward Ca current in 25 m*M* Ca solution (left axis). Vertical bars indicate standard deviations. Open circles indicate the immunoglobulin secretion per cell per day in picograms (right axis).

trations of 100 μM (37% reduction). D600 at this concentration suppressed immunoglobulin secretion significantly even when it did not cause any significant reduction in cell density. This, of course, could be an effect of D600 other than blocking the Ca channel. This line of research is still in progress, and we believe the results are very promising (Fukushima, Hagiwara, and Saxton, 1984).

K CURRENT IN CYTOTOXIC T-LYMPHOCYTES

Recently methods have been developed for cloning cytotoxic T-lymphocytes (Fathman and Fitch, 1982). We examined membrane electrical properties of the mouse cytotoxic T-lymphocyte (CTL) cell line. In this lymphocyte no inward currents were observed under any condition. Instead, voltage-activated outward currents were found as shown in the inset of Figure 6 (Fukushima et al., 1984). The current activated with a time constant of several milliseconds followed by inactivation with a time constant of several hundred milliseconds. The current-voltage relations at the maximum of the outward current are shown in Figure 6. Two relations were obtained from the same cell, one in saline containing 2.5 m*M* $CaCl_2$ and the other with $CaCl_2$ being replaced with $MgCl_2$, that is, Ca-free. The two relations are nearly identical; this indicates that the outward current was not produced as a result of the influx of Ca^{2+} from the outside. The examination of the reversal potential indicated that the current is predominantly carried by K^+ ions.

The current is suppressed both by quinidine and tetraethylammonium (TEA) chloride. Figure 7 shows the dose-response curves. Broken lines were drawn according to one-to-one binding. Quinidine is far more effec-

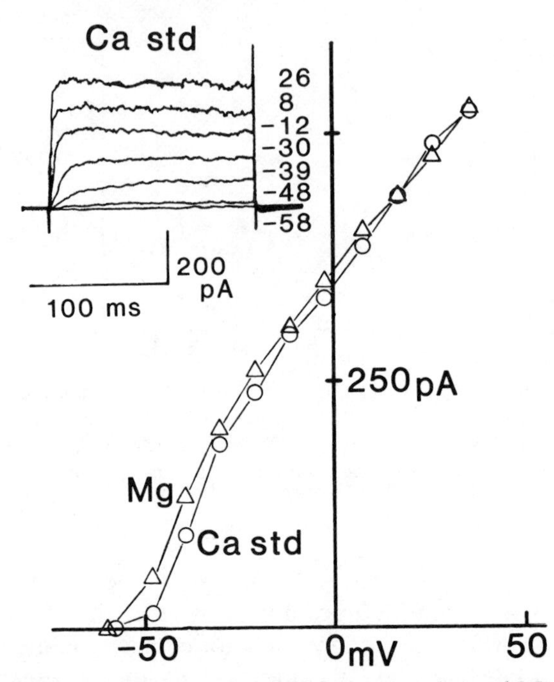

FIGURE 6. Current-voltage relation in 2.5 m*M* Ca (open circles) and 2.5 m*M* Mg (triangles) solutions at the peak of the outward current. I-V relations were measured 15–30 min after establishing the clamp. Inset, current traces recorded in 2.5 m*M* Ca solution. The figures on the right indicate the membrane potential (mV) during the voltage pulse. Holding potential was −90 mV.

tive than TEA in reducing the currents: 100 μ*M* quinidine eliminated the K current completely. The dissociation constant of quinidine was 23 μ*M*. Quinidine has been known to specifically block the Ca-induced K channel in a variety of tissues such as erythrocytes (Armandy-Hardy et al., 1975), Langerhans' islet cells (Atwater et al., 1979), barnacle photoreceptor cells (Hanani and Shaw, 1979), hepatocytes (Burgess et al., 1981), and fibroblast L cells (Okada et al., 1982). The K current of the CTL is substantially more sensitive to quinidine than the Ca-induced K currents of these preparations. Despite this fact we could not find any positive evidence that this K current is the Ca-induced one. In a few experiments the internal electrode contained 50 m*M* EGTA, and this should lower the internal free Ca^{2+} concentration. However, no significant changes were observed. It is necessary to make a further investigation to determine the character of this K channel. A similar K channel has been found in other types of T-lymphocytes (human helper and suppressor lymphocytes) by De Coursey, Chandy, Gupta, and Cahalan (1984) and Matteson and Deutsch (1984).

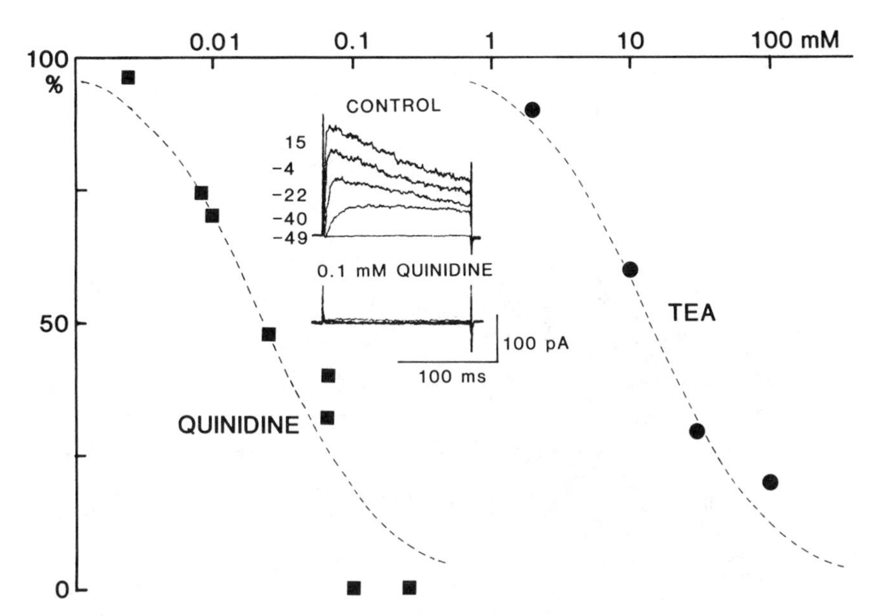

FIGURE 7. Dose-response relations for blocking by quinidine (closed squares) and TEA (closed circles). The ratio of the maximum chord conductance in the test solution to that in the control (no drug) solution is plotted against the drug concentration (logarithmic scale). Broken lines represent curves calculated for one-to-one binding. Inset: effect of 100 μM quinidine on the K currents in 10 mM solution.

CHANGES OF MEMBRANE CURRENT DURING CYTOTOXIC REACTION

CTL is an effector cell in cellular immunity. The CTL recognizes the difference of major histocompatibility antigens and kills nonself target cells. The CTL-mediated target cell lysis has been resolved into three successive stages: recognition-adhesion; the lethal hit or programming for lysis; and killer cell independent lysis (Martz, 1977; Martz et al., 1982). These three stages are characterized by requirements for certain divalent cations and temperatures. Mg^{2+} in the absence of Ca^{2+} is sufficient to support the adhesion formation, whereas Ca^{2+} is necessary for the lethal hit stage (Goldstein and Smith, 1976). Temperatures below 37°C inhibit the lethal hit but do not inhibit the adhesion formation (Wagner and Rollinghoff, 1974; Martz, 1975). After the target cell receives the lethal hit from the CTL, which occurs within several minutes, the lysis of the target cell occurs slowly during the next several hours without further participation of the CTL.

We attempted to observe the membrane changes of the CTL during the lethal hit in the following way. First CTL and target cell conjugates were

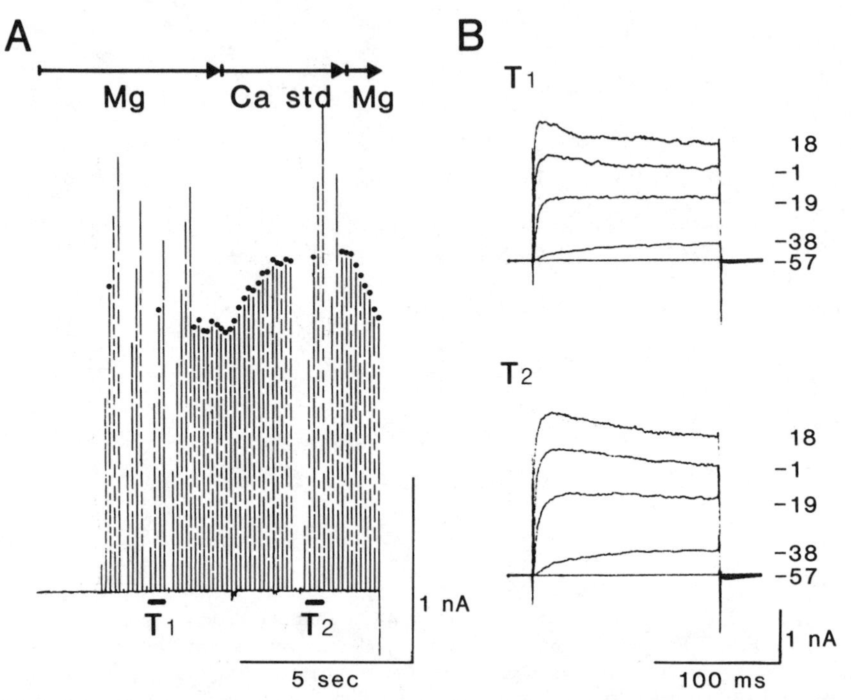

FIGURE 8. The chart record of a conjugated cell experiment. Current responses to a constant 150-msec voltage pulse to −1 mV are marked with dots. The pulses were applied at 8-sec intervals in this experiment. Currents were recorded first in 2.5 m*M* Mg solution, then in 2.5 m*M* Ca solution, and finally in Mg solution again. Time proceeds from left to right. Recordings were made at 37°C. The currents diminished and the cell appeared to disintegrate at the end of the chart record. (*B*) Current traces recorded at the two time periods indicated by thick lines in (*A*).

formed in Ca-free Mg medium at 20°C. Then preparations were warmed to 37°C; under these conditions lethal hit was not supposed to occur. Then the whole cell clamp was established in CTL, and the current-voltage relation was observed. This was followed by applying a solution at 37°C containing 2.5 m*M* Ca. This should lead to the lethal hit. In most cases recording conditions deteriorated rapidly at 37°C. However, on a few occasions we observed an increase of the outward K current when Ca solution was applied. In the case shown in Figure 8 the increase was about 20% over the entire range of the observed voltage range. In other words, the increase was not N shaped which is characteristic of the Ca-induced K current produced by the voltage-gated Ca^{2+} entry. At this time we have no conclusive evidence whether or not this is associated with the lethal hit since we did not observe the fate of the target cells. The phenomenon may be related to the finding that the lethal hit is associated with an enhanced $^{86}Rb^{+}$ efflux from preloaded CTL (Russel and Dobos, 1983).

CONCLUSION

The whole cell variation of the patch electrode voltge clamp technique gives us an excellent opportunity to investigate electrical properties of blood cells that play important roles in various immunological phenomena. There are a few disadvantages in using this method. The interior of the cell is rapidly exchanged with the solution in the electrode after the whole cell clamp is established. This may drastically alter the responsiveness of the cell. Some of the necessary enzymes may be depleted or some of the membrane proteins may undergo conformational changes. It is therefore necessary to add various factors in the internal solution. The recent finding (Doroshenko et al., 1982) that the reduction of the Ca current is prevented by adding APT, Mg^{2+}, cyclic AMP, and theophylline is a good example. To obtain a satisfactory seal between cell membrane and the patch electrode it is often necessary to treat the cell with enzymes to remove the structures on the cell membrane. As in the case of our studies, clonal tissue cultured cell lines are often used since the surface of cultured cells is usually very clean. There is no guarantee, however, that the treatment with enzymes does not alter the properties of the cell. The clonal cell lines may differ from normal cells. The properties of the cultured cell may change depending upon culture conditions. If we pay close attention to these points, this line of work may open a new approach to further elucidate immunological problems.

Acknowledgments

We express our appreciation to Dr. R. Irie of UCLA and Dr. J. Blustone of N.I.H. for kindly supplying cell culture lines and to Dr. M. Bosma for invaluable criticism during the preparation of the manuscript. The work was supported by USPHS Grant NS 09012 and a grant from the Muscular Dystrophy Association to Dr. S. Hagiwara and a Muscular Dystrophy Association Research Fellowship to Dr. Y. Fukushima.

REFERENCES

Armando-Hardy, M., J. C. Ellory, H. G. Ferreira, S. Fleminger, and V. L. Lew (1975) Inhibition of the calcium-induced increase in the potassium permeability of human red blood cells by quinine. *J. Physiol. (Lond.)* **205**:32–33P.

Atwater, I., C. M. Dawson, B. Ribalet, and E. Rojas (1979) Potassium permeability activated by intracellular calcium ion concentration in the pancreatic cell. *J. Physiol. (Lond.)* **288**:575–588.

DeCoursey, T., K. G. Chandy, S. Gupta, and M. D. Cahalan (1984) Voltage-gated K^+ channels in human T lymphocytes: a role in mitogenesis? *Nature* **307**:465–468.

Doroshenko, P. A., P. G. Kostyuk, and A. E. Martynyuk (1982) Intracellular metabolism of adenosine 3′,5′-cyclic monophosphate and calcium inward current in perfused neurones of *Helix pomatia*. *Neuroscience* **7**:2125–2134.

Fathman C. G. and F. W. Fitch (1982) *Isolation, Characterization and Utilization of T Lymphocyte Clones*. Academic Press, New York.

Fukushima, Y. and S. Hagiwara (1983) Voltage-gated Ca^{2+} channel in mouse myeloma cells. *Proc. Natl. Acad. Sci. U.S.A.* **80**:2240–2242.

Fukushima, Y., S. Hagiwara, and M. Henkart (1984) Potassium current of clonal cytotoxic T lymphocytes in the mouse. *J. Physiol. (Lond.)* **351**:645–656.

Fukushima, Y., S. Hagiwara, and R. E. Saxton (1984) Variation of calcium current during the cell growth cycle in moust hybridoma lines secreting immunoglobulins. *J. Physiol. (Lond.)* **355**:313–321.

Goldstein, P. and E. T. Smith (1976) The lethal hit stage of mouse T and non-T cell-mediated cytolysis: differences in cation requirements and characterization of an analytical "cation pulses" method. *Eur. J. Immunol.* **6**:31–37.

Golub, E. (1982) Connections between the nervous, haematopoietic and germ-cell systems. *Nature* **299**:483.

Hagiwara, S. (1983) *Membrane Potential Dependent Ion Channels in Cell Membrane: Phylogenetic and Developmental Approaches*. Distinguished lecture series of the Society of General Physiologists, Vol. 3. Raven Press, New York.

Hagiwara, S. and L. Byerly (1981) Calcium channel. *Ann. Rev. Neurosci.* **4**:69–125.

Hagiwara, S. and L. Byerly (1983) The calcium channel. *Trends NeuroSci.* **6**:189–193.

Hamill, O. P., A. Marty, E. Neher, B. Sakmann, and F. J. Sigworth (1981) Improved patch-clamp techniques for high-resolution current recording from cells and cell-free membrane patches. *Pflügers Arch.* **391**:85–100.

Hoyle, G. and T. Smyth, Jr. (1963) Neuromuscular physiology of giant muscle fibers of a barnacle, *Balanus nubilus* Darwin. *Comp. Biochem. Physiol.* **10**:291–314.

Martz, E. (1975) Early steps in specific tumor cell lysis by sensitized mouse T lymphocytes. I. Resolution and characterization. *J. Immunol.* **115**:261–267.

Martz, E. (1977) Mechanism of specific tumor lysis by alloimmune T-lymphocytes: resolution and characterization of discrete steps in the cellular interaction. *Contemp. Top. Immunobiol.* **7**:301–361.

Martz, E., W. L. Parker, M. K. Gately, and C. D. Tsoukas (1982) The role of calcium in the lethal hit of T lymphocyte-mediated cytolysis. *Adv. Exp. Med. Biol.* **146**:121–143.

Meech, R. W. (1974) The sensitivity of *Helix aspera* neurones to injected calcium ions. *J. Physiol. (Lond.)* **237**:259–277.

Okada, Y., W. Tsuchiya, and T. Yada (1982) Calcium channel and calcium pump involved in oscillatory hyperpolarizing responses of L-strain mouse fibroblasts. *J. Physiol. (Lond.)* **327**:449–461.

Okamoto, H., K. Takahahsi, and N. Yamashita (1977) Ionic currents through the membrane of the mammalian oocyte and their comparison with those in the tunicate and sea urchin. *J. Physiol. (Lond.)* **267**:665–695.

Russell, J. H. and C. B. Dobos (1983) Accelerated $^{86}Rb^{+}(K^{+})$ release from the cytotoxic T lymphocyte is a physiologic event associated with delivery of the lethal hit. *J. Immunol.* **131**:1138–1141.

Taki, M. (1970) Studies on blastogenesis of human lymphocytes by phytohemagglutinin, with special reference to changes of membrane potential during blastoid transformation. *Mie. Med. J.* **29**:245–262.

Matteson, D. R. and C. Deutsch (1984) K channels in T lymphocytes: a patch clamp study using monoclonal antibody adhesion. *Nature* **307**:468–471.

Tillotson, D. (1979) Inactivation of Ca conductance dependent on entry of Ca ions in molluscan neurons. *Proc. Natl. Acad. Sci. U.S.A.* **76:**1497–1500.

Wagner, H. and M. Rollinghoff (1974) T cell-mediated cytotoxicity: discrimination between antigen recognition, lethal hit, and cytolysis phase. *Eur. J. Immunol.* **4:**745–750.

Yamashita, N. (1982) Enhancement of ionic currents through voltage-gated channels in the mouse oocyte after fertilization. *J. Physiol. (Lond.)* **329:**263–280.

Chapter Eight

COMMUNICATION BY CONDUCTING EPITHELIA

ROBERT K. JOSEPHSON

School of Biological Sciences
University of California
Irvine, California

The topic to be considered here is epithelial conduction as a mode of communication between cells. In seeking a way to introduce the topic, to set the stage for the reader before the main exposition of experimental results, I turned to that usually reliable source of inspiration, Bullock and Horridge's *Structure and Function in the Nervous Systems of Invertebrates* (1965). The paucity of information about conducting epithelia therein reminded me rather forcefully that epithelial conduction as a mode of intercellular communication and as a mechanism of behavioral control has been recognized only recently. It has been known for a very long time, of course, that there can be spread of action potentials through collections of muscle cells, specifically cardiac and smooth muscle. There have also been hints of intercellular conduction in other tissues, in particular ciliated epithelia, that did not appear to involve neuronal activity. Nervelike conduction in cells other than nerve or muscle cells has been termed "neuroid" conduction following G. H. Parker who was an early investigator of such phenomena (e.g., Parker, 1919). Because the evidence for impulse conduction in cells other than nerve or muscle was incomplete and unconvincing at the time of writing, in the text of Bullock and Horridge it is generally assumed that nerve and muscle cells are the only cells that conduct excitation and that the spread of activity through a tissue, measured electrophysiologically or behaviorally, is a reflection of nerve or muscle activity. This is not said in criticism, in fact, I made the same assumptions in papers published at that time, but as an indication of the prevailing view, about 20 years ago, that conduction of excitation and neuronal activity were essentially synomous concepts.

Coincidentally, 1965 saw both the appearance of Bullock and Horridge's long-awaited monograph and the first truly convincing demonstration of epithelial conduction as a behavioral-coordinating mechanism; the latter in a paper by G. O. Mackie demonstrating electrical conduction of behaviorally meaningful signals in nerve-free epithelia of siphonophores. Subsequently epithelial conduction has been found to be moderately widespread. Impulse conduction has been convincingly demonstrated now in several dozen epithelia distributed among four different animal phyla (Cnidaria, Annelida, Mollusca, Chordata), and there are hints that it occurs in other animal groups as well (for review see Anderson, 1980).

Epithelia are sheets of cells that cover body surfaces or line cavities and that are separated from underlying structures by a basement membrane. Because epithelia are two dimensional, epithelial conduction is nicely intermediate in geometric complexity between axonal conduction, which is basically linear, and complex, three-dimensional excitation spread found, for example, in heart tissue. Electrical interactions between cells of different epithelia range from simple electrical coupling (e.g., Loewenstein, 1981 and Chapter 9 of this volume) to instances in which the coupled cells are electrically excitable but in which the action potentials are not fully regenerative and so die away with distance (Murakami and Machemer, 1982) to epithelia capable of generating essentially all-or-nothing action potentials that are propagated without decrement. The last, epithelia that produce propagated, all-or-nothing impulses, are considered here.

AN EXAMPLE OF EPITHELIAL CONDUCTION: THE EXUMBRELLAR EPITHELIUM OF *EUPHYSA JAPONICA*

The hydromedusa *Euphysa japonica* is a cup-shaped jellyfish about 1 cm high as an adult. Electrically or mechanically stimulating the upper (exumbrellar) surface of *E. japonica* or of any of a number of other hydromedusae evokes a protective response termed *crumpling* in which swimming pulsations cease and the margins of the swimming bell are drawn inward by contraction of radial and marginal muscles (Hyman, 1940; Mackie and Passano, 1968). Excitation is conducted in the exumbrellar epithelium from the point of stimulation, and its progress can be monitored with electrodes on the epithelium as an electrical event conducted at 5–50 cm/sec varying with the temperature and species. In all hydromedusae that have been carefully examined with light and electron microscopy, the exumbrellar epithelium has been found to lack nerves. The exumbrellar epithelia are structurally very simple tissues, consisting of only a single layer of flattened epithelial cells sitting on a thick, transparent mesoglea. If nerves were present in the epithelia, it is inconceivable that they would not have been found. Thus there can be no doubt that conduction in the exumbrellar epithelium does not invole nerves; it is purely epithelial conduction.

Intracellular recordings quite clearly show that the exumbrellar epithelial cells of *E. japonica* are electrically coupled and electrically excitable (unless otherwise specified, information in the remainder of this section is from Josephson and Schwab, 1979 or Schwab and Josephson, 1982). The cells of the epithelium are broad (about 70 μm) but quite thin (0.4–4 μm, average 1.4 μm), which poses technical problems for successful penetration of a cell with a microelectrode (Fig. 1). Nevertheless, with care it is possible to record from cells and simultaneously to impale separate cells in the epithelium. Passing electrical current into one cell, thereby changing its membrane potential, results in potential changes in nearby cells, thus demonstrating cell coupling (Fig. 2). The cell coupling is presumably mediated by the gap junctions found between adjacent epithelial cells.

The amplitude of the potential change in one epithelial cell resulting from the injection of a fixed current into another cell is a function of the distance between the current and recording electrodes. Because the epithelium is a two-dimensional sheet, the rules governing the decline in potential with distance are a bit different from those that apply to the more familiar one-dimensional cables such as axons. Fortunately steady-state solutions for the relations between current and voltage in a two-dimensional cable have been worked out (e.g., Jack et al., 1975; see Fig. 3). Cable analysis to determine the passive electrical properties of a two-dimensional epithelium is only a little more difficult than cable analysis of an extended cylinder such as an axon or a muscle cell. From the decay of potential with distance from the current source in the epithelium of *E. japonica*, and the average epithelial thickness, it is calculated that the electrical resistance of the cytoplasm including the resistance of the intercellular junctions is about 200 Ω-cm. This value is remarkable since it is similar to specific resistance values determined for cytoplasm alone in many cells. Seemingly, most of the internal longitudinal resistance within the epithelium can be attributed to the cytoplasm; the intercellular junctions must be very permeable indeed.

In contrast to the intercellular junctions, the resistance of nonjunctional cell boundaries in the epithelium is rather high. The calculated resistance between the cytoplasm and the outside of the epithelium is 23 KΩ-cm^2. This represents the combined resistance of the apical and basal membranes as parallel pathways. The resistance of neither apical nor basal membranes can be less than 23 KΩ-cm^2, but how the total resistance is apportioned between apical and basal portions of the cells is unknown. Somewhat paradoxically given the high membrane resistance, the transverse resistance across the epithelium, from its mesogleal to its outer surface, is quite

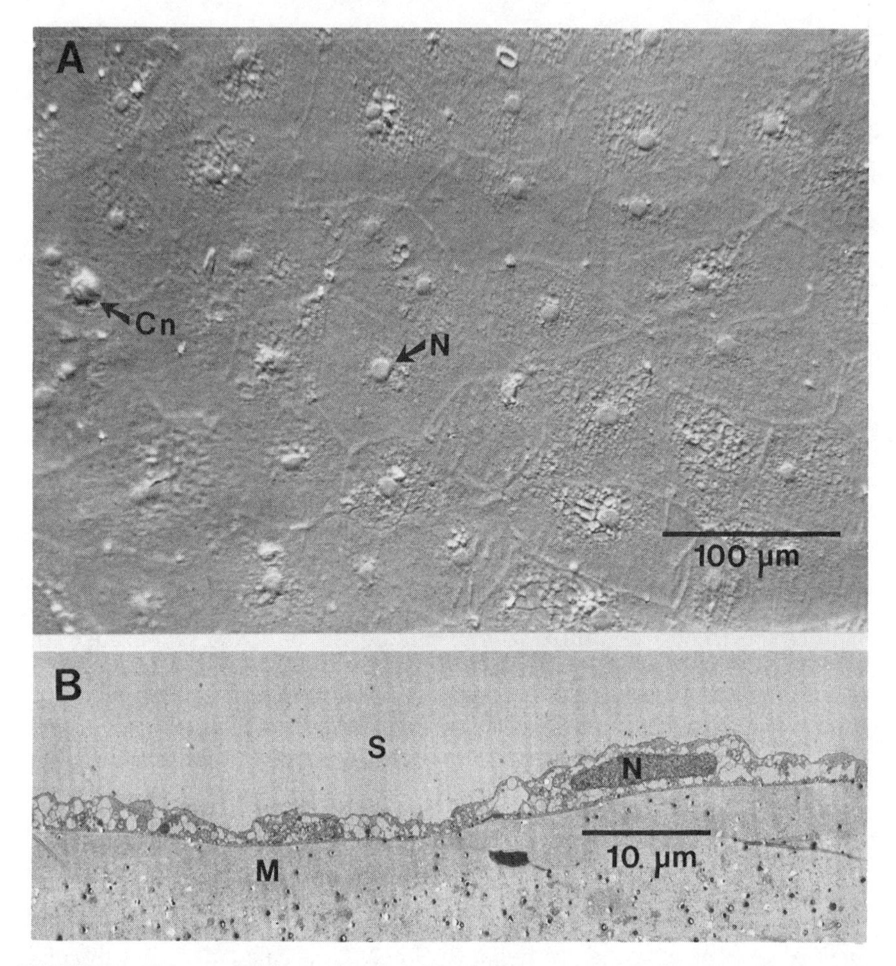

FIGURE 1. The exumbrellar epithelium of *Euphysa japonica*. (*a*) Surface view of living cells, Hoffman modulation contrast optics. (*b*) Transmission electron micrograph of section through the epithelium. Cn, nematocyst; N, nucleus; M, mesoglea; S, seawater. (Reproduced from *The Journal of General Physiology*, Vol. 74, 1979, by copyright permission of the Rockefeller University Press.)

low (7.5 Ω-cm^2), indicating that there are major low-resistance pathways transversely between epithelial cells. Morphological counterparts of these pathways have not been seen in the limited electron microscopical examination done with this tissue. A similar cable analysis of a myoepithelium in a siphonophore gave values of 110 Ω-cm^2 for the membrane resistance and 34 Ω-cm for the specific resistivity of cytoplasm and junctions (Chain et al., 1981). The value for internal resistivity seems questionable since it is much lower than values determined for cytoplasm alone in many other tissues (reviewed in Josephson and Schwab, 1979).

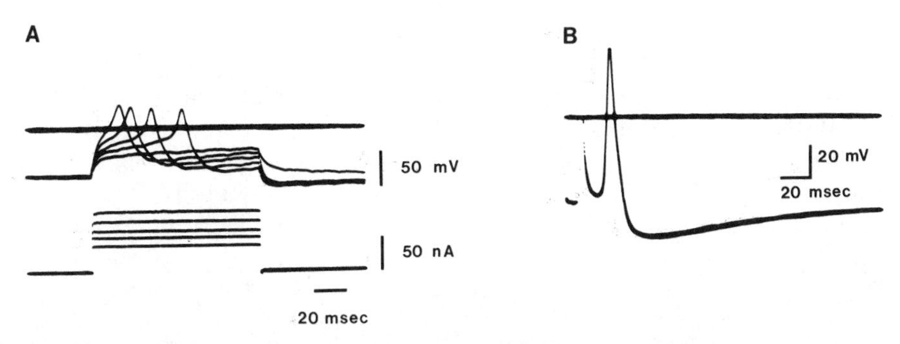

FIGURE 2. Electrical coupling and electrical excitability in the epithelium of *E. japonica*. (*A*) Current injected into one cell (monitored in lower set of traces) results in voltage change in adjacent cell (middle set of traces, upper trace marks the zero potential level) showing that the cells are electrically coupled. Large depolarizing currents evoke action potentials. (*B*) An action potential recorded with an intracellular electrode in an epithelial cell. (Reproduced from *The Journal of General Physiology*, Vol. 74, 1979, by copyright permission of the Rockefeller University Press.)

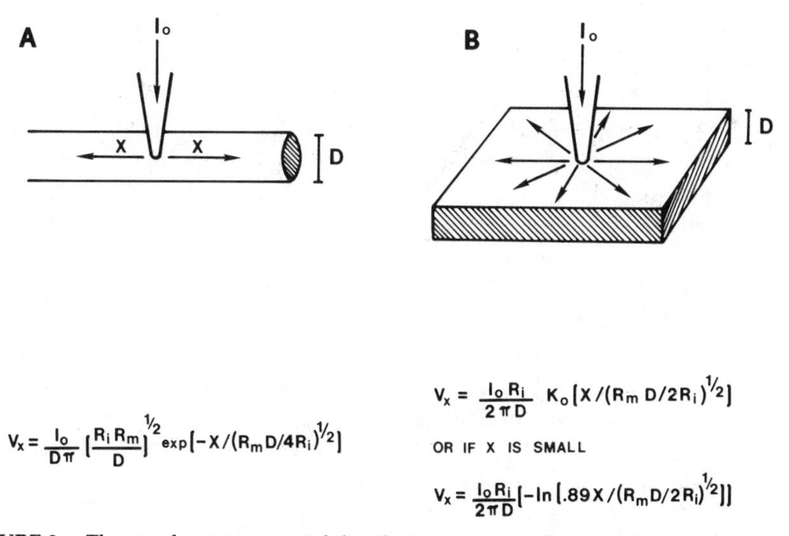

FIGURE 3. The steady-state potential distribution to imposed current in a one-dimensional cable such as an axon (*A*) and in a two-dimensional cable such as an epithelial sheet (*B*). The lower equation in (*B*) is a reasonable approximation if $X/(R_m D/2R_i)^{1/2}$ is less than about 0.5 (see Josephson and Schwab, 1979). D, cable diameter or epithelial sheet thickness (cm); I_o, amplitude of injected current (A); K_o, a modified Bessel function; R_i, the resistivity of cytoplasm and, for the epithelium, of intercellular junctions (Ω-cm); R_m, specific membrane resistivity (Ω-cm^2); X, distance from current source (cm); V_x, the potential at distance X (V).

Depolarizing current pulses evoke action potentials, thus demonstrating the electrical excitability of the tissue (Fig. 2). The depolarizing phase of the action potential overshoots zero and is about 11 msec long at 11°C. The initial depolarizing spike is followed by a hyperpolarizing after-potential lasting nearly a second. Available evidence suggests that the depolarizing current during an action potential may be carried by either Na^+ or Ca^{2+}. Action potentials were not blocked by tetrodotoxin (TTX, up to 3×10^{-3} *M*) or by bathing the epithelium in Na^+-free solution (Tris as substitute) or in Ca^{2+}-free solution (Na^+ as substitute, solution contained 2 m*M* EGTA), but action potentials were blocked by bathing the preparation in solutions lacking both Na^+ and Ca^{2+} (Schwab and Josephson, 1982; Schwab, personal communication). Ion substitution experiments with *E. japonica* and other hydromedusae must be interpreted cautiously since in most preparations the epithelium sits on a thick layer of mesoglea which may act as a source of ions, buffering the epithelium somewhat from changes in the ionic composition of the bathing solution.

It is of interest that action potentials in conducting epithelia are quite diverse in shape, in ionic dependence, and in pharmacological susceptibility (for details see Anderson, 1980). Some epithelial action potentials are short and spikelike and are followed by no afterpotentials, by depolarizing afterpotentials, or by hyperpolarizing afterpotentials (e.g., Kater et al., 1978; Mackie, 1976; Fig. 2); some action potentials are broad or with a pronounced plateau (Spencer and Satterlie, 1981; Roberts and Sterling, 1971); and some action potentials are initially short but develop a pronounced plateau during repetitive activation (Anderson, 1979; Chain et al., 1981). In most excitable epithelia both sodium and calcium participate in the inward, depolarizing current of an action potential. As indicated, in *Euphysa* either sodium or calcium seems adequate for action potential generation. In some epithelia action potentials are inhibited either by blocking inward sodium current with sodium-free bathing solution or TTX, or by blocking calcium current with antagonists such as cobalt or manganese (Anderson, 1979; Spencer and Satterlie, 1981). In some epithelia the action potential is basically a sodium spike with perhaps a small calcium component (Roberts and Sterling, 1971; Chain et al., 1981); in others the action potential is a calcium spike with a small and dispensable sodium component (Hadley et al., 1980). In the elytral epithelium of the worm *Hesperonoe complanata*, cells that produce a large calcium action potential share the epithelium with cells producing a sodium action potential (Herrera, 1979). TTX is effective in blocking sodium currents in some epithelia (Roberts and Sterling, 1971; Hadley et al., 1980); in others TTX at moderate concentration does not affect electrogenesis even when ion substitution experiments indicate that there is a significant sodium contribution to the action potential (Herrera, 1979; Chain et al., 1981). Action potentials in the excitable epithelium of young *Xenopus* tadpoles are blocked by bathing the basal side of the epithelium in sodium-free solution, whereas removing sodium from only the outer surface of the epithelium does not stop propagated activity (Roberts and Sterling, 1971). Thus in the amphibian skin, conduction seems to use voltage-sensitive sodium channels on only the inner face of the epithelial cells. I am uncertain as to whether tunicate hearts are more properly considered to be myoepithelia or thin muscle sheets. Like many epithelia, the heart tissue is a thin cell layer that lines a cavity, but, unlike conventional epithelia, the heart cells do not sit on a basement membrane that separates them from underlying tissue. The tunicate heart is a single cellular layer which does conduct action potentials and in which, as in larval amphibian skin, the action potentials are generated by just one of the two tissue surfaces—in the heart by the luminal surface (Kriebel, 1973; Weiss et al., 1976). It is not yet known if it is common for epithelial conduction to use voltage-sensitive channels on just one of the two epithelial surfaces, as in larval amphibian skin or tunicate hearts.

Conduction velocity in a neuronal pathway may either increase or decrease during repetitive activation (e.g., Bullock, 1951; Bliss and Rosenberg, 1979; Kocsis et al., 1979). In the conducting epithelia that have been examined, conduction velocity usually decreases, sometimes markedly, during repetitive activation (e.g., Spencer, 1975, 1978; Mackie, 1975; Bone and Mackie, 1975; for an exception see Mackie and Carre, 1983). Changes in conduction velocity in the epithelium of *E. japonica* are associated with a decrease in the positive potential reached during the action potential overshoot (Fig. 4). The second of two closely spaced action potentials begins during the hyperpolarizing after potential of the first, and the peak potential is diminished, as is the conduction velocity. During repetitive stimulation there is a further progressive decline in the amplitude of the action potential overshoot for impulses after the second action potential and a progressive decline in conduction velocity. The relation between action potential overshoot and conduction velocity is quite steep. A reduction of 4–5 mV in the amplitude of the overshoot out of a total action potential amplitude of about 70 mV is associated with a more than 20% decrease in the conduction velocity across the epithelium.

INPUTS AND OUTPUTS OF EPITHELIAL CONDUCTION SYSTEMS

The interactions between conducting epithelia and effectors, sensors, and neural coordinating mechanisms are diverse, occurring in almost every conceivable combination (Fig. 5).

Some coelenterate epithelia are simultaneously sensory systems, conducting systems, and effectors (Fig. 5*A*). Coelenterates are composed basically of two epithelial layers, an outer ectoderm and an inner endoderm. The effectors in coelenterates are generally epithelial. This is true even for musculature that typically occurs as sets of elongate, striated or nonstriated, contractile elements formed by basal processes of epithelial cells.

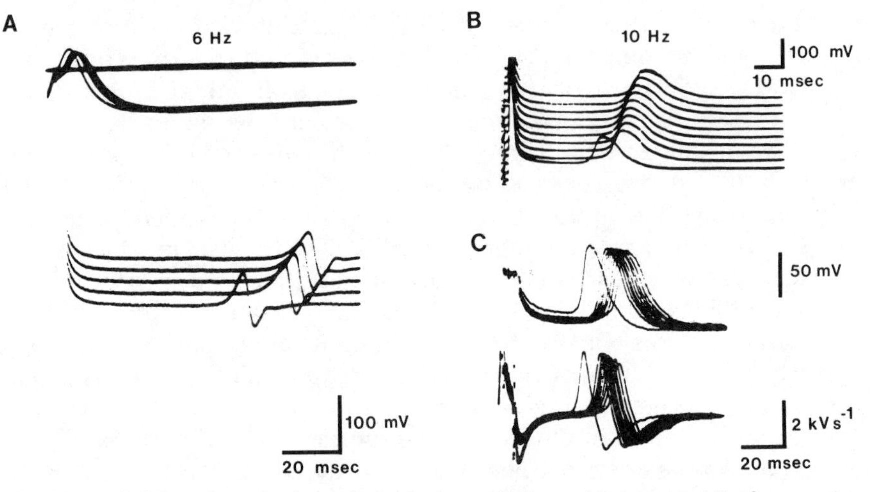

FIGURE 4. Lability of conduction velocity in the epithelium of *E. japonica*. (*A*) Above: action potentials during repetitive stimulation (6 Hz) recorded with an intracellular electrode about 0.2 mm from the stimulating electrode on the epithelial surface. The straight line marks the zero potential level. The action potential with the shortest latency is the first of the series. Below: The same actiion potentials recorded with an extracellular suction electrode 5.6 mm from the stimulating electrode. A raster generator was used to offset the traces; the first here is the lowest and the last the highest of the series. The amplitude calibration applies to the intracellular recordings only. (*B*) Intracellular action potentials during repetitive stimulation (10 Hz) recorded about 2.5 mm from the stimulating electrode. The traces were stepped upward sequentially so the first trace is the lowest of the series. (*C*) Upper: the action potentials in (*B*) without raster displacement. Lower: the time derivative of the action potentials above. (From Schwab and Josephson, 1982.)

Epithelia with well developed contractile components are sometimes termed myoepithelia. Sensitivity to stimuli, in particular to mechanical stimuli, and the capacity for effector responses occur in conducting epithelia. This is the case in the exumbrellar epithelium of nectophores from the siphonophore *Hippopodius*. Nectophores are medusalike individuals of the siphonophore colonies. Stimulating the exumbrellar surface of a nectophore initiates a propagated impulse in the exumbrellar epithelium. As the impulse traverses the epithelium, it initiates luminescence in the epithelial cells and an increase in opacity ("blanching") in the underlying mesoglea; the latter presumably is due to secretion into or uptake from the mesoglea of some component by the epithelium (Mackie and Mackie, 1967; Bassot et al., 1978). Luminescence and blanching are thought to be part of a protective response. There are other components of the protective response that are also controlled by the propagated epithelial activity, including marginal involution (Mackie, 1965; Bassot et al., 1978) and secretion from the rete mirabile, a specialized secretory epithelium which forms part of the ventral radial canal of the endoderm (Mackie, 1976). Interestingly,

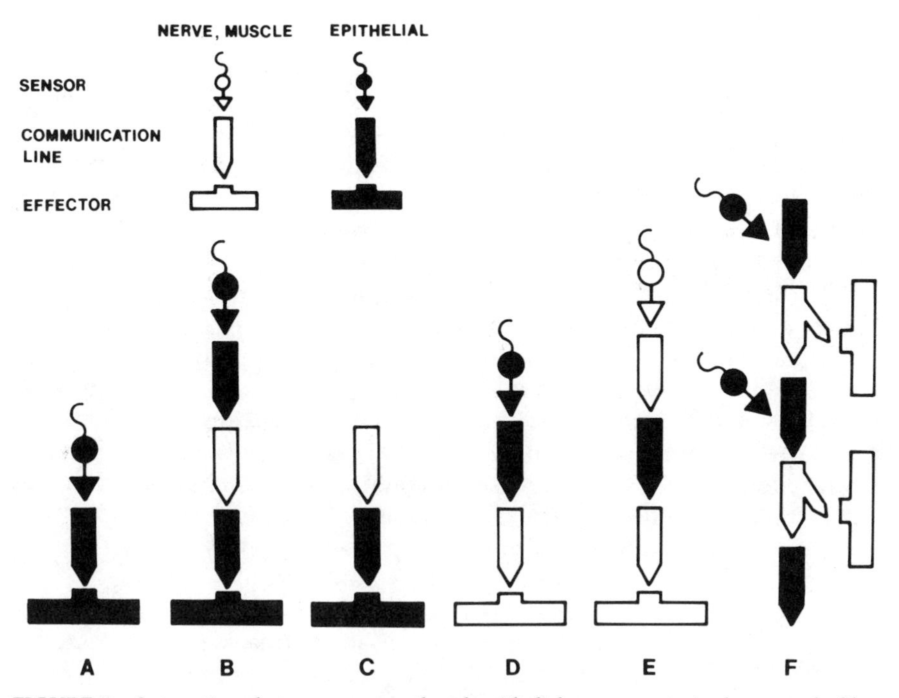

FIGURE 5. Interactions between neuronal and epithelial components in the control of behavior. See text for discussion.

epithelial excitation following exumbrellar stimulation must pass from the ectoderm to the endoderm to reach the rete and some of the marginal musculature participating in involution which, like the rete, is endodermal. Activity is thought to pass from the ectoderm to the endodermal epithelium along tissue bridges between the ectoderm and endoderm (Mackie and Passano, 1968; Mackie and Singla, 1975; Spencer, 1979).

The protective response following exumbrellar stimulation in *Hippopodius* and similar responses in other siphonophores are obviously related to the protective crumpling in hydromedusae described previously. Portions of the crumpling responses of hydromedusae probably utilize only epithelial components. However, part of the crumpling response of the hydromedusa *Polyorchis*, and likely of other hydromedusae as well, involves an interpolated neural pathway (Fig. 5*B*). In *Polyorchis* contraction of the radial subumbrellar muscles is a major component of the crumpling response. Activity in the exumbrellar epithelium reaches the radial muscles by way of marginal and possibly radial nerves (King and Spencer, 1981). Similarly, stimulation of the endodermal epithelium, which also evokes crumpling, activates marginal and radial nerves, which in turn activate the radial musculature.

In medusae the swimming pulsations are timed by neural pacemakers located in marginal ganglia (Scyphomedusae) or marginal nerve rings (Hydromedusae, medusoid individuals of siphonophore colonies; see reviews in Passano, 1982 and Spencer and Schwab, 1982). In hydromedusae and siphonophores the principal swimming musculature is circular muscle of the subumbrellar, ectodermal myoepithelium. In several cases examined the circular muscle sheet is nerve free. Activity spreads across the sheet by epithelial conduction from the margin or radii where the epithelium is innervated (Mackie and Singla, 1975; Spencer, 1979, 1982; Chain et al., 1981). The neuroepithelial synapses, from their ultrastructure, are conventional chemical synapses. The control of the swimming pulsations, then, consists of neuronal pacemakers and initial neuronal pathways which excite epithelial conduction and effectors (Fig. 5C). The scheme in Figure 5C would also apply to these instances in which a conducting epithelium and associated effectors are the only epithelial components of an otherwise neural control chain, as is the case for some snail salivary glands (Kater et al., 1978) and luminescent scales in an annelid (Herrera, 1979).

One component of the crumpling response evoked by exciting the exumbrellar epithelium in hydromedusae is inhibition of swimming. Exumbrellar epithelial conduction in medusoid individuals of siphonophore colonies may inhibit or evoke swimming pulsations (Mackie, 1965; Bassot et al., 1978; Mackie and Carre, 1983). The swimming pacemakers in these cases are neuronal and the initial activity epithelial. The scheme in Figure 5*B* would also apply to such instances in which activity in an excitable epithelium inhibits or excites neuronal pacemakers which in turn activate epithelial musculature.

In larval amphibians (Roberts, 1971; Roberts and Sterling, 1971) and some larval and adult urochordates (reviewed by Bone and Mackie, 1982) an excitable surface epithelium serves as an extension of neuronal sensory fields (Fig. 5*D*). Activity in the epithelium, initiated by mechanical stimulation, spreads across the epithelium exciting sensory neurons in route. Activity in the sensory neurons reaches the central nervous system where it eventually activates motoneurons and muscles or inhibits ongoing activity.

Colonial coordination in some compound ascidians involves excitation spread along the epithelium of the common circulatory system of the colony (Mackie and Singla, 1983). Here the sensors and the initial coordinating mechanisms are apparently neuronal, as are the motor mechanisms. The epithelial component, then, is interpolated between neuronal ones (Fig. 5*E*).

A remarkable example of interaction between epithelial and neuronal coordinating mechanisms is found in salp chains (Bone et al., 1980; Anderson and Bone, 1980). Stimulating the outer surface of one individual salp in the anterior chain causes momentary cessation of swimming in the chain or, with stronger stimuli, reversal in the direction of swimming by all the joined individuals making up the chain. Stimuli to the rear of the chain lead to acceleration of swimming. The initial pathway involved is the outer skin, which is an excitable epithelium. Impulses in the epithelium spread across the skin and reach specialized epithelial plaques which are the points of attachment between members of a chain. A junction between individuals is formed by two plaques, one from each of the individuals. The two plaques at a junction are asymmetrical. One plaque bears sensory cells with expanded, branching cilia, the opposing plaque has a group of "button cells" against which the ends of the cilia terminate. The button cell-cilia junction is functionally a polarized synapse, the button cells being presynaptic and the cilia postsynaptic. Epithelial impulses on the button cell side excite the sensory cells of the opposing plaque. An individual in a chain is joined to each of its neighbors by two plaques polarized such that the animal is presynaptic at one and postynaptic at the other. Activity in the sensory axons evoked by epithelial impulses in the neighboring individual is conducted to the brain where it modifies swimming behavior and where, through neuroepithelial synapses, epithelial impulses are initiated in the newly responding individual. This epithelial activity, in turn, excites new individuals in the chain. Thus propagation of activity along the chain utilizes an alternating sequence of neuroepithelial synapses; conducting epithelia; interanimal, epithelio-neural synapses of a quite novel sort; and neuronal pathways (Fig. 5*F*).

LIMITATIONS ON THE PERFORMANCE OF CONDUCTING EPITHELIA

Since the mid-1960s several dozen cases of demonstrated or presumed conducting epithelia have been reported in four unrelated phyla. It is inconceivable that the number of examples of excitable epithelia and the list of taxa in which they occur will not continue to grow. But it also seems likely that most epithelia will prove to be inexcitable and that major animal groups will lack epithelial conduction. Because of the large cellular surface areas involved, the ionic currents associated with epithelial conduction are rather large, and electrical correlates of epithelial conduction are correspondingly easy to detect. If conducting epithelia were common in such often-studied groups as insects, crustacea, or adult vertebrates, it seems likely that they would have been noticed long ago. But they have not. The ability to produce propagating action potentials appears to be a relatively uncommon attribute of epithelial tissues.

Why should epithelial conduction be rare? All animals, at least those more advanced than sponges or mesozoans, can synthesize voltage-activated ionic channels. Electrical coupling between the cells of animal epithelia is widespread (Loewenstein, 1981). The capacity for epithelial conduction requires only a sufficient density of appropriate, voltage-activated ionic channels in an epithelium of electrically coupled cells with suitable geometry. Why then are conducting epithelia not used more often as supplements to neuronal conduction and control? What disadvantages, if any, are inherent to conducting epithelia which limit their usefulness as coordinating pathways?

One obvious limitation to conducting epithelia as components in behavioral coordination is their lack of specificity. A neuron can conduct information between distant points without activating intervening cells. With epithelial conduction, in contrast, all the epithelial cells along the route are excited. Further, sheets of excitable cells are quite often extensive relative to the size of the organism containing them, and so conduction in the epithelium can be quite widespread. An epithelial impulse in the skin of a larval amphibian spreads across the entire surface of the animal including the corneas of its eyes (Roberts and Sterling, 1971). Similarly, an impulse in the exumbrellar epithelium of a hydromedusa spreads over the entire upper surface.

Sometimes there are boundaries of unknown nature in an epithelium that limit excitation spread. An impulse in the exumbrellar ectodermal epithelium of a hydromedusa does not pass to the subumbrellar ectoderm even though the ectoderm of the exumbrellar is confluent with that of the subumbrellar surface at the margin (Mackie and Passano, 1968). An impulse carried along the endodermal lamella of a hydromedusa, also an epithelial sheet, may be restricted to one quadrant of the animal even though the endodermal sheet is continuous around the animal with individual lamellar sections linked by the radial canals (Mackie, 1975). Spread of excitation in the stolon of the hydroid *Cordylophora* is thought to be along an excitable epithelium (Jha and Mackie, 1967). In *Cordylophora* a single impulse elicited by a stimulus to the stolon spreads several millimeters and stops (Josephson, 1961). The impulse leaves a temporary trace of its existence at the boundry reached, for the boundary is changed and another impulse following the first at a sufficiently short interval passes the former boundary and spreads further along the stolon. Thus in the stolon there is facilitation of conduction distance during repetitive firing. How this is achieved is quite unknown.

Some other limitations on the performance of epithelial conduction are a consequence of the two-dimensional geometry of epithelial sheets. In an elongate neural process, current flow about a source is linear and in two directions. In an epithelium, current flow about a source is radial and in many directions (Fig. 3). Because of the greater effective area of the cytoplasmic pathway for current flow in an epithelium, and the greater available membrane area, the input resistance of an epithelium is much lower than that of a neural process of equal width. Therefore the epithelium is much less sensitive to imposed current than is the neural process. Further, in a cylindrical neural process current density declines with distance from a source because of current leakage through the surface membrane. In a two-dimensional epithelial sheet, current density in the cytoplasm declines both because of leakage through the surface membrane and because the cross-sectional area of the current pathway increases with distance from the source. Because there is both an increasing area component and a membrane current component in the decline of internal longitudinal current density in an epithelium, the current density, and therefore the voltage change in the cytoplasm due to the imposed current, decays more rapidly with distance in an epithelium than in a neuronal process of similar width. The effective space constant of an epithelial sheet is short, which probably limits conduction velocity.

The differences in the passive electrical properties of an epithelium and a cylindrical cell process, for example an axon, can be illustrated by an example. Consider an axon 10 μm in diameter and a epithelial sheet 10 μm in height. Assume that in each the cytoplasm is bounded by membrane with specific electrical resistance of 1000 Ω-cm^2. In the epithelium there will actually be two membranes, an apical and a basal. Further, assume that the electrical resistivity of the cytoplasm in each is 100 Ω-cm. For the epithelium this resistance will be both that of the cytoplasm itself and that of the intercellular junctions. The input resistance of an axon is defined as the ratio between the voltage change at the site of current injection and the amplitude of the intracellularly injected current. From standard cable equations, the input resistance of our axon will be 3.2×10^6 Ω, and the evoked potential change will have decayed to one-half its initial value in a distance of 350 μm. The approach must be a little different with a two-dimensional cable such as an epithelial sheet. In the usual treatment of a cylindrical cable, current is assumed to originate from a plane perpendicular to the long axis of the cable. Since the plane has finite area, the current density is finite even at the origin. In the treatments available for a two-dimensional sheet, current is regarded as originating at a line perpendicular to the surface of the sheet. Since the line has no area, the theoretical current density at the origin is infinite, as is the resulting potential change. To avoid this singularity at the origin, I define the input resistance of an epithelium as the ratio between the evoked potential change measured one tissue width from the current source (here 10 μm) and the amplitude of injected current. The input resistance of the epithelium so defined is 7×10^4 Ω, and the evoked potential would decay to one-half its initial amplitude (i.e., that at 10 μm) at a distance of 90 μm from the current source. Thus the input resistance of the axon is more than 40 times greater than that of the epithelium, and the rate of potential decay with distance is much slower in the axon than the epithelium. It might be noted that in the axon described the evoked potential 10 μm from current source is about 98% as large as at the source, so the preceding argument would not be changed if the initial electrical potential in the axon were taken as that 10 μm away from the current source as was done for the epithelium.

Some recent considerations of the evolutionary origins of nervous systems have suggested that epithelial conduction preceeded neuronal coordination and that neurons evolved from excitable epithelial cells, the evolutionary advantages being greater specificity and an increase in the number of available pathways (Horridge, 1968; Mackie, 1970). The passive electrical properties of epithelial sheets suggest that one of the first steps in the evolution of neurons from epithelial cells might have been the electrical isolation of small areas within the epithelium, the advantage being a greater input resistance and therefore greater sensitivity to sensory inputs. If this line of reasoning is correct, the earliest components of nervous systems would have been sensory structures whose outputs were presumably to conducting epithelia and epithelial effectors.

Acknowledgments

Original research was supported by grants BNS-09530 and PCM-8201559 from the National Science Foundation. I acknowledge especially W. E. Schwab whose technical skill and intellectual daring made possible the work with Euphysa.

REFERENCES

Anderson, P. A. V. (1979) Epithelial conduction in salps. I. Properties of the outer skin pulse system of the stolon. *J. Exp. Biol.* **80**:231–239.

Anderson, P. A. V. (1980) Epithelial conduction: Its properties and functions. *Prog. Neurobiol.* **15**:161–203.

Anderson, P. A. V., and Q. Bone (1980) Communication between individuals in salp chains. II. Physiology. *Proc. R. Soc. Lond.* **B210**:559–574.

Bassot, J.-M., A. Bilbaut, G. O. Mackie, L. M. Passano, and M. Pavans de Ceccatty (1978) Bioluminescence and other responses spread by epithelial conduction in the siphonophore *Hippopodius. Biol. Bull. Mar. Biol. Lab., Woods Hole* **155**:473–498.

Bliss, T. V. P. and M. E. Rosenberg (1979) Activity-dependent changes in conduction velocity in the olfactory nerve of the tortoise. *Pflugers Arch.* **381**:209–216.

Bone, Q., P. A. V. Anderson, and A. Pulsford (1980) The communication between individuals in salp chains. I. Morphology of the system. *Proc. R. Soc. Lond.* **B210**:549–558.

Bone, Q. and G. O. Mackie (1975) Skin impulses and locomotion in *Oikopleura* (Tunicata: Larvacea). *Biol. Bull. Mar. Biol. Lab., Woods Hole* **149**:267–286.

Bone, Q. and G. O. Mackie (1982) Urochordata. In *Electrical Conduction and Behaviour in "Simple" Invertebrates,* G. A. B. Shelton, ed. Clarendon Press, Oxford, pp. 473–535.

Bullock, T. H. (1951) Facilitation of conduction rate in nerve fibres. *J. Physiol.* **114**:89–97.

Bullock, T. H. and G. A. Horridge (1965) *Structure and Function in the Nervous Systems of Invertebrates.* W. H. Freeman and Co., San Francisco.

Chain, B. (1981) A sodium-dependent "twitch" muscle in a coelenterate: the ectodermal myoepithelium of the gastrozooids in *Agalma sp.* (Siphonophora). *J. Exp. Biol.* **90**:101–108.

Chain, B. M., Q. Bone, and P. A. V. Anderson (1981) Electrophysiology of a myoid epithelium in *Chelophyes* (Coelenterata: Siphonophora). *J. Comp. Physiol.* **143**:329–338.

Hadley, R. D., A. D. Murphy, and S. B. Kater (1980) Ionic bases of resting and action potentials in salivary gland acinar cells of the snail *Helisoma. J. Exp. Biol.,* **84**:213–225.

Herrera, A. A. (1979) Electrophysiology of bioluminescent excitable epithelial cells in a polynoid polychaete worm. *J. Comp. Physiol.* **129:**67–78.

Horridge, G. A. (1968) The origins of the nervous system. In *The Structure and Function of Nervous Tissue, Vol. 1, Structure I,* G. H. Bourne, ed. Academic Press, New York, pp. 1–31.

Hyman, L. A. (1940) Observations and experiments in the physiology of medusae. *Biol. Bull. Mar. Biol. Lab., Woods Hole* **79:**282–296.

Jack, J. J. B., D. Noble, and R. W. Tsien (1975) *Electric Current Flow in Excitable Cells,* Clarendon Press, Oxford, 502 pp.

Jha, R. K. and G. O. Mackie (1967) The recognition, distribution and ultrastructure of hydrozoan nerve elements. *J. Morph.* **123:**43–62.

Josephson, R. K. (1961) Repetitive potentials following brief electric stimuli in a hydroid. *J. Exp. Biol.* **38:**579–593.

Josephson, R. K. and W. E. Schwab (1979) Electrical properties of an excitable epithelium. *J. Gen. Physiol.* **74:**213–236.

Kater, S. B., J. R. Rued, and A. D. Murphy (1978) Propagation of action potentials through electrotonic junctions in the salivary glands of the pulmonate mollusc, *Helisoma trivolvis. J. Exp. Biol.* **72:**77–90.

King, M. G. and A. N. Spencer (1981) The involvement of nerves in the epithelial control of crumpling behaviour in a hydrozoan jellyfish. *J. Exp. Biol.* **94:**203–218.

Kocsis, J. D., H. A. Swadlow, S. G. Waxman, and H. M. Brill (1979) Variation in conduction velocity during the relative refractory and supernormal periods: a mechanism for impulse entrainment in central axons. *Exp. Neurol.* **65:**230–236.

Kriebel, M. E. (1973) Action potentials occur only on lumen surface of tunicate myoendothelial cells. *Comp. Biochem. Physiol.* **46A:**463–468.

Loewenstein, W. R. (1981) Junctional intercellular communication: the cell-to-cell membrane channel. *Physiol. Rev.* **61:**829–913.

Mackie, G. O. (1965) Conduction in the nerve-free epithelia of siphonophores. *Am. Zool.* **5:**439–453.

Mackie, G. O. (1970) Neuroid conduction and the evolution of conducting tissue. *Q. Rev. Biol.* **45:**319–332.

Mackie, G. O. (1975) Neurobiology of *Stomotoca.* II. Pacemakers and conduction pathways. *J. Neurobiol.* **6:**357–378.

Mackie, G. O. (1976) Propagated spikes and secretion in a coelenterate glandular epithelium. *J. Gen. Physiol.* **68:**313–325.

Mackie, G. O. and D. Carre (1983) Coordination in a diphyid siphonophore. *Mar. Behav. Physiol.* **9:**139–170.

Mackie, G. O. and G. V. Mackie (1967) Mesogleal ultrastructure and reversible opacity in a transparent siphonophore. *Vie et Milieu, Ser. A. Biol. Mar.* **18:**47–71.

Mackie, G. O. and L. M. Passano (1968) Epithelial conduction in hydromedusae. *J. Gen. Physiol.* **52:**600–621.

Mackie, G. O. and C. L. Singla (1975) Neurobiology of *Stomotoca.* I. Action systems. *J. Neurobiol.* **6:**339–356.

Mackie, G. O. and C. L. Singla (1983) Coordination of compound ascidians by epithelial conduction in the colonial blood vessels. *Biol. Bull. Mar. Biol. Lab., Woods Hole* **165:**209–220.

Murakami, A. and H. Machemer (1982) Mechanoreception and signal transmission in the lateral ciliated cells on the gill of *Mytilus. J. Comp. Physiol. A* **145:**351–362.

Parker, G. H. (1919) *The Elementary Nervous System.* Lippincott, Philadelphia and London.

Passano, L. M. (1982) Scyphozoa and Cubozoa. In *Electrical Conduction and Behaviour in "Simple" Invertebrates,* G. A. B. Shelton, ed. Clarendon Press, Oxford, pp. 149–202.

Roberts, A. (1971) The role of propagated skin impulses in the sensory system of young tadpoles. *Z. Vgl. Physiol.* **75**:388–401.

Roberts, A. and C. A. Sterling (1971) The properties and propagation of cardiaclike impulses in the skin of young tadpoles. *Z. Vgl. Physiol.* **71**:295–310.

Schwab, W. E. and R. K. Josephson (1982) Lability of conduction velocity during repetitive activation of an excitable epithelium. *J. Exp. Biol.* **98**:175–193.

Spencer, A. N. (1975) Behavior and electrical activity in the hydrozoan *Proboscidactyla flavicirrata* (Brandt). II. The medusa. *Biol. Bull. Mar. Biol. Lab., Woods Hole* **149**:236–250.

Spencer, A. N. (1978) Neurobiology of *Polyorchis.* I. Function of effector systems. *J. Neurobiol.* **9**:143–157.

Spencer, A. N. (1979) Neurobiology of Polyorchis. II. Structure of effector systems. *J. Neurobiol.* **10**:95–117.

Spencer, A. N. (1982) The physiology of a coelenterate neuromuscular synapse. *J. Comp. Physiol.* **148**:353–363.

Spencer, A. N. and R. A. Satterlie (1981) The action potential and contraction in subumbrellar swimming muscle of *Polyorchis penicillatus* (Hydromedusae). *J. Comp. Physiol.* **144**:401–407.

Spencer, A. N. and W. E. Schwab (1982) Hydrozoa. In *Electrical Conduction and Behaviour in "Simple" Invertebrates,* G. A. B. Shelton, ed. Clarendon Press, Oxford, pp. 73–148.

Weiss, J., Y. Goldman, and M. Morad (1976) Electromechanical properties of the single cell-layered heart of the tunicate *Boltenia ovifera* (Sea Potato). *J. Gen. Physiol.* **68**:503–518.

Chapter Nine

CHANNELS IN THE JUNCTIONS BETWEEN CELLS

Formation and Permeability

WERNER R. LOEWENSTEIN

Department of Physiology and Biophysics
University of Miami School of Medicine
Miami, Florida

Cells in organized tissues are commonly interconnected at their junctions. The junctional elements are large membrane channels through which a range of hydrophilic cellular molecules can flow directly from one cell interior to another. Such cell-to-cell channels are present throughout the phylogenetic scale in tissues of mesenchymal and epithelial origin, adult and embryonic, from sponges to man. As a rule, a given cell in a tissue is connected with several neighbors, and so whole organs or large organ parts are continuous from within.

This connecting channel was an elegant solution of the problem that cells faced in their evolution to multicell ensembles, the problem of how to get communication between cytoplasms by hydrophilic molecules without losing cell circumscription and genetic individuality. Circumscription was preserved by making a channel of two leakproof abutting halves, one contributed by each cell membrane, and individuality was preserved by choosing a critical channel size. The evolutionary gain was a relatively free intercellular exchange of inorganic ions, metabolites, nucleotides, high-energy phosphates, and the like. In respect to those cytoplasmic molecules, the connected cell ensemble, and not the single cell, became the functional compartmental unit.

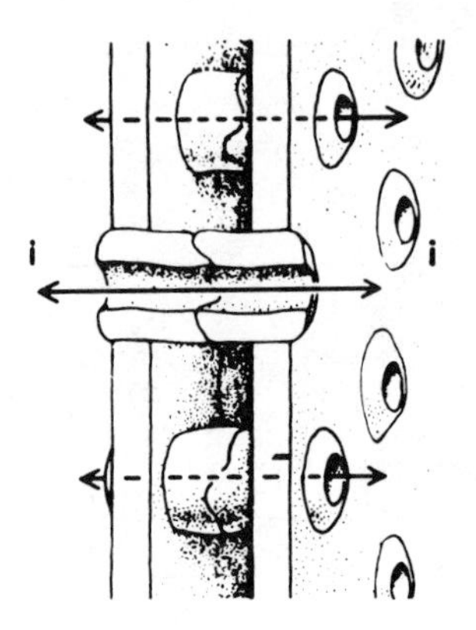

FIGURE 1. The Cell-to-cell channel. The channel unit (as inferred from electrophysiological and fluorescent tracer studies) made of a pair of matching channel halves traversing the membranes. The pair has a continuous aqueous bore and interlocking walls providing continuous insulation. (Reprinted with permission from Loewenstein, 1975a.)

I hit upon this general connectivity in 1963, about 10 years after leaving Ted Bullock's laboratory. But I know that, with its wide evolutionary ramifications, this is the sort of thing that interests him. The comparative aspects of the field, in fact, are still waiting to be plowed.

In the evolutionary sense, the most basic function of the channel is homeostatic, a coordination toward uniformity in the multicellular compartmental unit. In many instances this constitutes a true Gleischschaltung, an evening out of chemical and electrical potentials in the cell ensemble wherein individual cell variations are rapidly buffered.

THE CELL-TO-CELL CHANNEL AND ITS FORMATION

We imagine the cell-to-cell channel as made of two matching, interlocking halves, one contributed by each cell, forming a continuous and direct aqueous passageway between cells (Fig. 1). I proposed this idea originally based on electrical measurements of the whole junctional pathway and of probings with fluorescent tracer molecules (Loewenstein, 1966), but there is now evidence for the channel of the most direct kind. Electrically, the single channels were resolved as quantal events of conductance in a forming junction, and, structurally, they were traced to the membrane particles of gap junction.

Channel Openings Are Detected as Stable Quantal Steps of Conductance

Cell-to-cell channels can form spontaneously when cells come into contact. The first experimental inductions of channels were made by manipulating into contact cells isolated from sponges or early amphibian embryos, while monitoring the electrical coupling. A conductive pathway developed spontaneously within a few minutes of the cell contact (Loewenstein, 1967; Ito and Loewenstein, 1969). There was little chance that the cells selected the contact regions by active movement: we pushed the cells together at randomly chosen spots, and we blocked cell-generated movement by colchicine or low external Ca^{2+}. At the level of the light microscope at least, the conductive pathway seemed to form anywhere on the membrane, not just in certain predetermined regions—a dynamism particularly clear in the amphibian embryo cells. With these large cells (nearly 0.5 mm in diameter) we could make communicating junctions repeatedly by simply pulling the pair apart and putting it together again at a different spot (the pathway sealed upon disjunction).

These findings revealed that a large part of the cell membrane of these cells, perhaps all of it, can rapidly develop a communicating junction. They implied that channels, as open entitities, can form on membrane regions where before there were none and gave a basis for the idea that the channels are made of preformed halves, the *protochannels* in each membrane.

A further clue about the junction formation process came when we monitored cell-to-cell conductance continuously during the formation of a junction between the pairs: the conductance rose gradually over several minutes to a plateau. In terms of channels, the simple interpretation was that permeable junction develops by a progressive accretion of channel units (Ito et al., 1974).

Thus, with high enough resolution, one would expect the channel accretion to show itself as a series of quantal increments of cell-to-cell conductance. This, in fact, turned out to be so when we measured the conductance of nascent junction with a phase-sensitive technique against a background of very small, or zero, conductance. During the early phase of junction formation, while few channels were forming, the openings of the channels manifested themselves as discrete, quantal jumps of conductance, moreover as stable quantal jumps. The channels thus detected stayed open over minutes of observation, and closures were not seen unless the intracellular calcium concentration was experimentally elevated (Fig. 2; Loewenstein et al., 1978a).

This stability in the open state contrasts with the behavior of the membrane channels of ordinary nonjunctional cell membranes, such as the potassium channel, the sodium channel, the acetylcholine channel, all short-lived. The molecular stability, which this stability in the open state implies, we think, arises as a consequence of the paired nature of the channel, by an interlocking of the two channel halves (Loewenstein et al., 1978b). The gramicidin channel may provide an instructive analogy: it converts from short-lived to stable form on covalent dimer linkage (Bamberg et al., 1978; Urry et al., 1971).

The available data do not rule out the possibility of unstable openings of

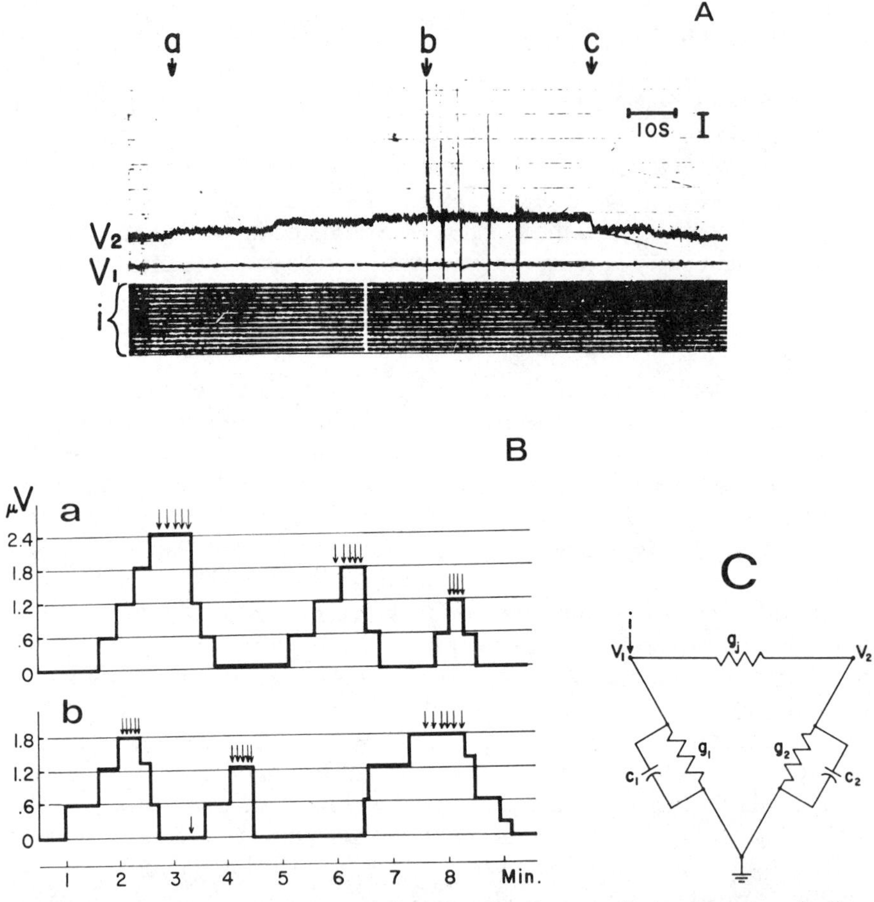

FIGURE 2. (*A*) Quantal development of junctional conductance during formation of cell-to-cell channels. Changes in junctional conductance monitored at high resolution. Current i is sinusoidal, and V_1 and V_2 are recorded by phase-sensitive detectors at selectable phase angle with respect to current and converted to proportional DC potentials. Oscillographic record displays V_2, V_1 (V_1 at lower gain), and sinusoidal current i. Junctional conductance is directly proportional to V_2 here. First of three quantal upsteps occurs at a. At b a series of 5 Ca^{2+} pulses is delivered into cell 1 (spikes on V_2 trace are capacitive artifacts caused by solenoid that controls injection), which leads at c to a series of quantal V_2 downsteps. Voltage calibrations: $V_1 = 100\ \mu V$; $V_2 = 3\ \mu V$; $i = 1.00 \times 10^{-8}$ A (root mean square); 25 Hz. (*B*) Time course of quantal conductance development. Cell contact at time 0. Upsteps of V_2 occur spontaneously; downsteps occur after Ca^{2+} injections (arrows) only. In a all changes are integer multiples of a unitary value, $V_2^\circ = 0.6080 \pm$ SE 0.0035 μV. In b downsteps are not integer multiples of upstep quantum. Upstep $V_2^\circ = 0.6106 \pm$ SE 0.0052 μV. (C) Equivalent circuit for coupling cell pair. Terminals represent cells 1 and 2 and extracellular ground in common for the i-passing and V-recording circuits. g_j, Junctional membrane conductance; g_1 and g_2, nonjunctional membrane conductances; c_1 and c_2, their corresponding capacitances. g_1 and g_2 are of similar orders of magnitude, and V_2 is several orders smaller than V_1; hence V_2 is directly proportional to g_j and serves as a direct index of nascent g_j. (From Loewenstein, Kanno, and Socolar, 1978a. Reprinted by permission from *Nature* **274:**133–136. Copyright 1978 Macmillan Journals Limited.)

the cell-to-cell channel in addition to the stable ones. The time constant of the phase detection system we used was such that had open states of 300 msec or less occurred, we would have missed them. There were, in fact, occasional conductance transients with half-times of about 1.3 sec, which may have reflected unstable channel openings and closings (Fig. 2*A*, c) (Loewenstein et al., 1978a).

A Channel Formation Hypothesis. A Self-Trap Model

We envision the cell-to-cell channel formation as a two-step process. First, the protochannels interlock with their counterparts from the opposite membrane. Second, they open up. Most simply, the second step may be a structural consequence of the first. Indeed, if the protochannels are made of symmetrical subunits (see following), the structural stabilization of the forming channel unit and its stabilization in the open state could result at once as molecular interactions of the same set of subunits (Loewenstein et al., 1978b).

How the protochannels get to the junction and within interaction range of each other—the necessary preliminaries of such channel formation—needs further elaboration. We may build on the premise that the protochannels cannot be transferred from one membrane to the other across the junction. There is no justification for this premise other than that the thermodynamic cost for such an exchange would be steep, and this is the reason why, since the inception of the cell-to-cell channel concept, I had envisaged the channel as a half-and-half contribution by two membranes. In principle, in agreement with the current ideas about biosynthesis and mobility of membrane proteins, the protochannel material could be directly inserted into the membrane junction, or, if inserted elsewhere, it could reach the junction by lateral movement in the membrane. Schemes of the first kind would require special mechanisms for recognition of the junctional membrane spots, which, given the dynamic behavior of such spots in experimental conditions of disjunction and junction, would imply that the recognition system be at least as dynamic. Lateral-movement schemes are simpler. I have toyed with the following one.

Suppose that the protochannnels (or subunits) move randomly in the fluid matrix and that they interlock with their opposite numbers of the other membranes whenever they get within range of Van der Waals forces of each other. The chances for such close encounter, nil where membranes are separate, increase steeply where the membranes are joined. There the protochannels would trap each other by their intercellular ends on encounter, and their outward drift is further opposed by the bending moment of the membranes at the margins of the junction (Fig. 3)—a perfect trap where the combined energies of attraction between the particles in the pairs and of the bending moment of the membranes by far exceed the particles' thermal energies.

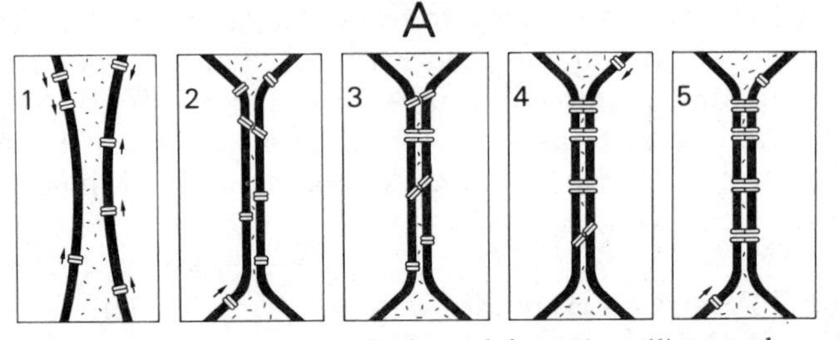

FIGURE 3. Self-trap model of cell-to-cell channel formation. (1) protochannels (channel halves or their precursor subunits) move randomly in membrane lipid matrix. At region of membrane apposition (2–5) they get within end-on attractive range of each other, interlock, and open up as complete cell-cell channels (3). Channels concentrate in region with time as protochannels get trapped there by their mutually attractive forces and the bending moment of the membranes (which by far exceed thermal energies of protochannel random movement). (Reprinted with permission from Loewenstein, 1981.)

This lateral-movement model (Loewenstein, 1981) relies entirely on particle self-interaction. It requires no special bell for channel congregation or special forces for drawing the protochannels to the junctional site. A humdrum membrane acquires a local organization on joining another membrane. The junction itself (by which I mean a close membrane apposition) brings the protochannels within range of molecular interaction and starts the sequence of events leading to protochannel concentration and channel self-assembly.

R. Skalak and I have analyzed the model quantitatively by Monte Carlo simulation, using particle sizes and membrane appositional areas of gap junctions and diffusion coefficients known for membrane proteins of similar sizes. The model did well: it produced channel clusters of gap-junction dimensions, and it did so fast enough even at low initial (dispersed) particle density (see Loewenstein, 1981).

Channel Formation and Cell Adhesion

Cell junction, in the stable sense used here, implies adhesion between cell membranes. The postulated hookup of the protochannel pairs, of course, is itself an adhesion, but there may be other stable molecular membrane bondings within the boundaries of a permeable junction or in its immediate vicinity. In terms of the foregoing model, such a bonding would be the prelude to cell-to-cell channel formation.

Experiments with cells isolated from the sponge *Haliclona* have shown that cell adhesion is indeed closely associated with channel formation. These cells need Ca^{2+} or Mg^{2+} and a glycoprotein present on their surface to adhere to each other (Moscona, 1968; Henkart et al., 1973; Burger et al., 1978). When pairs of cells that had been deprived of these factors were manipulated together *in vitro*, they did not develop electrical coupling until these factors were added to the bathing medium (Loewenstein, 1967). Cell adhesion and electrical coupling also correlate with respect to zoological-order specificity. When paired *in vitro*, cells from *Haliclona* fail to adhere (Galtsoff, 1925) and to couple electrically with cells from the sponge *Microciona* (Loewenstein, 1967), a specificity that also extends to their glycoproteins (Humphreys, 1965).

Recent experiments with mammalian liver cells treated with antibodies that block cell-cell adhesion speak further to this point. As do other higher cells (Edelman, 1983), cells from adult rat liver have specific glycoprotein surface molecules for mutual adhesion, called CAM (Ocklind and Obrink, 1982; Ocklind et al., 1983). The rat epithelial liver cell line RL expresses CAM in culture, and reaggregated cells make junctions that transfer carboxyfluorescein within 20 minutes. Treatment with antibodies against CAM prevents both cell-to-cell adhesion (but not cell-to-cell dish adhesion, which is mediated by another molecule) and permeable junction formation for as long as there are enough antibodies (fluorescent) on the cell membranes (K. Machida, B. Obrink, and W. R. Loewenstein, unpublished results).

All this indicates that membrane adhesion mediated by elements other than the cell-to-cell channels is necessary for channel formation, at least for multiple channel formation.

Channels between Cells of Different Type

Cells from different organs and different mammalian species, when paired in culture, will form communicating junctions with each other. Cells from rabbit lens couple to cells from rat liver, hamster kidney, or human skin, for example (Michalke and Loewenstein, 1971). In terms of the channel-formation hypothesis this means that the protochannels from these diverse cells can pair.

Such compatibility between cell types is widespread in the vertebrate phylum (Azarnia and Loewenstein, 1971; Pitts, 1972; Flagg-Newton and Loewenstein, 1980). Even cells from different zoological orders, such as birds and mammals, couple electrically to each other in culture. Among the arthropods, cells from different homopteran species can couple, but cells from three different orders (Homoptera, Lepidoptera, Diptera) probably cannot (Epstein and Gilula, 1977). Interphyla, arthropod and vertebrate cells do not couple (Epstein and Gilula, 1977); the channels of these phyla also are very different in size and permselectivity.

There is less information on heterologous junctions in organized tissues, but the indications are that such junctions are not infrequently communicating. In *Chironomus* salivary gland, where the coupling topology of the entire organ, consisting of 30–35 cells, has been worked out, all cells—including at least three different cell types—are interconnected (B. Rose, unpublished results). In rat ovarian follicle, cumulus oophorus and oocyte cells are connected (Gilula et al., 1978), and in embryonic tissue communicating heterologous junctions are quite common (Furshpan and Potter, 1968; Sheridan, 1968; Warner, 1973; Caveney, 1974; Lo and Gilula, 1979).

If the channels from various cell types are so broadly compatible, what then prevents the formation of unwanted communication? That there are limits to promiscuity is attested by several examples. In frog skin the mitochondria-rich cells are not coupled to the interconnected mass of epithelial transport cells with which they are interspersed (R. Rick and H. Ussing, personal communication). In leech ganglion, P sensory neurons couple to L motoneurons, but not to Retzius neurons—a specificity preserved in culture when the cells are placed in close apposition to each other (Fuchs et al., 1981). In mammalian mixed-cell cultures epithelioid cells from rat liver or human mammary gland make communicating junctions less frequently or more slowly than they make homologous ones (Fentiman et al., 1976; Pitts and Burk, 1976).

The self-trap model accounts for such selectivity simply in terms of cell approximation. Who makes channels with whom in mixed cell populations (with compatible protochannels) would depend on whether the cell membranes get within close (Van der Waals) range of each other. To get them that close, the electrostatic cell-repulsive forces must be overcome—a function presumably of the sugar groups of more or less specific adhesion molecules on the membrane surfaces. Then, whatever keeps cells apart, lack of adhesive forces or actual spacer material, would be sufficient for the cells not to form channels.

In this light, one may expect that molecular specificities in cell adhesion (that commonly go under the name *recognition*) and *basement membranes* or other spacer forms play a part in the selectivity of communication. The question relating to recognition is as yet untouched, but something has been learned about spacers. In experiments where the collagen coats of mammalian cells in culture were attacked with purified collagenases, several cell types made more communicating junctions as a result (C. Laurido and W. R. Loewenstein, unpublished results).

CHANNEL PERMEABILITY

The Channel is 16–20 Å Wide

Our knowledge of the permeability properties of the cell-to-cell channel comes mainly from probings with four kinds of fluorescent-labeled molecules. One set of probes consists of linear polyamino acids, negatively charged. They form a series of progressively increasing length and charge. Their width (the abaxial, permeation-limiting dimension) ranges from 14 to 16 Å. A second set is comprised of neutral linear oligosaccharides, polymers of glucose or mannose, alpha or beta linked, 16–20 Å wide. A third set is composed of neutral branched glycopeptides 20–30 Å wide. Finally, we make use of a set of peptides with tertiary structure too wide for permeation—fibrinopeptide, insulin A chain, polylysine. These various molecules are covalently labeled with the fluorophores, dansyl, lissamine rhodamine B, or fluorescein isothiocyanate, at an amine group (the sugars are reductively aminated). They are injected into the cells, one or two molecular species at a time, to probe junctional permeability; the three fluorophores have different excitation and emission spectra (Simpson et al., 1977; Schwartzmann et al., 1981).

From the widths of the largest permeants, the channel diameter was estimated: 16–20 Å for mammalian cells and 20–30 Å for arthropod cells (Simpson et al., 1977; Schwartzmann et al., 1981; Flagg-Newton et al., 1979). It is the widest membrane channel known. It is wide enough to admit probably most small-molecular, but not macromolecular, cytoplasmic solutes. The calculated equivalent conductance is of the order of 10^{-10} mho (Loewenstein, 1975b).

The channel behaves as if guarded by a fixed or induced charge. It discriminates against negatively charged molecules, a selectivity particularly prominent in the narrower mammalian channel (Flagg-Newton et al., 1979; Brink and Dewey, 1980).

Permeability Regulation

The permeability of a cell interface, the degree of intercellular communication, depends on the number of cell-to-cell channels that are in the open state. Two kinds of mechanisms are known to regulate the degree of communication. One a slow mechanism, controls the number of channels by a cAMP-promoted phosphorylation. This control presumably operates at the level of channel formation. The other, a fast mechanism, controls the channel aperture.

A cAMP-dependent Phosphorylation Regulates Permeability

The junctional permeability of various mammalian cell types in culture increases when the concentration of their intracellular cAMP $[cAMP]_i$) is elevated, and vice versa. The upregulation of permeability—it goes hand in hand with an increase in the number of gap-junction particles—was shown by supplying the cells with exogenous cAMP, by exposing them to phosphodiesterase inhibitor, or by stimulating their cAMP synthesis with choleragen (Flagg-Newton et al., 1981; Flagg-Newton and Loewenstein, 1981; Azarnia et al., 1981). Cells with normal expression of cell-to-cell channels respond by raising their junctional permeability above the base level, and this is the most general finding. But, more spectacularly, the rise can also occur from a zero level: in the mouse C1-1D cell line, an aberrant type that lacks detectable cell-to-cell channels in ordinary culture conditions, the channels develop on experimental elevation of $[cAMP]_i$ (Azarnia et al., 1981).

The upregulation by cAMP takes several hours to develop and, at least one component, depends on protein synthesis. It is blocked by the protein synthesis inhibitors cycloheximide and puromycin (but not by cytochalasin B) (Azarnia et al., 1981). All this, plus the fact that the upregulation involves a proliferation of gap-junction particles, leads us to suspect that cAMP promotes channel formation. However, a modulation of unit-channel permeability as an alternative or additional mechanism is not excluded.

Is this regulation mediated by phosphorylating protein kinase as it is in other cAMP-dependent cellular regulations (Kuo and Greengard, 1969; Krebs, 1972; Rosen and Krebs, 1981)? Work with mutant mammalian (CHO) cells (Chinese hamster kidney) deficient in cAMP-dependent protein kinase showed this to be the case. Mutant cells lacking kinase type I turned out to be deficient in permeable junctions, and this channel deficiency could be corrected by supplying the cells with exogenous catalytic subunit of the missing enzyme (Wiener and Loewenstein, 1983).

The permeability increase can be elicited by hormones that elevate $[cAMP]_i$. This point was made with the aid of two hormone-sensitive cultured mammalian cell types, the rat glioma cell C-6 that has β-adrenergic receptors and the human lung cell W1-38 that has prostaglandin receptors. The junctional permeability of these cells rose upon stimulation with catecholamine (isoproterenol) and prostaglandin E_1, respectively (Radu et al., 1982).

There is evidence for such a hormone action also in an organized tissue: gonadotropin produces increase of permeability at the oocyte/follicle cell junction (Browne and Wiley, 1979). There is further suggestive evidence of a morphological nature: thyroid hormone and estrogens cause proliferation of gap junctions in ependyma (Decker, 1976) and myometrium (Dahl and Berger, 1978; Garfield et al., 1978), respectively. Although such morphological data alone do not tell about junctional permeability (and it is not possible to infer that the gap-junction particles represent patent channels), a cAMP-mediated enhancement of junctional permeability seems a reasonable possibility in the light of the findings described in the preceding paragraph and of the knowledge that these hormones cause $[cAMP]_i$ elevation elsewhere.

Hormones may thus serve as regulators of junctional communication, that is, *slow* regulators for conditions calling for changes over hours in cellular states. Moreover, the interaction between the junctional and hormonal form of communication confers on the latter potential for coordination of cellular responses—simple homeostatic coordination or propagation of hormonal responses in tissues (hormonal cellular response amplification) where the hormones' second messengers are channel permeant (Loewenstein, 1981)

We know less about the downregulatory side of this cAMP-dependent mechanism. But, it is clear that it operates in this direction, too: when $[cAMP^{2+}]_i$ is lowered experimentally by treating cultured mammalian cells with serum or increasing their density, junctional permeability falls, hand in hand with a fall in the number of gap-junction particles. The downregulation is slow, on the order of 24–48 hr (Flagg-Newton and Loewenstein, 1981; Azarnia et al., 1981).

The available data fit a physiological regulation scheme of the following sort:

$$s_1 + r_1 \rightarrow \text{cAMP} \uparrow \rightarrow P_j \uparrow$$
$$s_2 + r_2 \rightarrow \text{cAMP} \downarrow \rightarrow P_j \downarrow$$

where *s* represents the stimulus and *r* its receptor, and where up- or downregulation of junctional permeability P_j depends on whether *r* activates or inhibits cAMP synthesis.

I use P_j rather than *number of channels*, to be safe. However, as already mentioned, the combined electrophysiological and morphological evidence suggests that the *number of channels* is the variable determining P_j here. If this is so, the scheme would demand that the channel be turned over fast enough for the observed times of permeability downregulation. The turnover of gap-junction protein, indeed, is fast (Revel et al., 1980; Dahl et al., 1981; Traub et al., 1983); in mammalian liver, its half-life is estimated at 5 hr (Fallon and Goodenough, 1981).

Another kind of downregulatory mechanism of junctional permeability has recently been found. This involves a protein kinase of a different sort, one that phosphorylates tyrosine rather than serine or threonine residues as in the case of cAMP-dependent phosphorylation (Loewenstein, 1985).

Elevation of Cytoplasmic $Ca^{2}+$ or H^+ Concentration Closes the Channel

The channel responds to changes of various sorts in the intracellular milieu by changing its open state. The free ionized calcium appears to play a pivotal role here.

The junctional permeability is high at the low Ca^{2+} concentrations normally prevailing in cytoplasm ($<10^{-8}$ *M*). It falls when the cytoplasmic Ca^{2+} concentration rises in the junctional locale.

This was originally shown by experiments in which a hole was made in the plasma membrane of a large *Chironomus* salivary gland cell, and the cell interior was equilibrated with known Ca^{2+} concentrations of the exterior through the hole. The conductance of the junctional membrane and its permeability for tracer molecules then fell steeply at exterior Ca^{2+} concentrations above 4–8 $\times$ 10^{-5} M (Loewenstein et al., 1967; Oliveira-Castro and Loewenstein, 1971). The most direct demonstration of this action of calcium was given by experiments in which the cytoplasmic free Ca^{2+} concentration $[Ca^{2+}]_i$ was monitored with the aid of aequorin. Microinjection of Ca^{2+} or elevation of $[Ca^{2+}]_i$ by treating cells with metabolic inhibitors or Ca-ionophores caused prompt and reversible channel closure whenever the $[Ca^{2+}]_i$ rose in the junctional locale (Rose and Loewenstein, 1976).

This channel closure mechanism by Ca^{2+} has been found to operate in a wide variety of animal cells [see Loewenstein and Rose (1978) for a review]. The regulation may be summarized by the scheme

$$\text{channel open state} \underset{[Ca^{2+}]_i \downarrow}{\overset{\uparrow [Ca^{2+}]_i}{\rightleftarrows}} \text{channel closed state}$$

The transitions between the states probably occur in the $[Ca^{2+}]_i$ range of 10^{-7} to 10^{-5} M. The threshold concentration for closure is estimated at 5 $\times$ 10^{-7} M (*Chironomus* salivary gland and mammalian heart) (Rose and Loewenstein, 1976; Dahl and Isenberg, 1980).

There is little as yet to guide us to the molecular mechanism of Ca^{2+}. From the aequorin experiments it is clear that Ca^{2+} acts in the immediate vicinity of the junction and that the channel reaction to Ca^{2+} is fast. This points to a direct membrane action, perhaps a binding reaction leading to a change in the channel's molecular (or fixed-charge) configuration (Loewenstein, 1966; 1975b).

Is the Ca^{2+} action mediated by a calmodulin? Several observations are suggestive: calmodulin binds to purified protein of gap junction of mammalian liver and lens (Hertzberg and Gilula, 1982) and to lens gap junction protein *in situ* (Welsh et al., 1982); calmodulin inhibitors (trifluorperazine, chlorpromazine, calmidazolium) cause electrical uncoupling of cells in insect epidermis (Lees-Miller and Caveney, 1982) or prevent uncoupling in *Xenopus* embryo (C. Peracchia, personal communication).

Channel closure can also be produced by cytoplasmic acidification in many cell types (Turin and Warner, 1977, 1980; Rose and Rick, 1978; Reber and Weingart, 1982). In *Xenopus* blastomere junction, for example, about half of the channels close on reducing the cytoplasmic pH by 0.6 unit (Spray et al., 1981). The effect is prompt and reversible.

Ca^{2+} Is Sufficient for Closing the Channel

Because acidification generally causes release of Ca^{2+} from intracellular stores (Rose and Rick, 1978; Lea and Ashley, 1978; Connor and Ahmed, 1979; Rick et al., 1980), the question naturally arises whether the effect is mediated by Ca^{2+}. But there is another side to that coin: elevation of $[Ca^{2+}]_i$ can cause release of H^+ (Meech and Thomas, 1977). Mitochondria, for example, release H^+ as they take up Ca^{2+} (Åkerman, 1978). Thus the question becomes which of the two ions mediates the channel closure—or in experimental terms—which of the two is sufficient?

There is a clear answer that Ca^{2+} is sufficient. Injection of pH-buffered Ca^{2+} solutions into *Chironomus* salivary gland cells causes channel closure when pH_i is constant and even when it is raised (pH_i, $[Ca^{2+}]_i$, and electrical coupling were monitored simultaneously in these experiments) (Rose and Rick, 1978).

As for the sufficiency of H^+, a definitive answer is still lacking. The most suggestive evidence comes from experiments by Spray et al. (1982) in which the pH at one side of a junction was varied by perfusing one cell of a pair of *Xenopus* blastomeres with solutions of known pH. Junctional conductance fell as the pH fell on that side (cis). It is unclear, however, whether the channel closure here was due to H^+ acting on the cis side or to Ca^{2+} acting on the trans side, that is, Ca^{2+} released by H^+ diffusing from cis to trans; H^+ does diffuse rapidly through junctions. Spray et al., (1982) considered such a trans mechanism unlikely because the Ca^{2+} sensitivity on the cis side was relatively low in their experiments. However, this argument is not cogent if calmodulin mediates the Ca^{2+} action. Calmodulin may have been partially washed out on the cis side by the perfusion, accounting for the low Ca^{2+} sensitivity. Precisely an effect of this sort is suggested by the results of Johnston and Ramon (1981) of perfusion of the crayfish (septate) nerve junction: upon perfusion of *both* sides of the junction, the cell-to-cell channels lost both Ca and pH sensitivity.

The Channel Closure by Ca at the Structural Level

Knowledge about the cell-to-cell channel came through transport studies long before intelligence about its structure or even its whereabouts in cell junctions. It is now widely thought that the channel is embodied by the intramembrane particles seen electron microscopically in clusters called *gap junction* or *nexus*. The particles are 60–80 Å in diameter in vertebrates and 110–150 Å in arthropods (Chalcroft and Bullivant, 1970; McNutt and Weinstein, 1973; Peracchia, 1973; Gilula, 1974; Revel, 1974; Makowski et al., 1977). They are in register on the two membranes, forming bonded pairs that span the 20–30-Å membrane gap (Caspar et al., 1977; Zampighi and Unwin, 1979; Baker et al., 1983). Each pair probably constitutes a channel unit; the members in a pair probably correspond to the protochannels.

By Fourier synthesis of electron microscopic views of particles at different tilts (mammalian liver), Unwin and Zampighi (1980) arrived at a structure consisting of six rodlike protein subunits, about 25 Å thick, 75 Å long, tilted with respect to the membrane plane and the sixfold axis of symmetry, giving the whole a left-handed twist. The subunits surround a

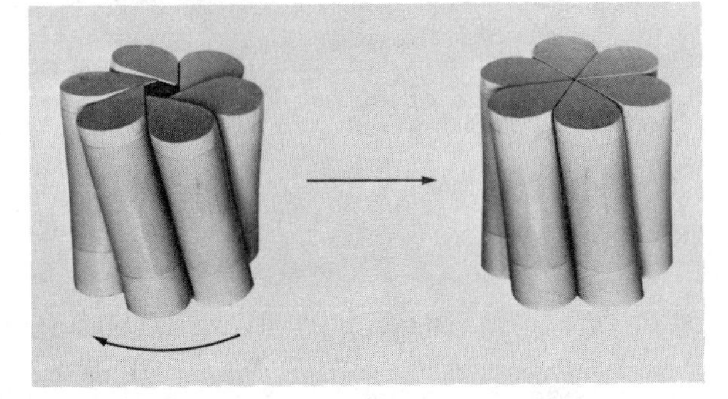

FIGURE 4. A structural model of the cell-to-cell channel as inferred from three-dimensional Fourier synthesis of electronmicroscopic views of membrane particles in gap junction: a tilted hexameric structure with sixfold symmetry (traversing the lipid bilayer). The structure corresponds to a channel half (protochannel) of Figure 1. Left: in the open state; right: closed state, on exposure to Ca^{2+}. (Reproduced with permission of Unwin and Zampighi (1980); see also Unwin and Ennis 1983.)

central opening of about 20-Å diameter, outlined by hydrophilic stain (Fig. 4). The opening has just been traced through the entire particle: it is a roughly cylindrical channel (P. N. T. Unwin, personal communication).

This structure has now been shown to go into a closed configuration by the action of Ca^{2+}. When the isolated gap-junction membrane preparation was exposed to 10^{-5} *M* Ca^{2+}, the particle underwent a conformational change constricting the central channel. Exposure to low pH (6.0) did not produce this change, neither was there a change seen when the pH was varied anywhere in the range of 6.0 to 8.0 (Unwin and Ennis, 1983). The structural transition by Ca^{2+} appears to be generated by a clockwise rotation of the subunit ensemble that straightens the subunits out with respect to the membrane plane and displaces them radially—an elegant iris diaphragm mechanism where the parts move largely tangentially to the container ring, the membrane lipid (Fig. 4).

The results of these structural studies are in pleasing agreement with those obtained by the electrophysiological probings of the channel in living cells.

The Role of Channel Closure as a Mechanism for Survival of Cell Ensembles

The channel-closure mechanism allows a connected cell ensemble to seal itself off from an unhealthy member. All elements of such a sealing reaction are built into the normal cell system and are critically poised: the steep chemical and electrical gradients drive Ca^{2+} inward, and the channels can rapidly close in the presence of high $[Ca^{2+}]_i$. All that is required to set the reaction into motion is a discontinuity in the cell membrane or a depression of cellular energy metabolism on which the intracellular sequestering and outward pumping of Ca depend. It is easy to see, in immediate connection with the Ca^{2+} mechanism, two general categories of functions for channel closure: (1) that of uncoupling a cell community from a cell member with defective Ca^{2+}-pumping or -buffering mechanisms, and (2) that of uncoupling it from a member with a damaged membrane. The first is likely to apply to cells reaching the end of their life span or when they get metabolically poisoned, and the second occurs upon injury to all sorts of tissues.

The channel closure mechanism probably arose early in phylogeny, allowing the first multicell interconnected organisms to survive local injury. The mechanism is kept throughout phylogeny. In fact, given the widespread connectivity, one can hardly imagine how our tissues could survive injury or death of even a single cell without some form of channel self-sealing.

REFERENCES

Akerman, K. E. O. (1978) Effect of pH and Ca^{2+} on the retention of Ca^{2+} by rat liver mitochondria. *Arch. Biochem. Biophys.* **189**:256–262.

Azarnia, R. and W. R. Loewenstein (1971) Intercellular communication and tissue growth. V. A cancer cell strain that fails to make permeable membrane junctions with normal cells. *J. Membrane Biol.* **6**:368–385.

Azarnia, R., G. Dahl, and W. R. Loewenstein (1981) Cell junction and cyclic AMP: III. Promotion of junctional membrane permeability and junctional membrane particles in a junction-deficient cell type. *J. Membrane Biol.* **63**:133–146.

Baker, T. S., D. L. D. Caspar, C. J. Hollingshead, and D. A. Goodenough (1983) Gap junction structures. IV. Asymmetric features revealed by low-irradiation microscopy. *J. Cell Biol.* **96**:204–216.

Bamberg, E., H. Apell, H. Alpes, E. Gross, J. L. Morrell, J. F. Harbaugh, K. Janko, and P. Lauger (1978) Ion channels formed by chemical analogs of gramicidin A. *Fed. Proc.* **37**:2633–2638.

Brink, P. R. and M. M. Dewey (1980) Evidence for fixed charge in the nexus. *Nature* **285**:101–102.

Browne, C. L. and H. S. Wiley (1979) Oocyte-follicle cell gap junctions in *Xenopus laevis* and the effects of gonadotropin on their permeability. *Science* **203**:182–183.

Burger, M. M., W. Burkart, G. Weinbaum, and J. Jumblatt (1978) Cell-cell recognition: molecular aspects. Recognition and its relation to morphogenetic processes in general. In *Cell-Cell Recognition*, A. S. G. Curtis, ed. Cambridge University Press, New York, pp. 1–24.

Caspar, D. L. D., D. A. Goodenough, L. Makowski, and W. C. Phillips (1977) Gap junction structures. I. Correlated electron microscopy and X-ray diffraction. *J. Cell Biol.* **74**:605–628.

Caveney, J. (1974) Intercellular communication in a positional field: Movement of small ions between insect epidermal cells. *Dev. Biol.* **40**:311–322.

Chalcroft, J. P. and S. Bullivant (1970) An interpretation of liver cell membrane and junction structure based on observation of freeze-fractured replicas of both sides of the fracture. *J. Cell Biol.* **47**:49–60.

Connor, J. A. and Z. Ahmed (1979) Intracellular calcium regulation by molluscan neurons. *Biophys. J.* **25:**265a.

Dahl, G. and W. Berger (1978) Nexus formation in the myometrium during parturition and induced by estrogen. *Cell. Biol. Int. Rep.* **2:**381–387.

Dahl, G. and G. Isenberg (1980) Decoupling of heart muscle cells: Correlation with increased cytoplasmic calcium activity and with changes of nexus ultrastructure. *J. Membrane Biol.* **53:**63–75.

Dahl, G., R. Azarnia, and R. Werner (1981) Induction of cell-cell channel formation by mRNA. *Nature* **289:**683–685.

Decker, R. J. (1976) Hormonal regulation of gap junction differentiation. *J. Cell Biol.* **69:**669–685.

Edelman, G. M. (1983) Cell adhesion molecules. *Science* **219:**450–457.

Epstein, M. L. and N. B. Gilula (1977) A study of communication specificity between cells in culture. *J. Cell Biol.* **75:**769–787.

Fallon, R. F. and D. A. Goodenough (1981) Five-hour half-life of mouse liver gap-junction protein. *J. Cell Biol.* **90:**521–526.

Fentiman, I. S., J. Taylor-Papadimitriou, and M. Stoker (1976) Selective contact-dependent cell communication. *Nature* **264:**760–762.

Flagg-Newton, J. L. and W. R. Loewenstein (1980) Asymmetrically permeable membrane channels in cell junction. *Science* **207:**771–773.

Flagg-Newton, J. L. and W. R. Loewenstein (1981) Cell junction and cyclic AMP: II. Modulations of junctional membrane permeability, dependent on serum and cell density. *J. Membrane Biol.* **63:**123–131.

Flagg-Newton, J. L., G. Dahl, and W. R. Loewenstein (1981) Cell junction and cyclic AMP: I. Upregulation of junctional membrane permeability and junctional membrane particles by administration of cyclic nucleotide or phosphodiesterase inhibitor. *J. Membrane Biol.* **63:**105–121.

Flagg-Newton, J. L., I. Simpson, and W. R. Loewenstein (1979) Permeability of the cell-to-cell membrane channels in mammalian cell junction. *Science* **205:**404–407.

Fuchs, P. A., J. G. Nicholls, and D. F. Ready (1981) Membrane properties and selective connexions of identified leech neurones in culture. *J. Physiol.* **316:**203–223.

Furshpan, E. J. and D. D. Potter (1968) Low resistance junctions between cells in embryos and tissue culture. *Curr. Top. Dev. Biol.* **3:**95–127.

Galtsoff, P. S. (1925) Regeneration after dissociation (an experimental study on sponges). I. Behavior of dissociated cells of *Microniona prolifera* under normal and altered conditions. *J. Exp. Zool.* **42:**183–221.

Garfield, R. E., S. M. Sims, M. S. Kannan, and E. E. Daniel (1978) Possible role of gap junctions in activation of myometrium during parturition. *Am. J. Physiol.* **235:**C168–C179.

Gilula, N. B. (1974) Junctions between cells. In *Cell Communication*, R. P. Cox, ed. Wiley, New York.

Gilula, N. B., M. Epstein, and W. H. Beers (1978) Cell-to-cell communication and ovulation. A study of the cumulus-oocyte complex. *J. Cell Biol.* **78:**58–75.

Henkart, P. S., J. Humphreys, and T. Humphreys (1973) Characterization of sponge aggregate factor; a unique proteoglycan complex. *Biochemistry* **12:**3045–3050.

Hertzberg, E. L. and N. B. Gilula (1982) Liver gap junctions and lens fiber junctions: Comparative analysis and calmodulin interaction. In *Organization of the Cytoplasm*, Cold Spring Harbor Symposium on Quantitative Biology, Vol. 46, pp. 639–645.

Humphreys, T. (1965) Cell surface components participating in aggregation: Evidence for a new cell particulate. *Exp. Cell Res.* **40:**539–543.

Ito, S. and W. R. Loewenstein (1969) Ionic communication between early embryonic cells. *Dev. Biol.* **19:**228–243.

Ito, S., E. Sato, and W. R. Loewenstein (1974) Studies on the formation of a permeable cell membrane junction. II. Evolving junctional conductance and junctional insulation. *J. Membrane Biol.* **19:**339–355.

Johnson, M. F. and F. Ramon (1981) Electrotonic coupling in internally perfused crayfish segmented axons. *J. Physiol.* **317:**509–518.

Krebs, E. G. (1972) Protein kinases. *Curr. Top. Cell. Regul.* **5:**99–133.

Kuo, J. F. and P. Greengard (1969) Cyclic nucleotide-dependent protein kinases. IV. Widespread occurrence of adenosine 3′,5′-monophosphate-dependent protein kinase in various tissues and phyla of the animal kingdom. *Proc. Natl. Acad. Sci. U.S.A.* **64:**1349–1355.

Lea, T. J. and C. C. Ashley (1978) Increase in free Ca^{2+} in muscle after exposure to Co_2. *Nature* **275:**236–238.

Lees-Miller, J. P. and S. Caveney (1982) Drugs that block calmodulin activity inhibit cell-to-cell coupling in the epidermis of *Tenebrio molitor*. *J. Membrane Biol.* **69:**233–245.

Lo, C. W. and N. B. Gilula (1979) Gap junctional communication in the preimplantation mouse embryo. *Cell* **18:**399–409.

Loewenstein, W. R. (1966) Permeability of membrane junctions. *Ann. N.Y. Acad. Sci.* **137:**441–472.

Loewenstein, W. R. (1967) On the genesis of cellular communication. *Dev. Biol.* **15:**503–520.

Loewenstein, W. R. (1975a). Cellular communication by permeable junctions. In *Cell Membranes: Biochemistry, Cell Biology and Pathology*, G. Weissmann and R. Claiborne, eds. H. P. Publishing, New York, pp. 105–114.

Loewenstein, W. R. (1975b) Permeable junctions. *Cold Spring Harbor Symp. Quant. Biol.* **40:**49–63.

Loewenstein, W. R. (1981) Junctional intercellular communication: The cell-to-cell membrane channel. *Physiol. Rev.* **61:**829–913.

Loewenstein, W. R. (1985) Regulation of cell-to-cell communication by phosphorylation. *Biochem. Soc. Symp.*, **50:**43–58.

Loewenstein, W. R. and B. Rose (1978) Calcium in (junctional) intercellular communication and a thought on its behavior in intracellular communication. *Ann. N.Y. Acad. Sci.* **307:**285–307.

Loewenstein, W. R., Y. Kanno, and S. J. Socolar (1978a) Quantum jumps of conductance during formation of membrane channels at cell-cell junction. *Nature* **274:**133–136.

Loewenstein, W. R., Y. Kanno, and S. J. Socolar (1978b) The cell-to-cell channel. *Fed. Proc. (Sympos.)* **37:**2645–2650.

Loewenstein, W. R., M. Nakas, and S. J. Socolar (1967) Junctional membrane uncoupling. Permeability transformations at a cell membrane junction. *J. Gen. Physiol.* **50:**1865–1891.

Makowski, L., D. L. D. Caspar, W. C. Phillips, and D. A. Goodenough (1977) Gap junction structure. II. Analysis of the X-ray diffraction data. *J. Cell Biol.* **74:**629–645.

McNutt, N., and R. S. Weinstein (1973) Membrane ultrastructure at mammalian intercellular junction. *Prog. Biophys. Mol. Biol.* **26:**45–101.

Meech, R. W. and R. C. Thomas (1977) The effect of calcium injection on the intracellular sodium and pH of snail neurones. *J. Physiol. (Lond.)* **265:**267–283.

Michalke, W. and W. R. Loewenstein (1971) Communication between cells of different types. *Nature* **232:**121–122.

Moscona, A. A. (1968) Cell aggregation: Properties of specific cell ligands and their role in the formation of multicellular systems. *Dev. Biol.* **18:**250–277.

Ocklind, C. and B. Obrink (1982) Intercellular adhesion of rat hepatocytes. Identification of a cell surface glycoprotein involved in the initial adhesion process. *J. Biol. Chem.* **257**:6788–6795.

Ocklind, C., U. Forsum, and B. Obrink (1983) Cell surface localization and tissue distribution of a hepatocyte cell-cell adhesion glycoprotein (Cell-CAM 105). *J. Cell Biol.* **96**:1168–1171.

Oliveira-Castro, G. M. and W. R. Loewenstein (1971) Junctional membrane permeability: Effects of divalent cations. *J. Membrane Biol.* **5**:51–77.

Peracchia, C. (1973) Low resistance junctions in crayfish. II. Structural details and further evidence for intercellular channels by freeze-fracture and negative staining. *J. Cell Biol.* **57**:66–80.

Pitts, J. D. (1972) Direct interaction between animal cells. In *Cell Interactions, Third Lepetit Colloquium*, L. G. Silvestri, ed. North-Holland, Amsterdam, pp. 277–285.

Pitts, J. D. and R. R. Burk (1976) Specificity of junctional communication between animal cells. *Nature* **264**:762–764.

Radu, A., G. Dahl, and W. R. Loewenstein (1982) Hormonal regulation of cell junction permeability: Upregulation by catecholamine and prostaglandin E_1. *J. Membrane Biol.* **70**:239–251.

Reber, W. R. and R. Weingart (1982) Ungulate cardiac Purkinje fibres: The influence of intracellular pH on the electrical cell-to-cell coupling. *J. Physiol.* **238**:87–104.

Revel, J-.P. (1974) Contacts and junctions between cells. In *Society for Experimental Biology*, Vol. 28. Cambridge University Press, London, pp. 447–467.

Revel, J.-P., S. B. Yancey, D. J. Meyer, and B. Nicholson (1980) Cell junctions and intercellular communication. *In Vitro* **16**(12):1010–1017.

Rink, T. J., R. Y. Tsien, and A. E. Warner (1980) Free calcium in *Xenopus* embryos measured with ion-selective microelectrodes. *Nature* **283**:658–660.

Rose, B. and W. R. Loewenstein (1976) Permeability of a cell junction and the local cytoplasmic free ionized calcium concentration. A study with aequorin. *J. Membrane Biol.* **28**:87–119.

Rose, B. and R. Rick (1978) Intracellular pH, intracellular free Ca, and junctional cell-cell coupling. *J. Membrane Biol.* **44**:377–415.

Rosen, O. M. and E. G. Krebs, eds. (1981) "Protein Phosphorylation," *Cold Spring Harbor Conference on Cell Proliferation*, Vol. 8 A and B.

Schwartzmann, G. O. H., H. Wiegandt, B. Rose, A. Zimmerman, D. Ben-Haim, and W. R. Loewenstein (1981) The diameter of the cell-to-cell junctional membrane channels, as probed with neutral molecules. *Science* **213**:551–553.

Sheridan, J. D. (1968) Electrophysiological evidence for low-resistance intercellular junctions in the early chick embryo. *J. Cell Biol.* **37**:650–659.

Simpson, I., B. Rose, and W. R. Loewenstein (1977) Size limit of molecules permeating the junctional membrane channels. *Science* **195**:294–296.

Spray, D. C., A. L. Harris, and M. V. L. Bennett (1981). Gap junctional conductance is a simple and sensitive function of intracellular pH. *Science* **211**:712–715.

Spray, D. C., J. H. Stern, A. L. Harris, and M. V. L. Bennett (1982) Gap junctional conductance: Comparison of sensitivities to H and Ca ions. *Proc. Natl. Acad. Sci. U.S.A.* **79**:441–455.

Traub, O., P. M. Druge, and K. Willecke (1983) Degradation and synthesis of gap junction protein in plasma membranes of regenerating liver after partial hepatectomy or cholestasis. *Proc. Natl. Acad. Sci. U.S.A.* **80**:755–759.

Turin, L. and A. Warner (1977) Carbon dioxide reversibly abolishes ionic communication between cells of early amphibian embryo. *Nature (Lond.)* **270**:56–69.

Turin, L. and A. Warner (1980) Intracellular pH in early *Xenopus* embryos: Its effect on current flow between blastomeres. *J. Physiol.* **300**:489–504.

Unwin, P. N. T. and P. D. Ennis (1983) Calcium-mediated changes in gap junction structure: Evidence from the low angle X-ray pattern. *J. Cell Biol.* **97:**1459–1466.

Unwin, P. N. T. and G. Zampighi (1980) Structure of the junction between communicating cells. *Nature,* **283:**545–549.

Urry, D. W., M. C. Goodall, J. D. Glickson, and D. F. Mayers (1971) The gramicidin A transmembrane channel: Characteristics of head-to-head dimerized $_{(L,D)}$helices. *Proc. Natl. Acad. Sci. U.S.A.* **68:**1907–1911.

Warner, A. (1973) The electrical properties of the ectoderm in the amphibian embryo during induction and early development of the nervous system. *J. Physiol.* **235:**267–286.

Welsh, M. J., J. C. Aster, M. Ireland, J. Alcala, and H. Maisel (1982) Calmodulin binds to chick lens gap junction protein in a calcium-independent manner. *Science* **216:**642–644.

Wiener, E. C. and W. R. Loewenstein (1983) Correction of cell-cell communication defect by introduction of a protein kinase into mutant cells. *Nature* **305:**433–435.

Zampighi, G. and P. N. T. Unwin (1979) Two forms of isolated gap junctions. *J. Mol. Biol.* **135:**451–464.

Chapter Ten

THE STRUCTURE OF THE NEUROENDOCRINE COMMANDS FOR EGG-LAYING BEHAVIOR IN *APLYSIA*

FELIX STRUMWASSER

Department of Physiology
Boston University
School of Medicine
Boston, Massachusetts
and
Marine Biological Laboratory
Woods Hole, Massachusetts

One mode of communication in the nervous system is the release into the circulation of a neurohormone that commands a goal-directed behavior. Among vertebrates, perhaps the best example is the discovery that angiotensin II, a natural octapeptide, induces a short-latency vigorous drinking response whether injected intracranially or systemically (Fitzsimons and Kucharczyk, 1978). Doses, intracranially, as low as femtomoles induce significant drinking. Among invertebrates, also, neuropeptides play an important role in initiating goal-directed behaviors. In moths the emergence of the adult from the old pupal cuticle involves writhing behaviors which are turned on by a brain peptide, eclosion hormone (Truman, 1971).

This chapter is concerned with the structure of the commands and the neural program for the innate reproductive act of egg laying in the opisthobranch mollusk *Aplysia*. Our laboratory has purified three peptides that initiate egg laying, and the primary structure of these peptides has been determined (Chiu et al., 1979; Heller et al., 1980). Recently the genes for these three peptides have been cloned by Scheller and colleagues (1983). The nucleotide sequences confirm the primary structure data obtained on the three peptides and show that there are relationships among the three genes.

The final command for egg laying is the neurohormone, egg-laying hormone (ELH). ELH is a basic peptide (pI 9.1) with molecular weight, determined from its primary structure, of 4385 daltons (Chiu et al., 1979). It is synthesized by the approximately 1000 bag cell neurons of the abdominal ganglion (Arch, 1972). The bag cell neurons are organized into two symmetrical clusters occurring at the junction of the abdominal ganglion and the two pleurovisceral connectives that connect this ganglion with the pleural ganglia in the head of the animal. Antibodies made in rabbits to ELH conjugated to thyroglobulin, after removal of the antithyroglobulin fraction, selectively stain each of the bag cells within the two clusters, using the peroxidase antiperoxidase second antibody technique (Chiu and Strumwasser, 1981).

The neural program for releasing ELH is a long-lasting synchronous pacemakerlike discharge of all the bag cells. This 30-minute discharge is easily evoked *in vitro* by brief electrical stimulation of either of the pleurovisceral connectives (Kupfermann and Kandel, 1970) and has been shown by *in vivo* recordings to always precede "spontaneous" egg laying in *Aplysia braziliana* (Dudek et al., 1979). Our studies have shown that this 30-minute program of pacemaker discharge is mediated by a cyclic AMP-dependent protein phosphorylation mechanism (Kaczmarek et al., 1978; Jennings et al., 1982; Strumwasser et al., 1982).

The precommand for egg laying is only partly worked out. Exactly what triggers the bag cells to discharge and release ELH in the intact *Aplysia* is not now known. Isolated *Aplysia* are known to lay eggs (MacGinitie, 1934), so that there is some mechanism, which we can term spontaneous egg laying, that does not appear to depend on any currently known external signal. Presumably mating, which occurs in these hermaphrodites, plays a role in reproductive function, but it is given a minor role as a causative factor in egg laying in *A. braziliana* (Blankenship et al., 1983). However, it is known that there are several peptides, extractable from a special region of the reproductive tract, the atrial gland, that can produce egg laying when injected into intact animals. Two of these peptides, termed peptides A and B, which bear no resemblance to ELH (Heller et al., 1980), mediate egg laying by acting through the abdominal ganglion since these peptides are inactive in the absence of the ganglion (Strumwasser et al., 1981). Furthermore, peptides A and B are known to initiate bag cell discharge *in vitro* at moderately low concentrations (Heller et al., 1980). A third peptide extracted from the atrial gland, called "egg-releasing hormone," resembles both ELH and peptides A and B and can induce egg laying in the absence of the abdominal ganglion (Schlesinger et al., 1981). When an *Aplysia* genomic library was repeatedly screened with a radioactive cDNA probe, coding for ELH-like sequences, a clone for this putative gene could not be found although the same probe turned up clones for the three other peptide genes (Scheller et al., 1983).

THE COMMAND FOR EGG LAYING

Purification and Structure of ELH

The command for egg laying is the release of ELH from the bag cells. ELH can be purified to homogeneity by a two-step procedure involving cation exchange chromatography on SP C25 (Sephadex) followed by gel filtration on a column of Bio-Rad P6 material (Chiu et al., 1979). The absorbance profile (at 215 nm) of fractions from the P6 column shows a major peak at Kav 0.2, the only region from which egg laying could be induced. This two-step procedure provides a 100-fold enrichment of ELH and on amino acid sequence analysis shows a single isoleucine NH2-terminus. When 2.5 nmol of this purified material is injected into *A. californica* it consistently induces egg laying (Chiu et al., 1979).

ELH is a basic peptide (pI 9.1) containing 36 amino acid residues (Chiu et al., 1979). The primary structure of ELH is shown in Figure 1. Isoleucine is the amino terminal residue and lysine the carboxy terminal residue. Since carboxypeptidases A and B were unable to cleave ELH, we suggested that the carboxy terminal lysine was probably blocked by amidation, but other possibilities were not excluded at the time.

5 10 15
A H-Ala -Val -Lys-Leu -Ser-Ser -Asp -Gly -Asn -Tyr -Pro-Phe -Asp -Leu -Ser
B H-Ala -Val -Lys-Ser -Ser-Ser -Tyr -Glu -Lys -Tyr -Pro-Phe -Asp -Leu -Ser
ELH H-Ile -Ser -Ile -Asn -Gln-Asp -Leu -Lys -Ala -Ile -Thr-Asp -Met -Leu -Leu-

20 25 30
A Lys -Glu -Asp -Gly -Ala -Gln-Pro -Tyr -Phe-Met -Thr -Pro-Arg -Leu -Arg
B Lys -Glu -Asp -Gly -Ala -Gln-Pro -Tyr -Phe-Met -Thr -Pro-Arg -Leu -Arg
ELH Thr -Glu -Gln -Ile -Arg -Glu-Arg -Gln -Arg -Tyr -Leu-Ala -Asp -Leu -Arg

34
A Phe-Tyr -Pro -Ile -NH_2
B Phe-Tyr -Pro -Ile -NH_2
ELH Gln -Arg -Leu -Leu-Glu -Lys-NH_2

FIGURE 1. Comparison of amino acid sequences of ELH and peptides A and B. Boxed areas include identical amino acids. Amidated amino-terminals are assumed from nucleotide sequences (Scheller et al., 1983).

Synthetic ELH

Recently, in collaborative experiments with Stephen B. H. Kent and Suzanna J. Horvath (of Caltech), various forms of ELH have been synthesized. Our preliminary tests of ELH–lysine–amide show that it induces egg laying as well as the normal behaviors associated with egg laying (inhibition of locomotor activity, head-weaving movements to wind the egg string, and mouth movements to attach the egg string to the substrate). Furthermore, the latencies for egg laying in the two test animals were normal (33 and 36 min) as well as the weight of the egg masses (1.1 and 1.8 g). The egg strings were examined under a dissecting microscope, and they both contained eggs, indicating that the synthetic material did indeed release eggs normally from the gonad into the small hermaphroditic duct. Packaging of the eggs within the egg string was also normal in both cases. Thus we can conclude that ELH–lysine–amide is sufficient to cause normal egg laying and associated behaviors in intact *Aplysia*.

Gene for ELH

The induction of egg laying with synthetic as well as native ELH is of great interest in view of the structure of the gene for ELH. Scheller and colleagues (1983) have sequenced a clone (ELH-1R) that codes for one complete copy of ELH as well as the 27 amino acid residue called acidic peptide (pI 4.8). This gene, if fully transcribed and translated, would code for a 41.8 Kilodaltons (Kd) precursor. Furthermore, from the presence of seven dibasic residues, excluding those related to the signal sequence, ELH and the acidic peptide, at least seven peptide fragments could be generated by action of endopeptidases. Rothman and colleagues (1983) have provided evidence that one of these peptide fragments, an octapeptide, alpha bag cell peptide, can be purified from bag cells and has inhibitory actions on certain autoactive identifiable neurons in the upper-left quadrant of the abdominal ganglion.

Role of the Abdominal Ganglion in Egg Laying

This latter result raises the question as to whether the abdominal ganglion itself plays any role in the overt behaviors of egg laying, as well as in the generation and extrusion of the egg string, except as a source of ELH. Some years ago we showed that surgical removal of the entire abdominal ganglion, including the two bag cell clusters, did not prevent the induction of egg laying and associated behaviors by crude extracts of the abdominal ganglion containing ELH (Strumwasser et al., 1972). Although these results should be repeated with synthetic ELH, they suggest that the abdominal ganglion does not play an essential role in the various aspects of egg laying referred to, including the associated behaviors, other than as a source of ELH. In conclusion, ELH by itself is a sufficient command to cause egg laying and associated behaviors in *Aplysia.*

THE NEURAL PROGRAM FOR RELEASING ELH

Bag Cell Discharge Precedes Egg Laying

It is known from *in vivo* recordings from the pleurovisceral connectives in *A. braziliana* that bag cells undergo a long-lasting electrical discharge (average of 21 min) prior to spontaneous bouts of egg laying (Dudek et al., 1979). It is also known that electrical stimulation of the pleurovisceral connectives either *in vivo* (Dudek et al., 1979) or *in vitro* (Kupfermann and Kandel, 1970) with the isolated abdominal ganglion causes an afterdischarge lasting about 30 min.

ELH, Acidic and Other Peptides are Released during Bag Cell Afterdischarge

By allowing isolated abdominal ganglia to incorporate radioactive amino acids into newly synthesized proteins it has been possible to study subsequently the release of radiolabeled peptides from a single isolated bag cell cluster during an electrically initiated afterdischarge. With this approach it has been demonstrated that ELH, the acidic peptide, and at least two other smaller peptides are released during an afterdischarge (Stuart et al., 1980). ELH was identified by its basic pI and comigration with purified radiolabeled marker ELH on isoelectric focusing (IEF) gels. The acidic peptide was identified by its pI of 4.8 on IEF gels. The two smaller peptides were not identified, but recent results by Rothman and colleagues (1983) would suggest that alpha bag cell peptide, an octapeptide, was one of the remaining two peptides released.

Afterdischarge Is a cAMP-Mediated Event

We have shown that the long-lasting afterdischarge in bag cells, after brief electrical stimulation of the pleurovisceral connectives, is a cAMP-dependent mechanism. cAMP rises within 1 min after initiation of an afterdischarge and peaks by 2 min at approximately 2.5 times the basal level before afterdischarge (Kaczmarek et al., 1978). This rise in cAMP appears to play a causal role in afterdischarge since membrane-permeant, phosphodiesterase-resistant, cAMP analogues such as 8-benzylthio or N6-*n*-butyl-8-(benzylthio)-cAMP initiate discharge of bag cells with similar time-varying spike-frequency characteristics as does electrically induced afterdischarge (Kaczmarek et al., 1978; Strumwasser et al., 1982). Forskolin, a diterpene activator of adenylate cyclase (Seamon et al.; 1981), when combined with theophylline, also initiates the characteristic bag cell discharge (Strumwasser et al., 1982), and phophodiesterase inhibitors, such as isobutylmethyl xanthine, caffeine, and papaverine significantly prolong electrically initiated afterdischarges (Kaczmarek et al., 1978).

Bag Cell Afterdischarge Is Associated with Phosphorlyation of Specific Proteins

Since bag cell afterdischarge is cAMP-dependent, we examined protein phosphorylation at 2 and 20 min into afterdischarge. Two different procedures were used (pre- and postlabeling). In the first procedure, abdominal ganglia were incubated in ^{32}P-orthophosphate for periods around 24 hours. Subsequently an afterdischarge was initiated in one isolated hemiganglion, while the other hemiganglion served as a control. The bag cell clusters were removed and homogenized at one of the two experimental time points into afterdischarge. Proteins were separated on SDS-polyacrylamide slab gels, and autoradiograms of the dried gels were obtained as described in the full publication (Jennings et al., 1982). With this procedure two phosphoproteins, in particular, showed consistent changes during afterdischarge. Both proteins showed enhanced phosphorylation: BC-1 (apparent mw of 33,000 daltons) was increased by 82 ± 14% at 2 min and 69 ± 43% at 20 min relative to controls. BC-2 (apparent mw of 21,000 daltons) was increased at 20 min (92 ± 23%) but not at 2 min (−19 ± 28%).

In the second procedure, bag cell clusters were homogenized at the same two time points into afterdischarge but phosphorylation was assayed in the cell-free preparation by using gamma-labeled ^{32}P-ATP and exogenous cAMP-dependent catalytic subunit of protein kinase. In this procedure the degree of phosphorylation observed in autoradiograms bears an inverse relationship to the endogenous phosphorylation since only the remaining available sites for phosphorylation can be occupied by radioactive ^{32}P. We found that there was a 73 ± 9% reduction in BC-2 at 20 min and no significant change at 2 min.

These results suggest that one or more proteins, such as BC-1 which shows an enhanced phosphorylation early in afterdischarge, could be playing a role in the enhanced membrane excitability of afterdischarge. Likewise, proteins such as BC-2, which show enhanced phosphorylation late in afterdischarge, may play some role in the termination of or recovery processes from afterdischarge. One approach to testing these hypotheses would be to raise antibodies, preferably monoclonal, to these proteins and use them as functional reagents to interfere with the normal processes of afterdischarge. There are currently examples of the use of monoclonal antibodies that interfere with various membrane enzymes, receptors, and channels, such as Na-K-ATPase (Schenk and Leffert, 1983), transferrin receptors and the growth of mammalian cells (Trowbridge and Lopez, 1982), and sodium channels and nerve impulse conduction (Meiri et. al.,

1983). We have started to develop monoclonal antibodies (MAb) against neural membranes of *Aplysia* (Viele et al., 1984). One MAb, 2BC4, blocks the postburst hyperpolarization when injected into the identified bursting pacemaker, R15, or isolated bag cells (Strumwasser et al., 1985).

The Program for Afterdischarge Is Stored in Individual Bag Cells

To further analyze the membrane mechanisms altered by cAMP-dependent protein phosphorylation, we developed techniques for dissociating bag cells and maintaining them in primary culture (Strumwasser et al., 1978; Kaczmarek et al., 1979). Such isolated bag cells primarily produce calcium spikes, on transmembrane stimulation, spikes that are insensitive to tetrodotoxin and sensitive to cobalt ions (Strumwasser et al., 1981). The amplitude and duration of these calcium spikes increase with repetitive transmembrane stimulation.

When cAMP analogues (see section on afterdischarge as a cAMP-mediated event) are added to such primary cultures the bag cells undergo spontaneous discharge (Kaczmarek and Strumwasser, 1981a). Interestingly, when spikes are added for a few seconds by transmembrane stimulation, there is an increase in the background rate of spike discharge for many seconds after the applied current (Kaczmarek and Strumwasser, 1981a). We tested for the afterdischarge property in a more stringent manner. After adding the cAMP analogue, an isolated bag cell was immediately hyperpolarized to prevent the spontaneous phase of afterdischarge. When the hyperpolarization was momentarily released, about 15 min later, to allow spike discharge for a few seconds, an afterdischarge lasting for 12–20 min occurred. These results demonstrate that the program of afterdischarge is intrinsic to each bag cell and furthermore that cAMP priming followed by calcium action potentials releases the program. Recently we have demonstrated, by measuring absorbance changes due to the calcium indicator dye, arsenazo III, that action potentials in bag cells do indeed increase calcium intracellularly (Woolum and Strumwasser, 1983).

Membrane Changes with Enhanced Protein Phosphorylation

Using isolated bag cells in primary culture, we have injected the catalytic subunit of cAMP-dependent protein kinase (PKC) purified from beef heart to further test the hypothesis that protein phosphorylation is responsible for the enhanced membrane excitability associated with afterdischarge (Kaczmarek et al., 1980). Intracellular PKC injections enhanced the rise time, amplitude, and duration of the calcium action potentials in bag cells. Associated with these changes was an increase in membrane resistance.

Voltage clamp studies of isolated bag cells show that net outward current is decreased during the action of external cAMP analogues (Kaczmarek and Strumwasser, 1981b), which is consistent with the enhancement of calcium action potentials seem with both PKC injections as well as with external cAMP analogues. In addition, it has been recently observed that a transient outward current, identified as the "A" current of Connor and Stevens (1971), is diminished during cAMP action (Kaczmarek and Strumwasser, 1984). Since there are at least three different potassium channels in molluskan neurons, one approach to further studies of the action of protein phosphorylation on the membrane of bag cells is the recording of single ion channels by the gigaseal patch recording technique (Hamill et al., 1981). We have successfully recorded single-channel activity in isolated bag cells and are currently classifying the nature of these channels (Strumwasser and McIntyre, unpublished). Siegelbaum and colleagues (1982) have recorded single-channel activity in sensory neurons of the intact abdominal ganglion and have found serotonin-sensitive outward current channels that are turned off by external cAMP analogues.

THE PRECOMMAND FOR EGG LAYING

Peptides A and B of the Atrial Gland

As indicated in the introduction to this chapter, there is incomplete information on the factors that initiate bag cell discharge *in vivo*. Arch and colleagues (1978) first showed that extracts of the atrial gland, a part of the large hermaphroditic duct near the gonopore, could induce egg laying. We purified two factors from the atrial gland of *A. californica*, peptides A and B, that induced egg laying. These peptides each contain 34 amino acid residues and differ from one another in only four residues (Heller et al., 1980). Peptide A has a calculated mw of 3924 daltons and a pI of 8.0; peptide B has a calculated mw of 4032 and a pI of 9.1. The amino acid sequences of peptides A and B are shown in Figure 1. Alanine is the amino terminal residue; isoleucine is the carboxy terminal residue.

Peptides A and B induce egg laying in *Aplysia* only when the abdominal ganglion containing the bag cells is left intact (Strumwasser et al., 1981). Peptides A and B also induce bag cell discharge in the *in vitro* abdominal ganglion preparation at concentrations as low as 0.1 μm (Heller et al., 1980).

The Nature of the Atrial Gland

These findings raise the issue as to the nature of the atrial gland. In recent studies Beard and colleagues (1982) conclude that the atrial gland is an exocrine gland. They find that the gland is a highly infolded stratified epithelium consisting of two cell types. The "capping" cell has cilia that border on the lumen of the hermaphroditic duct, whereas the "gobletlike" exocrine cell has 1–2 μm electron-dense granules. It is known that neighboring regions of the large hermaphroditic duct do not contain factors that induce egg laying. The structure of cells in this latter region is rather similar, but the large electron-dense granules have cristalike infoldings, and the granule is bounded by a double-limiting membrane, whereas the atrial gland granule has a single-limiting membrane and no crista (Beard et al., 1982).

Hence there is an enigma. The atrial gland contains at least three peptides that can induce egg laying (peptides A, B, and egg-releasing "hormone"). All three peptides activate the bag cells, and in addition ERH can induce egg laying in the absence of bag cells. However, the contents of the secretory cells in the atrial gland presumably enter the lumen of the large hermaphroditic duct in the intact animal close to the gonopore and the salty ocean just external to this orifice. It is possible then that the three peptides of the atrial gland never get into the circulation and serve some other function, presumably reproductive, yet to be discovered. On the other hand, it is possible that these reproductive tract peptides are similar to or identical with the putative transmitters that turn on the bag cells. Thus antibodies to these three peptides might lead us to the group of neurons that trigger the bag cells to discharge. It is noteworthy that antibodies to ELH, besides staining the bag cells in the abdominal ganglion (Chiu and Strumwasser, 1981), also stain discrete and symmetrical groups of neurons in the pleural and cerebral ganglia (Chiu and Strumwasser, 1983).

REFERENCES

Arch, S. (1972) Biosynthesis of the egg-laying hormone (ELH) in the bag cell neurons of *Aplysia californica*. *J. Gen. Physiol.* **60:**102–119.

Arch, S., T. Smock, R. Gurvis, and C. McCarthy (1978) Atrial gland induction of the egg-laying response in *Aplysia californica*. *J. Comp. Physiol.* **128:**67–70.

Beard, M., L. Millecchia, C. Masuoka, and S. Arch (1982) *Tissue Cell* **14:**297–308.

Blankenship, J. E., M. K. Rock, L. C. Robbins, C. A. Livingston, and H. K. Lehman (1983) Aspects of copulatory behavior and peptide control of egg-laying in *Aplysia*. *Fed. Proc.* **42:**96–100.

Chiu, A. Y., M. W. Hunkapiller, E. Heller, D. K. Stuart, L. E. Hood, and F. Strumwasser (1979) Purification and primary structure of the neuropeptide egg-laying hormone of *Aplysia californica*. *Proc. Natl. Acad. Sci. U.S.A.* **76:**6656–6660.

Chiu, A. Y. and F. Strumwasser (1981) An immunohistochemical study of the neuropeptidergic bag cells of *Aplysia*. *J. Neurosci.* **1:**812–826.

Chiu, A. Y. and F. Strumwasser (1984) Two neuronal populations in the head ganglia of *Aplysia californica* with egg-laying hormone-like immunoreactivity. *Brain Res.* **294:**83–93.

Connor, J. A. and C. F. Stevens (1971) Voltage clamp studies of a transient outward current in gastropod neural somata. *J. Physiol.* **213:**21–30.

Dudek, F. E., J. S. Cobbs, and H. M. Pinsker (1979) Bag cell electrical activity underlying spontaneous egg laying in freely behaving *Aplysia brasiliana*. *J. Neurophysiol.* **42:**804–817.

Fitzsimons, J. T. and J. Kucharczyk (1978) Drinking and haemodynamic changes induced in the dog by intracranial injection of components of the renin-angiotensin system. *J. Physiol. (Lond.)* **276:**419–434.

Hamill, O. P., A. Marty, E. Neher, B. Sakmann, and F. J. Sigworth (1981) Improved patch-clamp techniques for high resolution current recording from cells and cell-free membrane patches. *Pflügers Arch.* **391:**85–100.

Heller, E., L. K. Kaczmarek, M. W. Hunkapiller, L. E. Hood, and F. Strumwasser (1980) Purification and primary structure of two neuroactive peptides that cause bag cell after discharge and egg-laying in *Aplysia. Proc. Natl. Acad. Sci. U.S.A.* **77:**2328–2332.

Jennings, K. R., L. K. Kaczmarek, R. M. Hewick, W. J. Dreyer, and F. Strumwasser (1982) Protein phosphorylation during afterdischarge in peptidergic neurons of *Aplysia. J. Neurosci.* **2:**158–168.

Kaczmarek L. K., M. Finbow, J. P. Revel, and F. Strumwasser (1979) The morphology and coupling of *Aplysia* bag cells within the abdominal ganglion and in cell culture. *J. Neurobiol.* **10:**535–550.

Kaczmarek, L. K., K. R. Jennings, and F. Strumwasser (1978) Neurotransmitter modulation, phosphodiesterase inhibitor effects, and cyclic AMP correlates of afterdischarge in peptidergic neurites. *Proc. Natl. Acad. Sci. U.S.A.* **75:**5200–5204.

Kaczmarek, L. K., K. R. Jennings, F. Strumwasser, A. C. Nairn, U. Walter, F. D. Wilson, and P. Greengard (1980) Microinjection of catalytic subunit of cyclic AMP-dependent protein kinase enhances calcium action potentials of bag cell neurons in cell culture. *Proc. Natl. Acad. Sci. U.S.A.* **77:**7487–7491.

Kaczmarek, L. K. and F. Strumwasser (1981a) The expression of long lasting afterdischarge by isolated *Aplysia* neurons. *J. Neurosci.* **1:**626–634.

Kaczmarek, L. K. and F. Strumwasser (1981b) Net outward currents of bag cell neurons are diminished by a cAMP analogue. *Abstr. Soc. Neurosci.* **7:**932.

Kaczmarek, L. K. and F. Strumwasser (1984) A voltage-clamp analysis of currents underlying cyclic AMP-induced membrane modulation in isolated peptidergic neurons of *Aplysia. J. Neurophysiol.* **52:**340–349.

Kupfermann, I. and E. R. Kandel (1970) Electrophysiological properties and functional interconnections of two symmetrical neurosecretory clusters (bag cells) in abdominal ganglion of *Aplysia. J. Nerophysiol.* **33:**865–876.

MacGinitie, G. E. (1934) The egg-laying activities of the sea hare, *Tethys californicus. Biol. Bull.* **67:**300–303.

Meiri, H., I. Zeitoun, H. H. Grunhagen, V. Lev-Ram, Y. Cohen, Z. Eshhar, and J. Schlessinger (1983) Monoclonal antibodies against sodium channel. *Abstr. Soc. Neurosci.* **9:**20.

Rothman, B. S., E. Mayeri, R. O. Brown, P. M. Yuan, and J. E. Shively (1983) Primary structure and neuronal effects of bag cell peptide, a second candidate neurotransmitter encoded by a single gene in bag cell neurons of *Aplysia. Proc. Natl. Acad. Sci. U.S.A.* **80:**

Scheller, R. H., J. F. Jackson, L. B. McAllister, B. S. Rothman, E. Mayeri, and R. Axel (1983) A single gene encodes multiple neuropeptides mediating a stereotyped behavior. *Cell* **32:**7–22.

Schenk, D. B. and H. L. Leffert (1983) Monoclonal antibodies to rat Na, K-ATPase block enzymatic activity. *Proc. Natl. Acad. Sci. U.S.A.* **80:**5281–5285.

Schlesinger, D. H., S. P. Babirak, and J. E. Blankenship (1981) Primary structure of an egg releasing peptide from the atrial gland of *Aplysia californica.* In Symposium on Neurohypophyseal Peptide Hormones and Other Biologically Active Peptides, D. H. Schlesinger, ed. Elsevier/North Holland, pp. 137–150.

Seamon, K. B., W. Padgett, and J. W. Daly (1981) Forskolin: unique diterpene activator of adenylate cyclase in membranes and in intact cells. *Proc. Natl. Acad. Sci. U.S.A.* **78:**3363–3367.

Siegelbaum, S. A., J. S. Camardo, and E. R. Kandel (1982) Serotonin and cyclic AMP close single K channels in *Aplysia* sensory neurones. *Nature* **299**:413–417.

Strumwasser, F., L. K. Kaczmarek, and K. R. Jennings (1982) Intracellular modulation of membrane channels by cyclic AMP-mediated protein phosphorylation in the peptidergic bag cell neurons of *Aplysia*. *Fed. Proc.* **41**:2933–2939.

Strumwasser, F., L. K. Kaczmarek, K. Jennings, and A. Y. Chiu (1981) Studies of a model peptidergic neuronal system, the bag cells of *Aplysia*. In Farner, *Neurosecretion: Molecules, Cells, Systems*, D. S. Farner and K. Lederis, eds. Plenum, New York, pp. 249–268.

Strumwasser, F., L. K. Kaczmarek, and D. Viele (1978) The peptidergic bag cell neurons of *Aplysia*: morphological and electrophysiological studies of dissociated cells in tissue culture. *Soc. Neurosci. Abstr.* **4**:207.

Strumwasser, F., F. R. Schlechte, and S. Bower (1972) Distributed circadian oscillators in the nervous system of *Aplysia*. *Fed. Proc.* **31**:405.

Strumwasser, F., D. P. Viele, and K. D. Lovely (1985) Monoclonal antibody blocks the post burst hyperpolarization in R15 and bag cell neurons. *Biophys. J.* **47**:52a.

Stuart, D. K., A. Y. Chiu, and F. Strumwasser (1980) Neurosecretion of egg-laying hormone and other peptides from electrically active bag cell neurons of *Aplysia*. *J. Neurophysiol.* **43**:488–498.

Trowbridge, I. S. and F. Lopez (1982) Monoclonal antibody to transferrin receptor blocks transferrin binding and inhibits human tumor cell growth *in vitro*. *Proc. Natl. Acad. Sci. U.S.A.* **79**:1175–1179.

Truman, J. W. (1971) Physiology of insect ecdysis. I. The eclosion behavior of saturniid moths and its hormonal release. *J. Exp. Biol.* **53**:805–814.

Woolum, J. C. and F. Strumwasser (1983) Dynamic Ca measurements in isolated *Aplysia* bag cells. *Biophys. J.* **41**:59a.

Viele, D. P., F. Strumwasser, and K. D. Lovely (1984) Monoclonal antibodies against neural membranes of *Aplysia*. *Abstr. Soc. Neurosci.* **10**:759.

Chapter Eleven

A COMPARISON OF SIMPLE DEFENSIVE REFLEXES IN *APLYSIA*

Implications for General Mechanisms of Integration and Plasticity

E. T. WALTERS
J. H. BYRNE

Department of Physiology and Cell Biology
University of Texas Medical School
Houston, Texas

T. J. CAREW

Department of Psychology
Yale University,
New Haven, Connecticut

E. R. KANDEL

Center for Neurobiology and Behavior
Departments of Physiology and Psychiatry
College of Physicians and Surgeons
Columbia University, New York

Reflexes have attracted the interest of neuroscientists since Sherrington's insight that these simple behaviors offer a fine vantage point for studying the integrative actions of the nervous system. The interest in reflexes increased further when Pavlov discovered that even these simple behaviors could nonetheless be modified by experience and therefore could be used to explore general principles of learning.

Modern neurobiologists have continued to work within the two traditions established by Sherrington and by Pavlov. By contrast, the study of reflexes has benefitted little from a third tradition—the comparative tradition—that has been highly influential in the analysis of other types of behavior. The lack of a comparative approach to the study of reflex actions has several sources. Perhaps most important has been the traditional focus of ethology (and of other comparative approaches to behavior) on complex activities—such as reproduction, feeding, communication, and social interaction—rather than on simple behavior.

Despite the relative neglect of simple reflexes by students of comparative behavior, it seemed to us that a comparative approach may be useful in placing the principles of reflex action into a broader perspective. At the very least, a comparative approach could help distinguish general features of simple reflex functions from the particular features that apply only to a given case. We therefore wondered: Why should the analysis of simple reflex behavior be deprived of the benefits of an approach that has brought such a harvest of insights to complex behavior? So, having in mind Ted Bullock's pioneering efforts in bridging comparative physiology and neurobiology, we thought we could express our intellectual debt to him here by viewing two simple defensive withdrawal reflexes of *Aplysia* from a comparative perspective.

COMMON FUNCTIONAL CONSIDERATIONS IN THE ORGANIZATION OF DEFENSIVE REFLEXES: THE GILL AND TAIL WITHDRAWAL REFLEXES

During the last 10 years it has been possible in *Aplysia* to delineate several defensive reflexes and to analyze the neural circuit underlying each of them. These include siphon and gill withdrawal, inking, mucus and opaline release, and withdrawal of the tail. In this chapter we focus on only two reflexes: (1) siphon and gill withdrawal in response to siphon stimulation, mediated by the abdominal ganglion, and (2) tail withdrawal in response to tail stimulation, mediated by the pleuropedal ganglia.

As is the case with many flexion reflexes in vertebrates and invertebrates, both the gill withdrawal reflex (Fig. 1*A*) and tail withdrawal reflex (Fig. 1*B*) serve primarily to remove an exposed appendage from possible threat. Although this function seems straightforward, at least two addi-

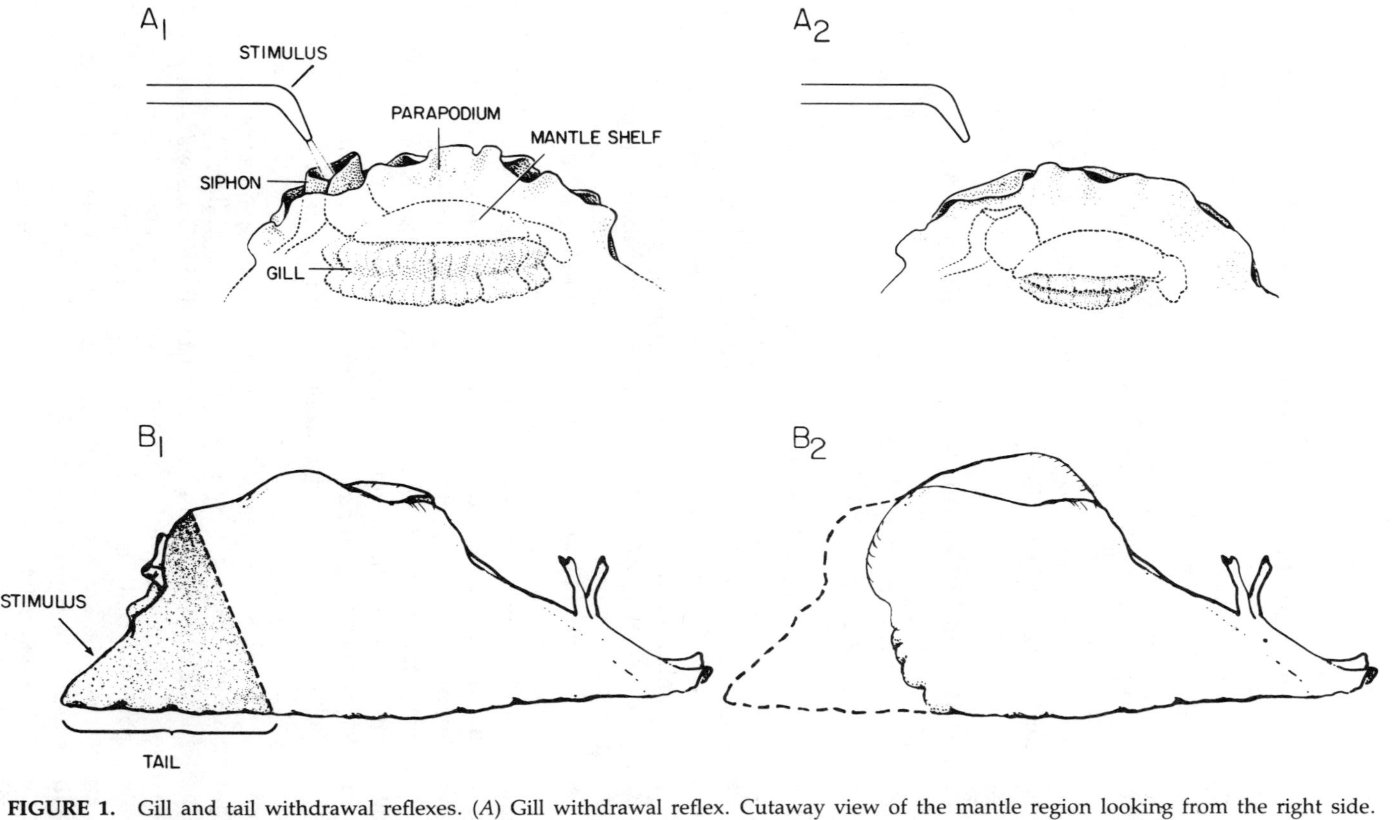

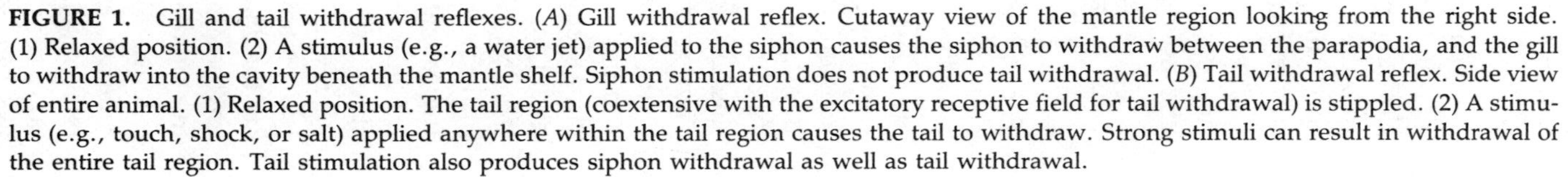

FIGURE 1. Gill and tail withdrawal reflexes. (*A*) Gill withdrawal reflex. Cutaway view of the mantle region looking from the right side. (1) Relaxed position. (2) A stimulus (e.g., a water jet) applied to the siphon causes the siphon to withdraw between the parapodia, and the gill to withdraw into the cavity beneath the mantle shelf. Siphon stimulation does not produce tail withdrawal. (*B*) Tail withdrawal reflex. Side view of entire animal. (1) Relaxed position. The tail region (coextensive with the excitatory receptive field for tail withdrawal) is stippled. (2) A stimulus (e.g., touch, shock, or salt) applied anywhere within the tail region causes the tail to withdraw. Strong stimuli can result in withdrawal of the entire tail region. Tail stimulation also produces siphon withdrawal as well as tail withdrawal.

tional considerations of behavioral flexibility suggest that even these simple reflexes possess interesting control features.

Even Simple Reflexes Involve Response Selection and Coordination

Both the gill and the tail are used for more than one response. In addition to withdrawing for defensive purposes, the gill contracts during respiratory pumping and mantle cleansing movements, and the tail contracts at the end of each step during pedal locomotion. The ability to select one among several responses raises the question: How is the underlying neuronal circuitry organized to subserve more than one kind of behavior, and how are competing responses integrated?

Even Simple Reflexes Are Modifiable by Experience

For energetic efficiency the strength and duration of a reflex response should match as closely as possible the "expected" importance of the stimulus. Most, if not all, reflexes show a profound sensitivity to prior stimulation; they show a capacity for learned modification in response to previous experience. Even simple reflexes in *Aplysia* display three forms of learning observed with reflexes in other animals: habituation, sensitization, and classical conditioning. Repetition of an innocuous mechanical stimulus applied to the siphon or tail results in habituation, a progressive depression of the gill or tail withdrawal reflex as the animal learns to "ignore" the unimportant stimulus (Pinsker et al., 1970; Carew et al., 1972; Walters and Byrne, unpublished observations). An unsignaled noxious stimulus applied elsewhere on the animal causes sensitization, a large enhancement of the gill and tail withdrawal reflexes in response to subsequent innocuous stimuli (Pinsker et al., 1970, 1973; Walters et al., 1983b), possibly because the animal "expects" additional threats and is primed for optimal defensive responsiveness. The gill withdrawal reflex (and possibly the tail withdrawal reflex, discussed later) is capable of more complex modification—classical conditioning—when the animal learns that a weak stimulus is a reliable signal predicting a strong stimulus for gill withdrawal (Carew et al., 1981, 1983).

At what level (or levels) in the neural circuitry subserving these reflexes are the control mechanisms for these common forms of modifiability? How are the modifications expressed, and in what part of the reflex apparatus is the memory stored? Although these questions have not yet been subject to extensive comparative analysis in the *Aplysia* gill and tail withdrawal reflexes, enough is now known to allow preliminary suggestions of the relationships between the physiology and function of each system. In addition, commonalities between the two systems allow some preliminary thoughts on general principles of integration and plasticity in defensive reflexes. The following sections discuss six points of commonality in the gill and tail withdrawal reflexes.

RAPID DEFENSIVE REFLEXES HAVE RESTRICTED EXCITATORY RECEPTIVE FIELDS

This point, so obvious that it is hardly worth stating, has important implications for the neural analysis of reflex mechanisms. It is therefore surprising that so few preparations have allowed direct study of the neural organization underlying topographic variation in reflex responsiveness. Defensive reflexes would be expected to have retricted excitatory fields since withdrawal of an appendage in response to stimulation of distant sites on the body would usually be defensively ineffective, and in addition would sometimes be incompatible with the local defensive responses in the stimulated region. For example, if a dog withdrew *both* forelegs in response to having *one* foreleg bitten by a snake, the dog could fall and be more likely to receive additional bites. Similarly, because of the constant volume of the hydrostatic skeleton of *Aplysia*, withdrawal of the head reduces the degree to which the tail can withdraw. Thus if the head withdrew as well as the tail, following a predator's contact with the tail, the tail would be less able to withdraw from the threat.

Both the gill and tail withdrawal reflexes have restricted excitatory receptive fields (Fig. 1). The gill withdrawal reflex is elicited best by siphon or tail stimulation, and the tail withdrawal reflex is only initiated by stimulation of the tail region. Stimulation of the head does not produce tail withdrawal (and, conversely, stimulation of the tail does not produce head withdrawal). The fields of each reflex are coextensive with the excitatory receptive fields of each population of identified mechanosensory neurons (in the abdominal ganglion Fig. 2*A* and pleural ganglia Fig. 2*B*). Figure 3 shows excitatory receptive fields from individual siphon and tail sensory neurons. In each case the receptive fields show considerable overlap (with stimulation of a single point expected to activate 5–10 sensory neurons) and variation in size. There is also somatotopic organization. In the case of the siphon and mantle sensory neurons the somatotopic organization appears to be limited to the restriction of the somata to particular clusters (Byrne et al., 1974). The tail sensory neurons show a more distinctive somatotopic pattern within each sensory (VC) cluster itself (Walters et al., 1983a). Thus far, monosynaptic connections to motor neurons (discussed later) involved in each reflex have only been found from sensory neurons innervating the excitatory receptive field of each reflex.

An interesting feature of these and other identified mechanosensory populations is the existence of an inhibitory surround. There is in each case a hyperpolarizing response to stimulation outside of the excitatory receptive field (Carew et al., 1971; Walters et al., 1983a). Although this hyper-

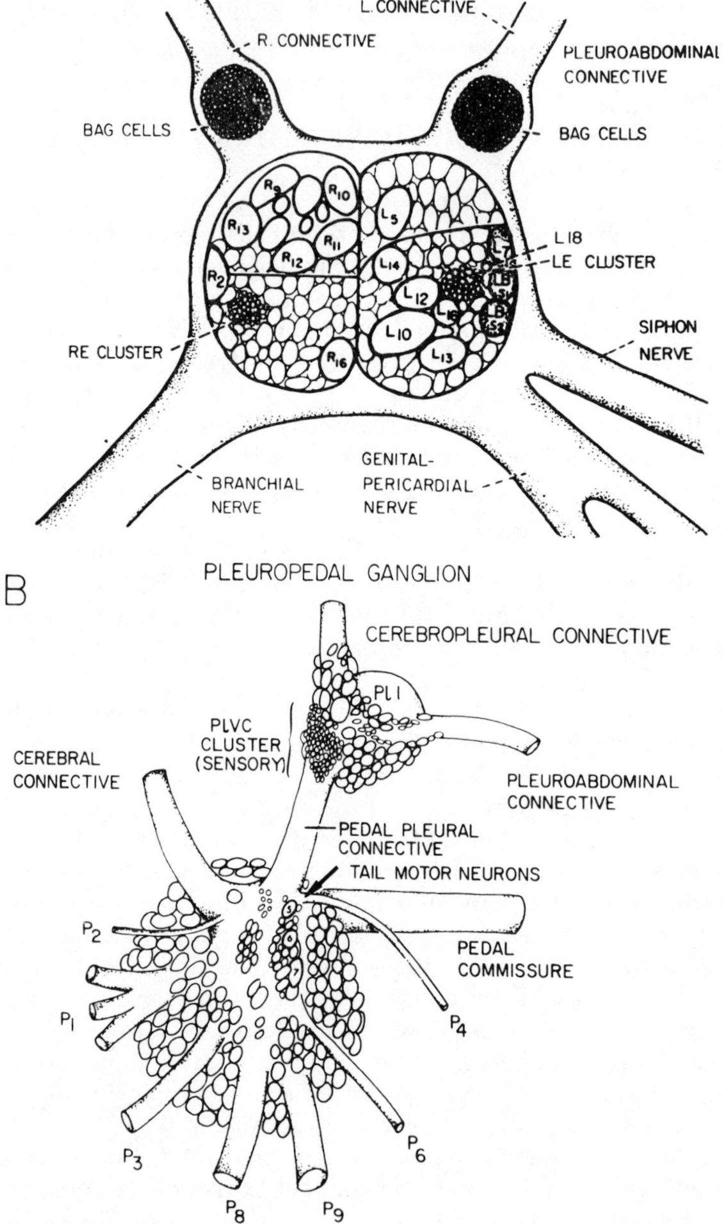

FIGURE 2. Ganglia involved in gill and tail withdrawal reflexes. (*A*) Abdominal ganglion (ventral surface). The sensory neurons innervating the siphon are in the LE cluster near the right margin of the drawing lying between L12 and L7. The RE cluster is a second population of sensory neurons that innervate the mantle shelf. L7 is one of the identified motor neurons involved in gill withdrawal. (From Byrne et al., 1974). (*B*) Pedal and pleural ganglia (left pair). The sensory neurons innervating the tail are located in the PlVC cluster in each pleural ganglion. Identified tail motor neurons P5, P6, and P7 are shown in the pedal ganglion. (From Walters et al., 1983a.)

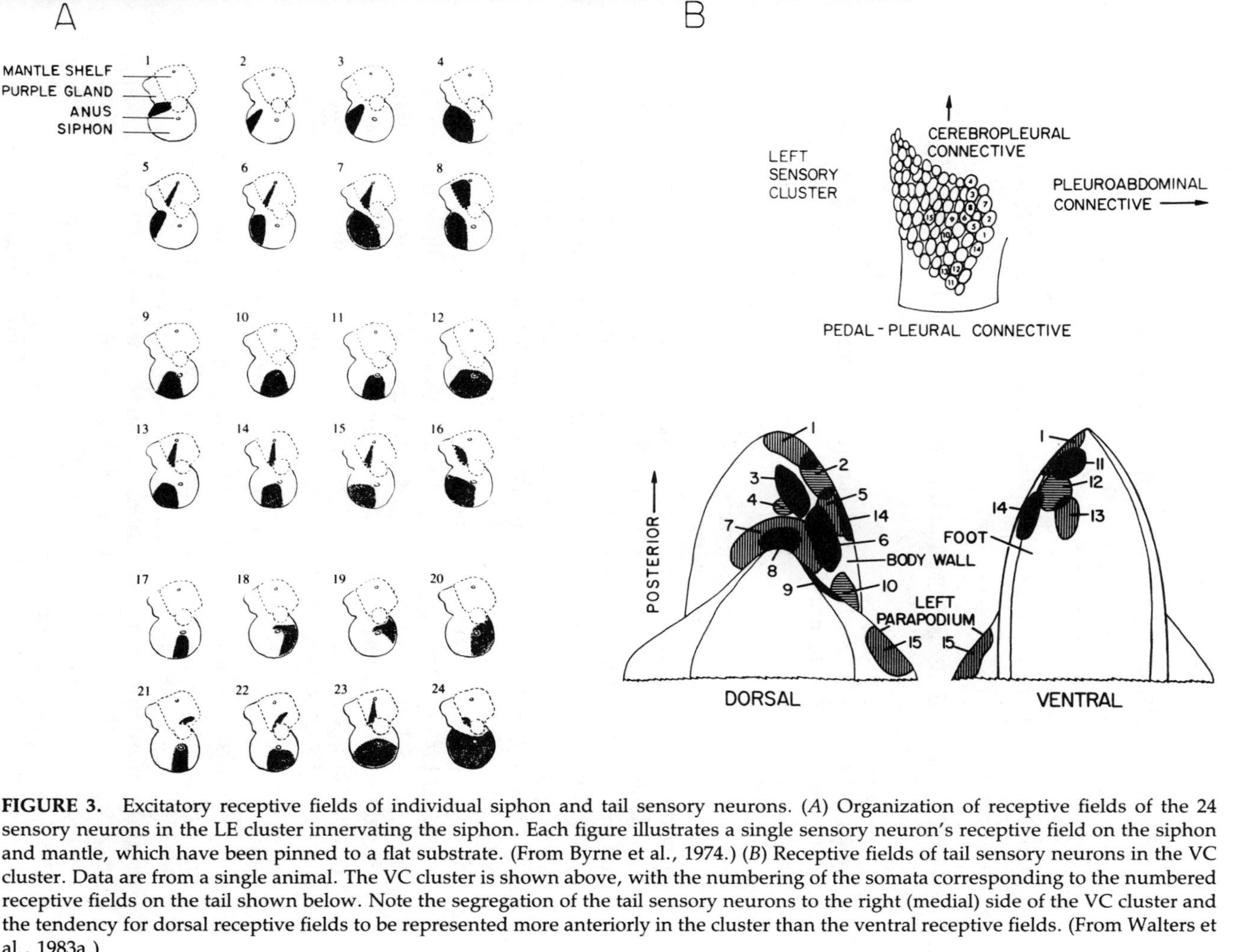

FIGURE 3. Excitatory receptive fields of individual siphon and tail sensory neurons. (*A*) Organization of receptive fields of the 24 sensory neurons in the LE cluster innervating the siphon. Each figure illustrates a single sensory neuron's receptive field on the siphon and mantle, which have been pinned to a flat substrate. (From Byrne et al., 1974.) (*B*) Receptive fields of tail sensory neurons in the VC cluster. Data are from a single animal. The VC cluster is shown above, with the numbering of the somata corresponding to the numbered receptive fields on the tail shown below. Note the segregation of the tail sensory neurons to the right (medial) side of the VC cluster and the tendency for dorsal receptive fields to be represented more anteriorly in the cluster than the ventral receptive fields. (From Walters et al., 1983a.)

polarizing response has not yet been shown to be functionally important (see Walters et al., 1983a), it seems likely that such diffuse inhibition outside of a stimulated field (surround inhibition) would sharpen the contrast between the sites of maximal stimulation and other stimulated sites, and thus reduce the likelihood of unnecessary or competing responses outside of the region of maximal stimulation. In addition, diffuse hyperpolarizing responses might help to reduce nonassociative modifications of defensive reflexes by reducing voltage-sensitive Ca^2 influx (see the section on afferent modifiability and common cell regulation, and Walters and Byrne, 1983b).

RAPID DEFENSIVE REFLEXES DEPEND ON BOTH SHORT- AND LONG-LATENCY PATHWAYS

Speed is important for effective defensive responses, and one means for increasing speed is to minimize the number of synaptic relays between stimulus and response. The gill and tail withdrawal reflexes rely heavily upon simple monosynaptic pathways between the sensory neurons and motor neurons in each reflexive circuit (Fig. 4). Figure 5 illustrates evidence for the monosynaptic nature of connections from sensory to motor neurons in each reflex. Monosynaptic excitatory postsynaptic potentials (EPSPs) are often large and show temporal summation, so that monosynaptic excitation produced by selective activation of a *single* sensory neuron can bring the motor neurons above threshold in each reflex (even in high-divalent cation solutions that reduce the possible contributions of interneurons). The direct monosynaptic connections from individual sensory neurons to motor neurons can be very potent. Selective activation of a single siphon or tail sensory neuron can sometimes produce not only short-latency activation of motor neurons but also rapid withdrawal of the gill or tail (Fig. 6; Byrne et al., 1978; Walters et al., 1983a). Disynaptic pathways involving several excitatory interneurons also contribute to the early phase of each withdrawal reflex (Fig. 4; Byrne, 1981; Hawkins et al., 1981; Walters and Byrne, unpublished observations). Thus in both reflexes the short latency (mono- and polysynaptic) connections from the mechanosensory neurons make large contributions to the early phase of the withdrawal response.

By contrast, the monosynaptic connections from the sensory to motor neurons seem unlikely to account directly for the *late* phase of the defensive withdrawal reflexes. The sensory neurons rarely respond to cutaneous stimulation by firing for more than a second, whereas the motor neuron activation and withdrawal responses in each reflex can last for several minutes. Thus it seems probable that the sensory neurons cause parallel activation of interneuronal pathways that can produce sustained activation of the motor neurons. In the case of the gill withdrawal reflex there are suggestions that the interneuronal network responsible for spontaneous respiratory pumping movements of the gill can be recruited by siphon

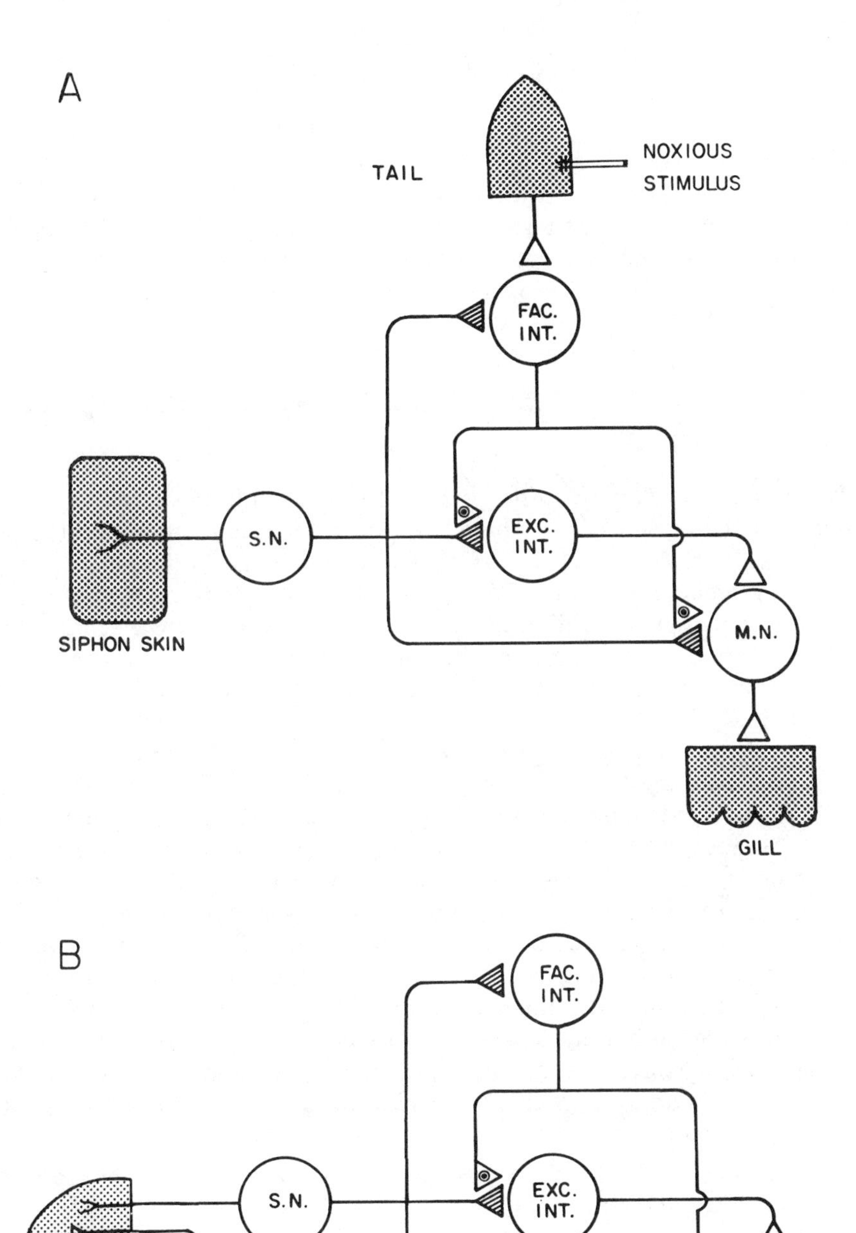
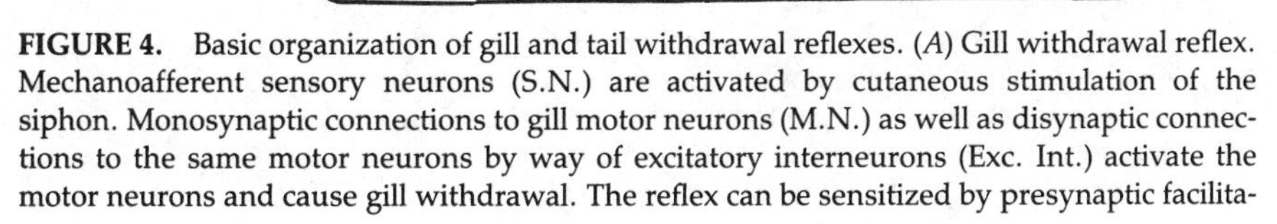

FIGURE 4. Basic organization of gill and tail withdrawal reflexes. (*A*) Gill withdrawal reflex. Mechanoafferent sensory neurons (S.N.) are activated by cutaneous stimulation of the siphon. Monosynaptic connections to gill motor neurons (M.N.) as well as disynaptic connections to the same motor neurons by way of excitatory interneurons (Exc. Int.) activate the motor neurons and cause gill withdrawal. The reflex can be sensitized by presynaptic facilita-

stimulation, prolonging the reflexive withdrawal of the gill (Kupfermann and Kandel, 1969; Kupfermann et al., 1974; Kanz et al., 1979; Byrne, 1983). Thus sensory neuron activation may cause defensive withdrawal by two parallel pathways: (1) a short-latency mono- and polysynaptic pathway for the rapid phase of the response, and (2) a longer-latency polysynaptic pathway for the prolonged phase of the response.

AFFERENT NEURONS FOR DEFENSIVE REFLEXES HAVE A WIDE DYNAMIC RANGE

The mechanosensory neurons for the gill and tail withdrawal reflexes show a graded response to a very broad range of stimulus intensities (Byrne et al., 1974, 1978; Walters et al., 1983a). These sensory neurons display slowly adapting responses to mechanical stimuli ranging from light pressure to intense pinching. In addition, sensory neurons respond to electric shock and to tissue injury. However, since they do not respond to movements of their innervated regions, these sensory neurons do not function as stretch receptors.

The responsiveness of the mechanosensory neurons to both innocuous and noxious stimuli contrasts with the response properties of first-order neurons in mammalian somatosensory systems, but resembles the properties of higher-order "wide dynamic range" spinothalamic sensory interneurons in pain pathways of mammals (Mountcastle, 1980; Chung et al., 1979). Another similarity of these mechanosensory neurons in *Aplysia* to the spinothalamic tract interneurons in mammals is that in both cases there is a limited excitatory receptive field and an extensive inhibitory surround—which can cover the entire body (Gerhart et al., 1981). These comparisons suggest that mechanosensory neurons in the simple nervous system of *Aplysia* combine some of the integrative functions that are found in mammals in both first-order sensory neurons and in higher-order sensory interneurons.

tion of the sensory neuron terminals. Facilitatory interneurons (Fac. Int.) are typically activated by tail stimulation in our experiments but can also be activated by stimulating other sites. The dot ("dense core vesicle") in the modulatory terminal from the facilitatory interneuron denotes the possibility that some of these modulatory neurons are serotonergic (see Fig. 11). Neither interneurons mediating longer latency polysynaptic excitation from the sensory neurons to the motorneurons nor interneurons mediating surround inhibition of the sensory neurons are depicted. (*B*) Tail withdrawal reflex. The same basic organization seen in the gill withdrawal reflex is displayed in this circuit. Tail stimulation and activation of the tail sensory neurons are known to produce facilitation of the sensory neuron terminals, but the facilitatory interneurons have not yet been identified. Interneurons mediating longer latency excitation and surround inhibition are not depicted.

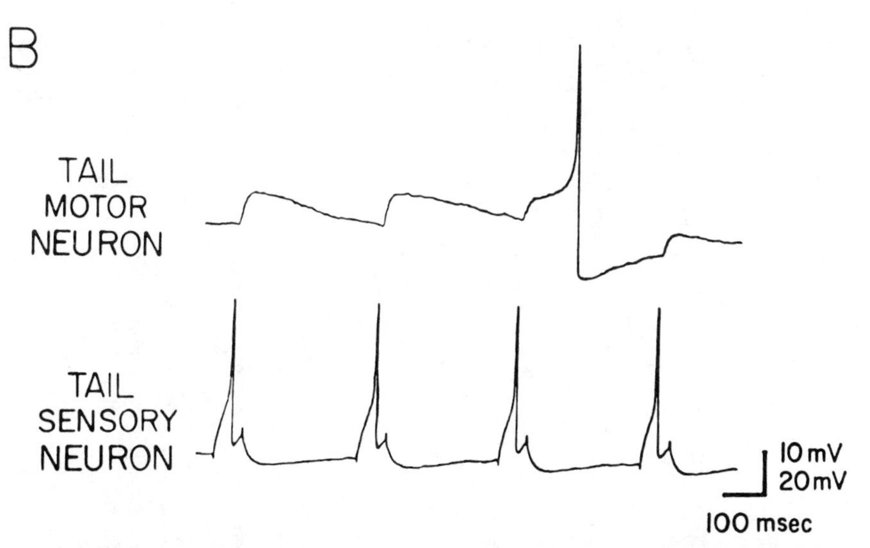

FIGURE 5. Monosynaptic components of gill and tail withdrawal reflexes. (*A*) Gill withdrawal reflex. Unchanged synaptic latency under low-release conditions (high Mg^{2+}, low Ca^{2+} seawater) indicates monosynaptic connection between LE sensory neuron and gill motor neuron L7. Unchanged latency with increased release (produced by intracellular injection of TEA) provides additional evidence for monosynapticity. (From Castellucci and Kandel, 1974.) (*B*) Tail withdrawal reflex. Constant, short latency EPSPs that follow high rates of sensory neuron activation in solutions that reduce the probability of interneuronal activation (high Mg^{2+}, high Ca^{2+} seawater) indicate that the connections between tail sensory neurons in the VC cluster and identified tail motor neurons in the pedal ganglion are monosynaptic. (From Walters et al., 1983a.)

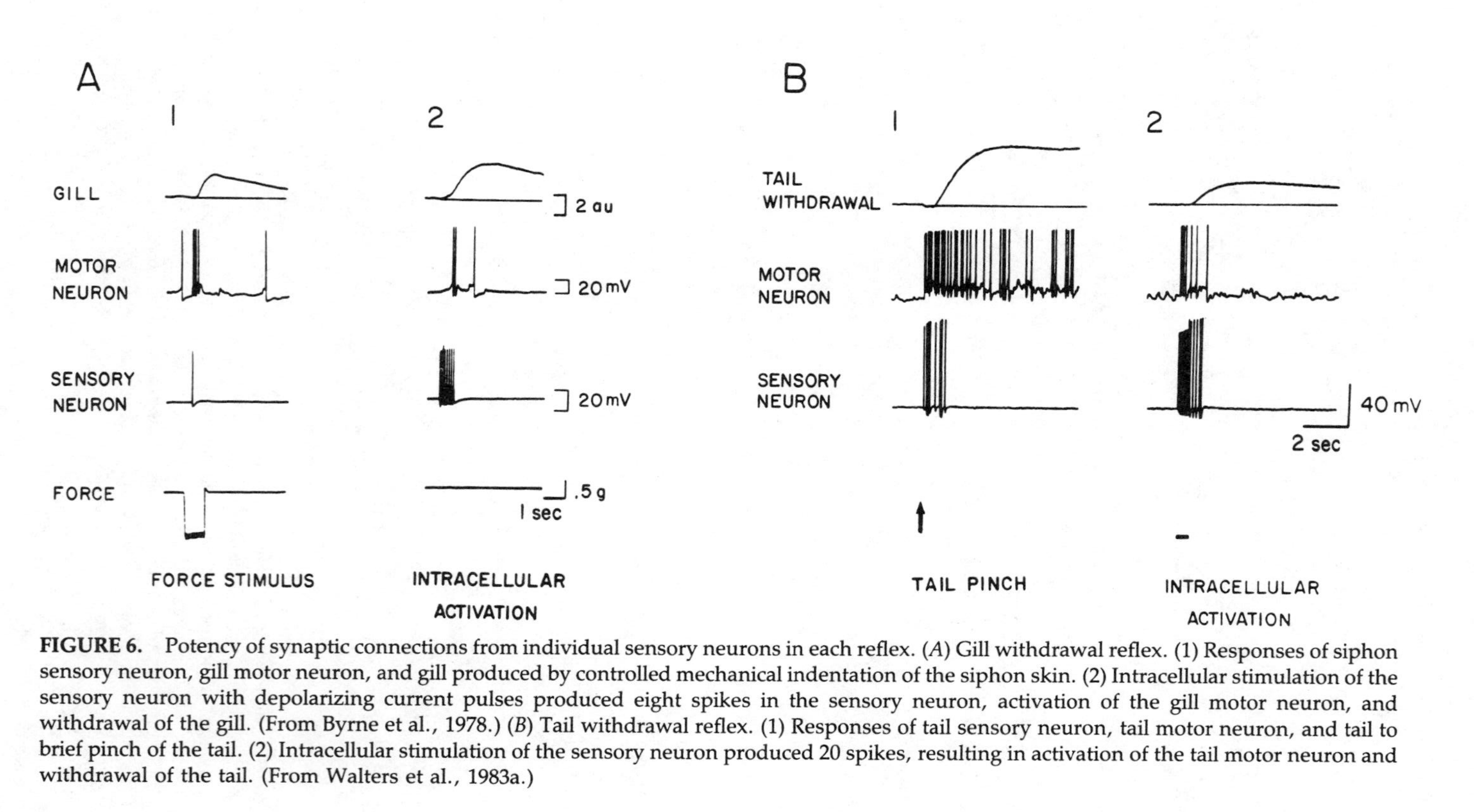

FIGURE 6. Potency of synaptic connections from individual sensory neurons in each reflex. (*A*) Gill withdrawal reflex. (1) Responses of siphon sensory neuron, gill motor neuron, and gill produced by controlled mechanical indentation of the siphon skin. (2) Intracellular stimulation of the sensory neuron with depolarizing current pulses produced eight spikes in the sensory neuron, activation of the gill motor neuron, and withdrawal of the gill. (From Byrne et al., 1978.) (*B*) Tail withdrawal reflex. (1) Responses of tail sensory neuron, tail motor neuron, and tail to brief pinch of the tail. (2) Intracellular stimulation of the sensory neuron produced 20 spikes, resulting in activation of the tail motor neuron and withdrawal of the tail. (From Walters et al., 1983a.)

AFFERENT SYNAPSES SHOW A WIDE RANGE OF PLASTICITY THAT CONTRIBUTES TO REFLEX MODIFIABILITY

The gill withdrawal reflex shows short- and long-term habituation, sensitization, and classical conditioning (Pinsker et al., 1970; Carew et al., 1972; Pinsker et al., 1973; Carew et al., 1981). The tail withdrawal reflex shows sensitization produced by stimulating the test site itself or by stimulating other sites. We are now exploring the degree to which it also is capable of habituation and classical conditioning (Walters et al., 1983b; Walters and Byrne, 1983a; and unpublished observations). All of these forms of reflex modification have been associated with changes in the monosynaptic connections between the sensory and motor neurons in each reflex, and each appears to contribute to reflex modifiability. In addition, a fourth class of synaptic plasticity has been examined in these synapses: posttetanic potentiation. Its function for reflex modifiability is not yet known.

The first form of plasticity is homosynaptic depression (Fig. 7). Progressive depression of the monosynaptic EPSP occurs at the activated synapse over a wide range of interstimulus intervals from 0.1 sec to 600 sec (Byrne, 1982; Castellucci et al., 1970; Walters et al., 1983a; Walters and Byrne, 1985). In the gill withdrawal reflex this depression has been shown to be presynaptic and has been correlated with habituation of the reflex (Kupfermann et al., 1970; Castellucci et al., 1970; Castellucci and Kandel, 1974). Habituation has not yet been examined systematically in the tail withdrawal reflex.

The second form of plasticity—heterosynaptic facilitation—is the only form that is expressed diffusely throughout the population of sensory neurons following focal cutaneous stimulation. Moderate to intense tactile stimulation of practically any point on the body seems to facilitate all of the central synapses from the sensory neurons innervating the siphon, mantle, and tail (Walters et al., 1983b and unpublished observations; Hawkins et al., 1983). General heterosynaptic facilitation is also produced by electrical stimulation of nerves and connectives (Fig. 8*A*; Castellucci et al., 1970; Walters et al., 1983b). Heterosynaptic facilitation has been shown to be presynaptic (Castellucci and Kandel, 1976) and to be associated with sensitization of both the gill and tail withdrawal reflexes (Fig. 8*B*; Castellucci et al., 1970; Walters et al., 1983b). Despite its diffuse nature, heterosynaptic facilitation need not occur at all branches of a given sensory neuron. For example, heterosynaptic facilitation can occur at central synapses without affecting peripheral sensory to motor neuron synapses (Clark and Kandel, 1984).

Third, we have found activity-dependent enhancement of presynaptic facilitation in the monosynaptic connections from each population of sensory neurons. Walters and Byrne (1983a) using the tail withdrawal circuit, and Hawkins et al., (1983) using the gill withdrawal circuit, independently found that activation of single sensory neurons immediately before mod-

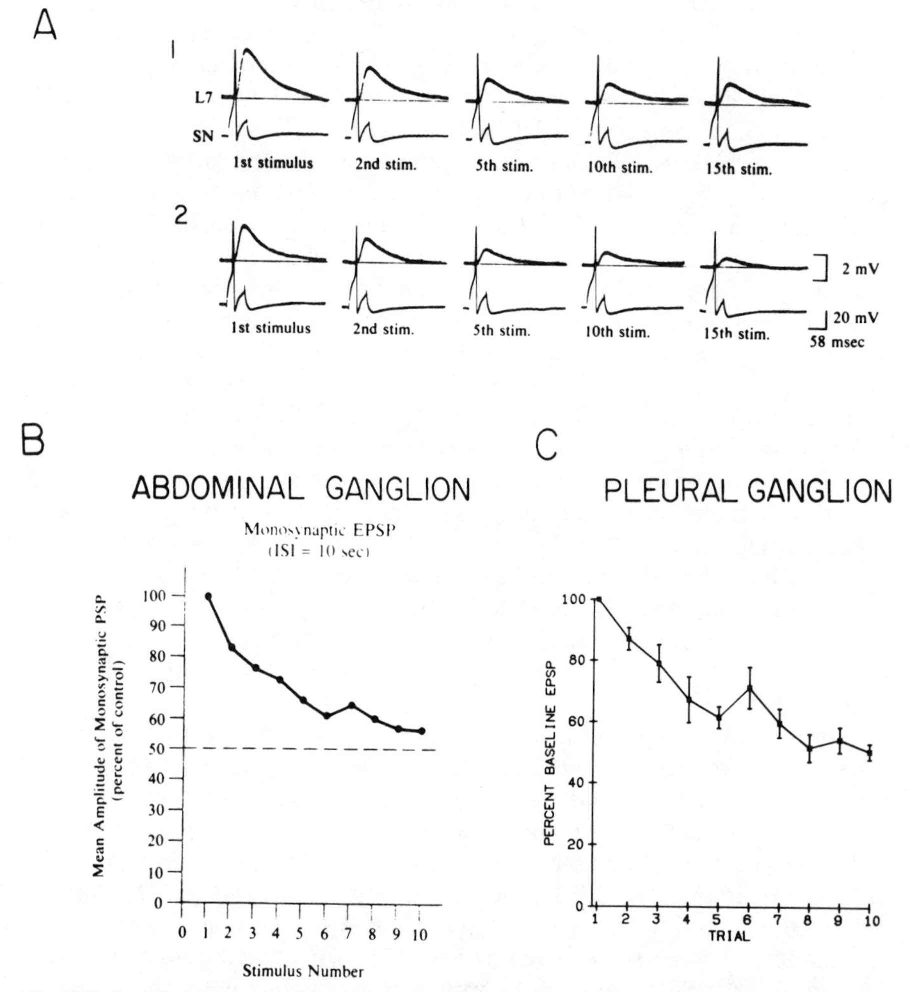

FIGURE 7. Homosynaptic depression in gill and tail withdrawal circuits. (*A*) Gill withdrawal circuit. (1) The amplitude of the monosynaptic EPSP in a gill motor neuron progressively declined with repeated activation of single spikes in the siphon sensory neuron at 10-sec intervals. (2) After a 15 min rest the EPSP showed partial recovery, but continued testing then showed that the depression had increased. (From Castellucci and Kandel, 1974.) (*B*) Gill withdrawal circuit: Plot of synaptic depression with 10-sec interstimulus interval. (*C*) Tail withdrawal circuit. Plot of depression of the EPSP from tail sensory neurons to tail motor neurons (N = 7). (From Walters et al., 1983a.)

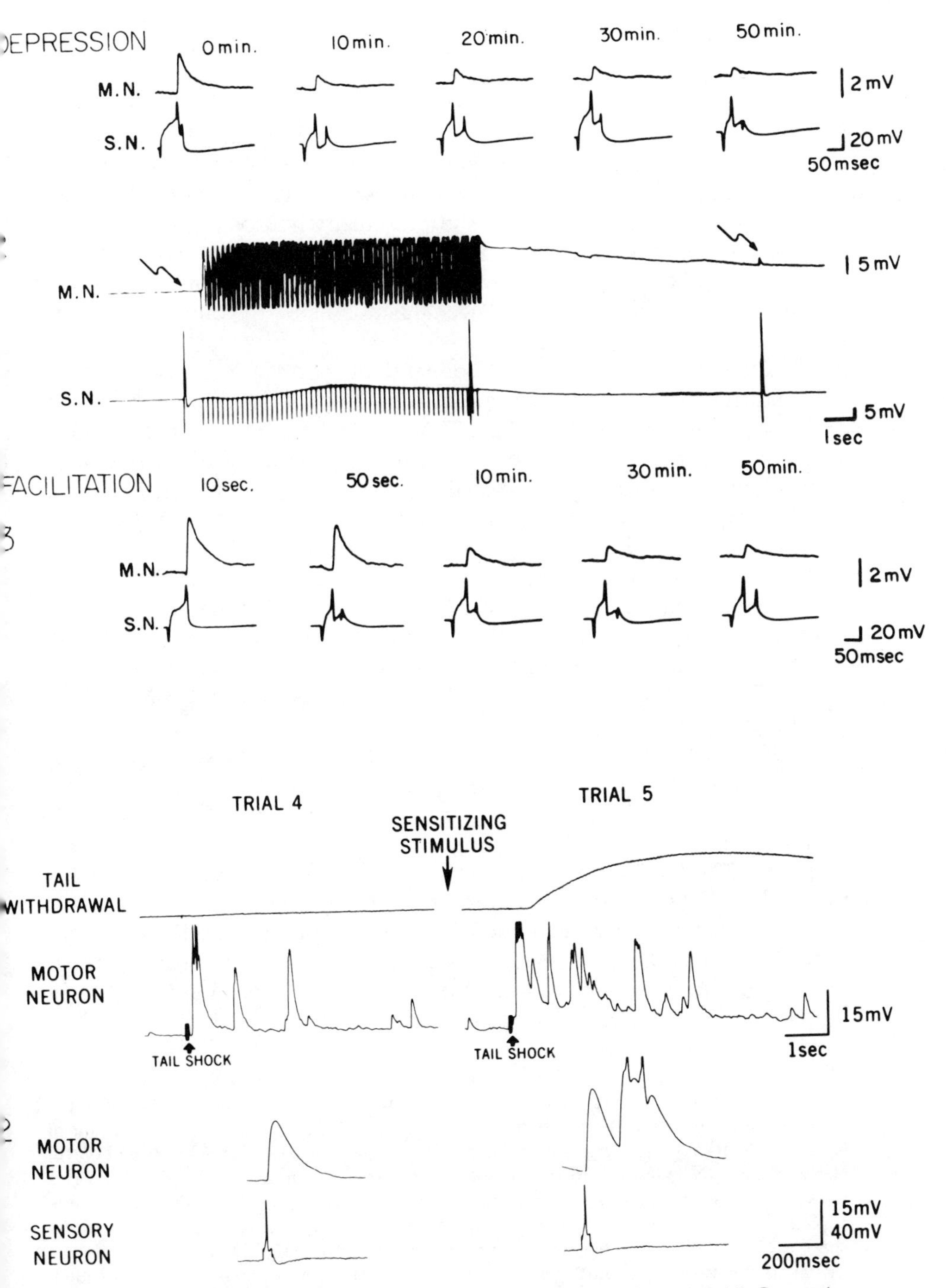

FIGURE 8. Heterosynaptic facilitation in gill and tail withdrawal circuits. (*A*) Connection between a siphon sensory neuron and gill motor neuron. (1) Depression with repeated testing at 10-sec intervals for 50 min. (2) Effects of the facilitating stimulus (6 Hz shock to the left

ulatory input from the sensitizing heterosynaptic stimulation produced a large amplification of the amount of general presynaptic facilitation shown by control cells (Fig. 9). Since temporally specific, activity-dependent modulation occurs at the initial stage of each reflex, it might be expected to underlie temporally specific changes in the reflex. Hawkins et al. (1983) provided support for this possibility by showing striking parallels between activity-dependent modulation of siphon sensory neurons and differential classical conditioning of gill withdrawal using siphon stimulation as a conditioned stimulus (see also Carew et al., 1983). Since this change occurs in the first relay of the reflex, this mechanism is likely to contribute importantly to associative conditioning.

An attractive feature of activity-dependent modulation is that it selectively addresses a diffuse modulatory signal to only those target cells that are active at the time of the signal. Thus cells whose activity coincides (or predicts) the strong unconditioned stimulus (causing release of the modulatory signal) show the largest plastic changes. This is consistent with and seems to explain the occurrence of differential conditioning of the gill withdrawal reflex elicited by siphon or mantle inputs using a tail stimulus as an unconditioned stimulus (Carew et al., 1983). If siphon stimulation is used as a conditioned stimulus (CS^+) and paired with tail stimulation (US) while mantle stimulation is unpaired (CS^-), only the paired CS^+ will show the effects of classical conditioning and only the sensory neurons of the paired CS^+ will show the consequences of activity-dependent enhancement of presynaptic facilitation.

A final form of heterosynaptic plasticity—posttetanic potentiation (PTP)—also occurs at the monosynaptic connections in each reflex (Fig. 10). PTP occurs with higher frequencies of stimulation and, like depression, is specific to the activated synapses (Walters and Byrne, 1983a, 1984; Clark and Kandel, 1984). Although the actual contribution of PTP to reflex modifiability has not yet been determined, it would be expected to interact closely with the activity-dependent modulation previously described. Indeed, Walters and Byrne (1985) have now found that long-term potentiation (lasting at least several hours) of tail sensory neuron synapses may involve an interaction of the mechanism for PTP with that of activity-dependent modulation.

---

connective of the abdominal ganglion for 10 sec) given after the 50-min test. Arrows show the last EPSP before the facilitating stimulus and the first EPSP afterwards. (3) Gradual decline of facilitation with continued testing for 50 more minutes. (From Kandel and Schwartz, 1982.) (*B*) Heterosynaptic facilitation correlated with sensitization of the tail withdrawal reflex. (1) Test responses of the tail and a tail motor neuron 3 min before (trial 4) and 2 min after (trial 5) brief application of a mechanical sensitizing stimulus to the tail. The weak test stimulus was a brief shock delivered at 5-min intervals through an electrode in the tail. The sensitizing stimulus was applied between the fourth and fifth tests, and caused an enhancement of subsequent motor neuron and tail withdrawal responses. (2) Monosynaptic EPSPs produced by tail sensory neuron activation before and after delivery of the sensitizing stimulus. (From Walters et al., 1983b.)

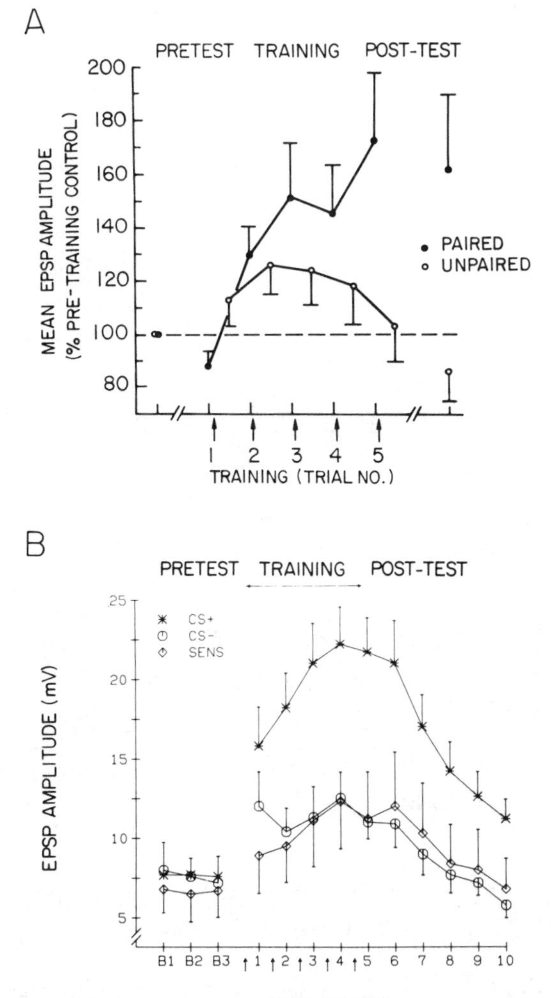

FIGURE 9. Activity-dependent modulation of monosynaptic connections in gill and tail withdrawal circuits. (*A*) Gill withdrawal circuit. Progressive increase in monosynaptic EPSPs from siphon sensory neurons trained with activity (five spikes) paired with modulatory input compared to cells examined simultaneously receiving unpaired training. The posttest was conducted either 5 or 15 min after the fifth training trial. (*B*) Tail withdrawal circuit. Test responses before (B1 to B3), during, and after training. The modulatory stimulus (tail shock) was applied at each arrow. Monosynaptic responses from tail sensory neurons trained with intracellular activation (nine spikes) paired with modulatory input (CS+) were larger than responses of cells receiving unpaired training (CS−) or modulatory input alone (SENS) during training and for at least 30 min after training (test 10). The modulatory tail shock did not itself activate the recorded sensory neurons. (From Walters and Byrne, 1983a.)

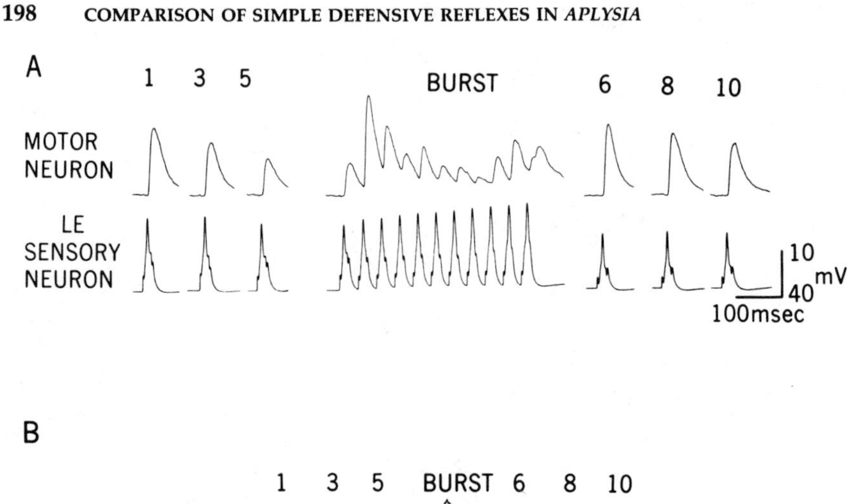

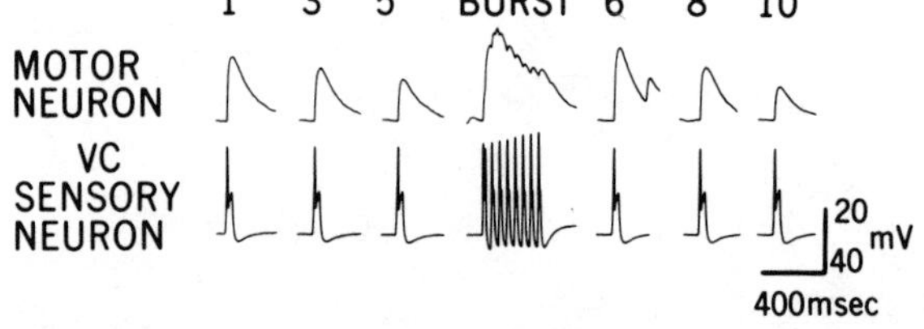

FIGURE 10. Posttetanic potentiation (PTP) in gill and tail withdrawal circuits. In each case 10 test stimuli were delivered at 60-sec intervals. Thirty seconds after the fifth test stimulus a high frequency (20 Hz) burst of spikes was produced by intracellular stimulation of the sensory neuron. The numbers in the figure indicate the number of each test. (*A*) PTP of monosynaptic connection between a siphon sensory neuron and a siphon motor neuron. (*B*) PTP of monosynaptic connection between a tail sensory neuron and a tail motor neuron. (From Walters and Byrne, 1984.)

PRESYNAPTIC AFFERENT MODIFIABILITY PROVIDES STIMULUS SPECIFICITY FOR REFLEX MODIFICATION

As already mentioned, the motor neurons (and possibly the interneurons) involved in defensive withdrawal reflexes in *Aplysia* participate in other behaviors as well. If the plastic changes underlying reflex modification involved these cells, interference could occur with other behaviors. For example, a persistent change in a few of the pedal motor neurons could interfere with the smooth stepping sequence during locomotion.

Aplysia seems to have avoided such problems by putting the primary loci of reflex modifiability in the afferent limb of the reflex—in the sensory neuron terminals (Fig. 4). The specificity of the change then depends on the source of the signal to the sensory neuron for plastic change. During homosynaptic forms of plasticity (homosynaptic depression, PTP, and activity-dependent modulation) the plastic change is dependent on activity within the sensory neuron synapse, and thus the plasticity is specific to activated synapses. Since the synapses of these mechanosensory neurons are normally only activated by discrete, well-defined stimuli (effective stimulation of the appropriate excitatory receptive field), these changes will be specific to those sensory stimuli. By contrast, the one form of activity-independent plasticity, heterosynaptic facilitation, is not stimulus specific. Intense focal stimulation of a point on the skin (activating only a few sensory neurons) causes the diffuse modulation of almost all central afferent synapses for the defensive reflexes. However, all forms of plasticity, whether specific or diffuse, appear to involve changes in the sensory neurons themselves, with little or no persistent changes in the motor neurons (Castellucci and Kandel, 1974, 1976; Walters et al., 1983b). In the case of habituation and classical conditioning, the reflex can show stimulus specificity as well, showing changes to particular stimuli without alteration of responses to other stimuli (Carew et al., 1971, 1983).

AFFERENT MODIFIABILITY INVOLVES COMMON CELL REGULATORY PROCESSES

We have discussed features common to two different defensive reflexes in *Aplysia*. This raises a basic question: Are these common features in the reflexive control in *Aplysia* and in other animals due to analogous but genetically unrelated evolutionary responses to similar pressures, or are there deep homologies in the underlying regulatory processes? This fundamental question, which lies at the heart of all comparative physiology, is difficult to answer, and it is too early to see the answers in the case of *reflex organization* per se. However, some answers are beginning to emerge in the analysis of *reflex modifiability*.

This evidence comes from our initial identification of subcellular processes involved in modification of the gill and tail withdrawal reflexes and the similarity of these processes to those in several distantly related organisms. Heterosynaptic facilitation is the best-studied form of plasticity. The facilitatory neuromodulator (probably serotonin but quite likely one or more other transmitters as well; Fig. 11) activates an adenylate cyclase in the sensory neuron soma and terminals (Brunelli et al., 1976; Klein and Kandel, 1978; Castellucci et al., 1980, 1982; Bernier et al., 1982; Ocorr et al., 1985; Pollock et al., 1982). This causes increased synthesis of cAMP which, through a protein kinase, results in the phosphorylation of substrate proteins, leading to closure of a novel K^+ channel. The closure of this K^+ channel prolongs the action potential, allowing more Ca^{2+} influx, thus facilitating transmitter release (Klein and Kandel, 1978, 1980).

In activity-dependent modulation it is thought that Ca^{2+} influx shortly before or during the heterosynaptic facilitation causes an amplification of the cAMP effects, perhaps by a synergistic activation of the adenylate cyclase (Walters and Byrne, 1983a, 1983b; Hawkins et al., 1983; Abrams et al., 1983). From the point of view of a comparative physiologist, the inter-

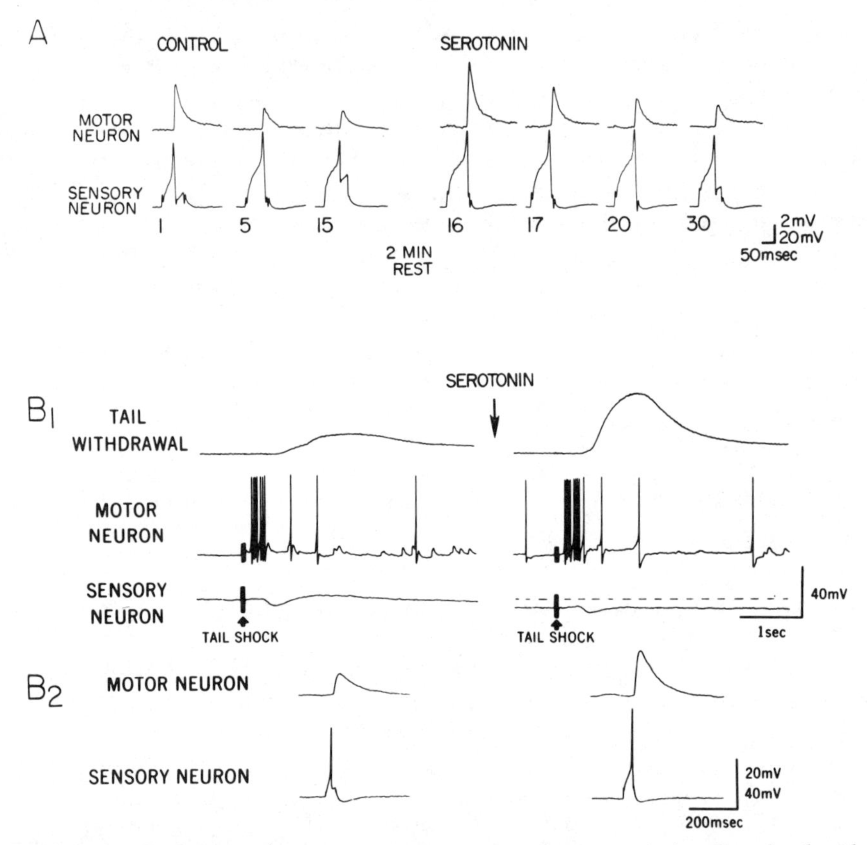

FIGURE 11. Facilitation of monosynaptic connections by serotonin in gill and tail withdrawal circuits. (*A*) Facilitation of connection from siphon sensory neuron to gill motor neuron. (From Kandel and Schwartz, 1982.) (*B*) Facilitation of tail sensory neuron connection correlated with enhancement of tail withdrawal reflex. (1) A weak test shock to the tail caused larger motor neuron and tail withdrawal responses following selective application of serotonin (5×10^{-5} *M*) to the central nervous system. (2) The monosynaptic EPSP tested 30 sec before the tests shown in part 1 was also facilitated. (From Walters and Byrne, 1983b.)

esting observation is that cyclic nucleotides and Ca^{2+} are probably the best known and most ubiquitous of the intracellular regulators. Thus reflex modifiability by way of changes in afferent neuron transmitter release may be closely linked to some of the most fundamental, highly conserved cell regulatory processes.

This possibility is further strengthened by recent evidence obtained from *Drosophila* and from mammals indicating that cyclic nucleotides and Ca^{2+} may be critical regulatory signals underlying plasticity and learning in these animals as well. For example, single gene mutants of *Drosophila* lacking cAMP phosphodiesterase are deficient in their capability for both associative learning and sensitization (Dudai et al., 1976; Byers et al., 1981;

Duerr and Quinn, 1982). In addition, Dudai (personal communication) and Livingstone et al. (1984) have found evidence that Ca^{2+}-dependent adenylate cyclase activity present in normal flies is deficient in mutants deficient in associative learning. Although mammals have not yet provided direct evidence for a link between nonassociative and associative learning, on the one hand, and cyclic nucleotides and Ca^{2+} on the other, a number of observations suggest that similar forms of neuronal plasticity occur in the mammalian brain and that these may involve regulatory processes similar to those seen in invertebrates. For example, activity-dependent modulation has been shown in cerebral cortex, and may involve activity-dependent cyclic nucleotide effects (Woody et al., 1978). In the hippocampus activity-dependent modulation may be involved in long-term synaptic potentiation (Hopkins and Johnson, 1984). Taken together with the prominent occurrence of Ca^{2+}-dependent adenylate cyclase in the mammalian brain (e.g., Brostrom et al., 1975), the synergistic activation of adenylate cyclase by Ca^{2+} and neuromodulators (Malnoe et al., 1983), and the growing evidence for the involvement of neuromodulation in memory and learning (e.g., Gold and Zornetzer, 1983), one wonders if at least some aspects of learning and reflex modification involve cellular regulatory processes common to most species.

CONCLUSIONS

To what degree do properties common to two different defensive reflexes in *Aplysia* represent general mechanisms of reflex control? It is interesting that many of these properties have also been found in other species. For example, many defensive reflexes including vertebrate flexion responses (Sherrington, 1906) and crayfish tail flip responses (Wine and Krasne, 1983), have restricted excitatory receptive fields. Other defensive reflexes also depend at least in part upon pathways having only a few synaptic relays: for example, monosynaptic pathways contributing to withdrawal reflexes of the leech (Kristan, 1982) and nudibranch mollusc *Tritonia* (Getting, 1977), trisynaptic pathways underlying the lateral giant tail flip response in the crayfish (Zucker, 1972), and four-neuron reflex arcs involved in flexion responses in mammals (Henneman, 1980). In most reflexes examined the dynamic range of individual sensory neurons is not known, but one can infer that the graded responsiveness of defensive reflexes reflects the wide dynamic range of at least part of the underlying circuitry. However, the breadth of the dynamic range is likely to be determined by interneurons as well as by primary sensory neurons, and the dynamic range of the reflex need not be represented completely in individual cells (e.g., Chung et al., 1979).

A striking commonality among reflexes subjected to cellular analysis is the investment of plastic capabilities in the sensory side of the reflex circuit.

As in *Aplysia*, habituation of defensive responses in the crayfish and cockroach is paralleled by depression of central synapses of mechanoreceptor neurons (Zucker, 1972; Zilber-Gachelin and Chartier, 1973). There is also indirect evidence for changes in first-order sensory neurons during long-lasting plasticity of mammalian stretch reflexes (Wolpaw et al., 1983). Changes in first-order sensory neurons have also been shown in the visual system of *Hermissenda* to be correlated with behavioral modifiability (Crow and Alkon, 1980; Farley and Alkon, 1982), and changes during conditioning have been observed in the afferent side of the pigeon's visual system (Gibbs and Cohen, 1980). Such afferent modifiability provides an efficient means for achieving stimulus specificity in either nonassociative or associative modification of reflexes. Perhaps most encouraging for the reductionists is the finding that the cellular mechanisms of reflex modifiability and neuronal plasticity show similarities in diverse species, supporting the notion that the fundamental regulatory processes underlying reflex regulation may be highly conserved. Not only is there evidence for the involvement of Ca^{2+}, cyclic nucleotides, and neuromodulators (as discussed before), but also many forms of plasticity in both *Aplysia* and other animals appear to involve the regulation of resting K^{+} conductances (Klein and Kandel, 1980; Crow and Alkon, 1980; Hoyle, 1982).

These apparent commonalities demand a more rigorous explanation than is yet available. Coupling of the comparative approach championed by Bullock with a molecular approach should provide these rigorous explanations.

Acknowledgments

This research was sponsored by NIH fellowship FN506455 to E. T. Walters; NIH Career Development Award K04NS00200 and NIH grant RO1NS19895 to J. H. Byrne; Research Career Development Award to T. J. Carew; a contract from the Office of Naval Research, "Molecular Mechanisms of Learning and Memory," 903-6015B, and a grant from the Systems Development Foundation to E. R. Kandel.

REFERENCES

Abrams, T. W., T. J. Carew, R. D. Hawkins, and E. R. Kandel (1983) Aspects of the cellular mechanism of temporal specificity: Preliminary evidence for Ca^{2+} influx as a signal of activity. *Abstr. Soc. Neurosci.* **9**:168.

Bernier, L., V. F. Castellucci, E. R. Kandel, and J. H. Schwartz (1982) Facilitatory transmitter causes a selective and prolonged increase in adenosine 3′:5′-monophosphate in sensory neurons mediating the gill and siphon withdrawal reflex in *Aplysia*. *J. Neurosci.* **2**:1682–1691.

Brostrom, C. D., Y. C. Huang, B. M. Breckenridge, and D. J. Wolff (1975) Identification of a calcium binding protein as a calcium-dependent regulator of brain adenylate cyclase. *Proc. Natl. Acad. Sci. U.S.A.* **72**:64–68.

Brunelli, M., V. Castellucci, and E. R. Kandel (1976) Presynaptic facilitation as a mechanism for behavioral sensitization in *Aplysia. Science* **194**:1176–1181.

Byers, D., R. L. Davis, and J. A. Kiger (1981) Defect in cyclic AMP phosphodiesterase due to the dunce mutation of learning in *Drosophila melanogaster. Nature* **289**:79–81.

Byrne, J. H. (1981) Comparative aspects of neural circuits for inking behavior and gill withdrawal in *Aplysia californica. J. Neurophysiol.* **48**:431–438.

Byrne, J. H. (1982) Analysis of synaptic depression contributing to habituation of gill-withdrawal reflex in *Aplysia californica. J. Neurophysiol.* **48**:431–438.

Byrne, J. H. (1983) Identification and initial characterization of a cluster of command and pattern-generating neurons underlying respiratory pumping in *Aplysia californica. J. Neurophysiol.* **49**:491–508.

Byrne, J., V. Castellucci, and E. R. Kandel (1974) Receptive fields and response properties of mechanoreceptor neurons innervating siphon and mantle shelf of *Aplysia. J. Neurophysiol.* **37**:1040–1064.

Byrne, J. H., V. F. Castellucci, and E. R. Kandel (1978) Contribution of individual mechanoreceptor sensory neurons to defensive gill-withdrawal reflex in *Aplysia. J. Neurophysiol.* **41**:418–431.

Carew, T. J., V. F. Castellucci, and E. R. Kandel (1971) An analysis of dishabituation and sensitization of the gill-withdrawal reflex in *Aplysia. Int. J. Neurosci.* **2**:79–98.

Carew, T. J., R. D. Hawkins, and E. R. Kandel, 1983. Differential classical conditioning of a defensive withdrawal reflex in *Aplysia californica. Science* **219**:397–400.

Carew, T. J., H. M. Pinsker, and E. R. Kandel (1972) Long-term habituation of a defensive withdrawal reflex in *Aplysia. Science* **175**:451–454.

Carew, T. J., E. T. Walters, and E. R. Kandel (1981) Classical conditioning in a simple withdrawal reflex in *Aplysia californica. J. Neurosci.* **1**:1426–1437.

Castellucci, V. and E. R. Kandel (1974) A quantal analysis of the synaptic depression underlying habituation of the gill-withdrawal reflex in *Aplysia. Proc. Natl. Acad. Sci. U.S.A.* **71**:5004–5008.

Castellucci, V. and E. R. Kandel (1976) Presynaptic facilitation as a mechanism for behavioral sensitization in *Aplysia. Science* **194**:1176–1181.

Castellucci, V., H. Pinsker, I. Kupfermann, and E. R. Kandel (1970) Neuronal mechanisms of habituation and dishabituation of the gill-withdrawal reflex in *Aplysia. Science* **167**:1745–1748.

Chung, J. M., D. R. Kenshalo, K. D. Gerhart, and W. D. Willis (1979) Excitation of primate spinothalamic neurons by cutaneous C-fiber volleys. *J. Neurophysiol.* **42**:1354–1369.

Clark, G. A. and E. R. Kandel (1984) Branch-specific heterosynaptic facilitation in *Aplysia* siphon sensory cells. *Proc. Natl. Acad. Sci. U.S.A.* **81**:2577–2581.

Crow, T. J. and D. L. Alkon (1980) Associative behavioral modification in *Hermissenda*: Cellular correlates. *Science* **209**:412–414.

Dudai, Y., Y.-N. Jan, D. Byers, W. G. Quinn, and S. Benzer (1976) Dunce, a mutant of *Drosophila* deficient in learning. *Proc. Natl. Acad. Sci. U.S.A.* **73**:1684–1688.

Duerr J. S. and W. G. Quinn (1982) Three *Drosophila* mutants that block associative learning also affect sensitization and habituation. *Proc. Natl. Acad. Sci. U.S.A.* **79**:3646–3650.

Farley, J. and D. L. Alkon (1982) Associative neural and behavioral change in *Hermissenda*: Consequences of nervous system orientation for light- and pairing-specificity. *J. Neurophysiol.* **48**:785–807.

Gerhart, K. D., R. P. Yezierski, G. J. Giesler, and W. D. Willis (1981) Inhibitory receptive fields of primate spinothalamic tract cells. *J. Neurophysiol.* **46**:1309–1325.

Getting, P. A. (1977) Neuronal organization of escape swimming in *Tritonia*. *J. Comp. Physiol.* **121:**325–342.

Gibbs, C. M. and D. H. Cohen (1980) Plasticity of the thalamofugal pathway during visual conditioning. *Soc. Neurosci. Abstr.* **6:**424.

Gold, P. E. and S. F. Zornetzer (1983) The memnon and its juices: Neuromodulation of memory processes. *Behav. Neural Biol.* **38:**151–189.

Hawkins, R. D., T. W. Abrams, T. J. Carew, and E. R. Kandel (1983) A cellular mechanism of classical conditioning in *Aplysia*: Activity-dependent amplification of presynaptic facilitation. *Science* **219:**400–405.

Hawkins, R. D., V. F. Castellucci, and E. R. Kandel (1981) Interneurons involved in mediation and modulation of gill-withdrawal reflex in *Aplysia*. I. Identification and characterization. *J. Neurophysiol.* **45:**304–314.

Henneman, E. (1980) Organization of the spinal cord and its reflexes. In *Medical Physiology*, V. B. Mountcastle, ed. Mosby, St. Louis, pp. 762–786.

Hopkins, W. F. and D. Johnston (1984) Frequency-dependent noradrenergic modulation of long-term potentiation in the hippocampus. *Science* **226:**350–352.

Hoyle, G. (1982) Pacemaker change in a learning paradigm. In *Cellular Pacemakers*, Vol. 2, D. O. Carpenter, ed. Wiley, New York, pp. 3–26.

Kandel, E. R. and J. H. Schwartz (1982) Molecular biology of an elementary form of learning: Modulation of transmitter release by cyclic AMP. *Science* **218:**433–443.

Kanz, J. E., L. B. Eberly, J. S. Cobbs, and H. M. Pinsker (1979) Neuronal correlates of siphon withdrawal in freely behaving *Aplysia*. *J. Neurophysiol.* **42:**1538–1556.

Klein, M. and E. R. Kandel (1978) Presynaptic modulation of voltage-dependent Ca^{2+} current: Mechanism for behavioral sensitization in *Aplysia californica*. *Proc. Natl. Acad. Sci. U.S.A.* **75:**3512–3516.

Klein, M. and E. R. Kandel (1980) Mechanism of calcium current modulation underlying presynaptic facilitation and behavioral sensitization in *Aplysia*. *Proc. Natl. Acad. Sci. U.S.A.* **77:**6912–6916.

Kristan, W. B. (1982) Sensory and motor neurones responsible for the local bending response in leeches. *J. Exp. Biol.* **96:**161–180.

Kupfermann, I., T. J. Carew, and E. R. Kandel (1974) Local, reflex and central commands controlling gill and siphon movements in *Aplysia*. *J. Neurophysiol.* **37:**966–1019.

Kupfermann, I., V. Castellucci, H. Pinsker, and E. Kandel (1970) Neuronal correlates of habituation and dishabituation of the gill-withdrawal reflex in *Aplysia*. *Science* **167:**1743–1745.

Kupfermann, I. and E. R. Kandel (1969) Neuronal controls of a behavioral response mediated by the abdominal ganglion of *Aplysia*. *Science* **164:**847–850.

Malnoe, A., E. A. Stein and J. A. Cox (1983) Synergistic activation of bovine cerebellum adenylate cyclase by calmodulin and beta-adrenergic agonists. *Neurochem. Int.* **5:**65–72.

Mountcastle, V. B. (1980) Neural mechanisms in somesthesis. In *Medical Physiology*, Vol. 1, V. B. Mountcastle, ed. Mosby, St. Louis, pp. 348–390.

Ocorr, K. A., E. T. Walters, and J. H. Byrne (1985) Associative conditioning analog selectively increases cAMP levels of tail sensory neurons in *Aplysia Proc. Natl. Acad. Sci. USA* **82:**2548–2552.

Paris, C. G., V. F. Castellucci, E. R. Kandel, and J. H. Schwartz (1981) Protein phosphorylation, presynaptic facilitation, and behavioral sensitization in *Aplysia*. *Cold Spring Harbor Conf. Cell Proliferation* **8:**1361–1375.

Pinsker, H., W. A. Hening, T. J. Carew, and E. R. Kandel (1973) Long-term sensitization of a defensive withdrawal reflex in *Aplysia*. *Science* **182:**1039–1042.

Pinsker, H., I. Kupfermann, V. Castellucci, and E. R. Kandel (1970) Habituation and dishabituation of the gill-withdrawal reflex in *Aplysia. Science* **167:**1740–1742.

Pollock, J. D., J. S. Camardo, L. Bernier, J. H. Schwartz, and E. R. Kandel (1982) Pleural sensory neurons in *Aplysia*: A new preparation for studying the biochemistry and biophysics of serotonin modulation of K^+ currents. *Abstr. Soc. Neurosci.* **8:**523.

Sherrington, C. S. (1906) *The Integrative Action of the Nervous System.* Yale University Press, New Haven.

Walters, E. T. and J. H. Byrne (1983a) Associative conditioning of single sensory neurons suggests a cellular mechanism for learning. *Science* **219:**405–408.

Walters, E. T. and J. H. Byrne (1983b) Slow depolarization produced by associative conditioning of *Aplysia* sensory neurons may enhance Ca^{2+} entry. *Brain Res.* **280:**165–168.

Walters, E. T. and J. H. Byrne (1984) Post-tetanic potentiation in *Aplysia* sensory neurons. *Brain Res.* **293:**377–380.

Walters, E. T. and J. H. Byrne (1985) Long-term enhancement produced by activity-dependent modulation of *Aplysia* sensory neurons. *J. Neurosci.* **5:**662–672.

Walters, E. T., J. H. Byrne, T. J. Carew, and E. R. Kandel (1983a) Mechanoefferent neurons innervating the tail of *Aplysia*: I. Response properties and synaptic connections. *J. Neurophysiol.* **50:**1522–1542.

Walters, E. T., J. H. Byrne, T. J. Carew, and E. R. Kandel (1983b) Mechanoafferent neurons innervating the tail of *Aplysia*: II. Modulation by sensitizing stimulation. *J. Neurophysiol.* **50:**1543–1559.

Wine, J. J. and F. B. Krasne (1983) The cellular organization of crayfish escape behavior. In *The Biology of Crustacea: Neural Integration and Behavior,* D. C. Sandeman and H. L. Atwood, eds. Academic, New York, pp. 241–292.

Wolpaw, J. R., D. J. Braitman, and R. F. Seegal (1983) Adaptive plasticity in primate spinal stretch reflex: Initial development. *J. Neurophysiol.* **50:**1296–1311.

Woody, C. D., B. E. Swartz, and E. Gruen 1978. Effects of acetylcholine and cyclic GMP on input resistance of cortical neurons in awake cats. *Brain Res.* **158:**373–395.

Zilber-Gachelin, N. F. and M. P. Chartier (1973) Modification of the motor reflex responses due to repetition of the peripheral stimulus in the cockroach. I. Habituation at the level of an isolated abdominal ganglion. *J. Exp. Biol.* **59:**359–381.

Zucker, R. S. (1972) Crayfish escape behavior and central synapses. II. Physiological mechanisms underlying behavioral habituation. *J. Neurophysiol.* **35:**621–637.

Part Three

INTEGRATIVE PROCESSES: NETWORKS

Commentary

JEROME Y. LETTVIN

Department of Electrical Engineering and Computer Science and Department of Biology
Massachusetts Institute of Technology
Cambridge, Massachusetts

Ted Bullock couples a broad knowledge of animals with an equally broad range of experiments on their nervous systems. We became friends about 25 years ago and have had long but infrequent conversations. I always learned more from him than he from me, and, although time has made us contemporaries, this asymmetry remains. It is an honor to chair a section of this colloquium in his honor.

Much of this festschrift concerns Ted's special interest, the processing of information by the nervous system. Given a wide enough spectrum of fauna, almost any imaginable operation on signals can be found and studied. The difficulties lie in relating what a system does (the process) to how it works (the mechanism). Such problems are not unique to biology. There is a modern thrust to ignore animal tissues and to confect alternative computational engines that purport to "do the same things" as living systems. Chiefing this movement is the discipline known as artificial intelligence, which, in the last few decades, has advanced most startlingly in its promises. The operational approach it uses, however grandly unsuccessful, has contributed a new mode of thought to the handling of nervous action. Such an engineering attitude, replacing the inspired mysticism of half a century ago, yields an interesting escape—if you cannot describe it, synthesize it. For some reason, it has spurred the analytic study of both nervous activity and animal action in terms of measurables connected with change rather than with steady state. Elsewhere in biology, for example, in the physiology of circulation, respiration, and excretion, this workmanlike approach has led to profound insight.

I do not think that either Ted or I will live to see the developed fruits of the new form of engineering biology as applied to brain and behavior. It looks pregnant, and Ted certainly qualifies as a suspected father, but the gestation has been uncommonly long, and every year there is a false alarm which sustains the hope but exhausts the patience of the waiting family of nervous physiologists.

The papers at this session are excellent in that they offer neither promises nor minor results that are inflated by pretense to depth or couched in an impenetrable formalism. They are pastiches that show, instead, clever experiments clearly interpreted and rich in meaning. In short, they have that lucidity that Ted has always required of his own work and of his students. Nevertheless, they also have the aspect not of final accounts but of interesting beginnings, and, in that they raise far more questions than they answer, they celebrate Bullock the Adventurer, rather than Bullock the Authority.

Pearson's approach to nervous circuitry introduces the modern flavor. Once it is realized that, from an electrical point of view, the membrane is a sheet of transactors—that is, three terminal devices—in which the charge carriers are specific ions and the ion-specific channels are independently switchable, anything that can be built electronically can also be built ionically. The only limitation is the bandwidth of the ionic devices which is, at best, little more than 1 kHz. Under the heritage of Hodgkin and Huxley, we know something about the nature of such transactors which port between two aqueous phases. They can be switched by Ca^{2+} as mediator for the value of free energy between the two aqueous solutions, or by neurotransmitters transducing the signals from another patch of membrane on another cell, or by mechanical free energy, or by configurational changes in a photosensitive molecule, and so on. In short, one can construct a cell with many such switches in different densities to be a linear or nonlinear amplifier, a two-state device, a monostable vibrator, a regular oscillator, a blocking oscillator, in a word, to be any function generator imaginable in electronics. And one can interconnect cells, as one interconnects integrated chips, to do specific tasks.

What makes the circuitry so interesting is the great flexibility to be had in the controls, or switches, of these transactors. The receptor, that is, the switch lead, can be tailored for substance specificity, for time of keeping the switch open or closed, and for species of channel to be switched. Combinations are almost unbounded. Thus a control system for a motor or glandular action, or a processor of sensory information can be readily devised.

Pearson's examples are a nice demonstration of the sorts of single cells and cell arrays that appeal to engineering intuition. As he hints in the text, it is not hard to model what he shows, and not hard to group some of the uses of the specific arrays. I wish he had expanded his note.

Llinas' chapter develops the theme with respect to certain sets of neurons as individually oscillatory, and, in a connected set, a group of mutually locked oscillators not only strongly fixed in period, after the theory of coupled oscillators, but also in specific phase relations with each other, as in a ring of vibrators. In this way, Llinas conceives that specific movement or processing sequences can be properly timed in phase relations with each other. He uses tissue *in vitro*, specifically the inferior olive, to demonstrate that the component neurons are separately oscillators and, connected, phase each other. Such a system is not easy to perturb; indeed, not only is it fixed in frequency, but also it tends to assume fixed phase relations of the components, depending on their interconnections, as I understand Llinas himself holds. But then, given such a circuit that resists not only change in frequency but also change in mutual phasing, what can possibly be the function of its input from elsewhere in the nervous system?

I am quite sure that Llinas has a clever and interesting answer, and I wish he had included it in the chapter. Left to dream on his most compelling figures, I can imagine what that answer might be. Such an interconnected set, locked in frequency and phase, is, nevertheless, capable of multiple phase sequences of the components, all equally stable under the same ensemble frequency. That is because they are not connected one dimensionally as in a line or ring, but two, or even three dimensionally. The ensemble can then be switched between alternative phase sequences by its input, thereby altering the phase clocking of what it controls. This kind of result of a randomly connected net of monostable elements was much investigated in the 1950s. I remember Wiener holding forth on it and Walter Pitts actually showing in a massive essay (now lost) that locked ensembles occur in a three dimensionally connected net.

Two chapters deal with the problem of movement of animals, which is no less complex than the problem of perception. One is by Stein, the other by Bizzi and Abend. Movement, as much as perception, couples the representation of the world (had by the animal) to the world itself. The transformation from central nervous state sequences to the muscle-and-joint state sequences is not easily treated. Those who have worked on vertebrates in which the brains have been disconnected from the spinal cord, know how much remarkable multipotence of coordinated action lies in the cord. Walking, swimming forward or backward, scratching, righting, presenting (for females in heat), stepping, are all different complex actions involving much the same motor nuclei but in different patterns of interconnected activation. One has the distinct impression that so many subprograms of motion may exist in the cord that higher level control is not so much a matter of detailing an action, muscle by muscle, but of switching sequences and combinations of subprograms.

On the other hand, there is the exquisite adaptive control of movement among mammals. Whether for mountain goat, spider monkey, or concert pianist, the precise actions of these animals bespeak so good a matching of body and limb trajectory to a contingently given task that, in some not easily defined way, one feels the choreography to express a kind of optimum determined by higher level process.

Obviously, these two views are not exclusive but only complementary. Stein explores some spinal programs that are almost jukeboxlike expressions in the sense that, once initiated, they go to completion or iteration independent of sensory control. Some can be studied in terms of ventral root patterns for a cord completely removed from the body. The phenomena, as he shows, are clear and distinct. But now the question is, what sort of connection patterns generate such complex stereotypes? For although an action, for example, a scratch reflex, is simple to name in conventional language, it is extremely difficult to describe in terms of a set of coupled pendula giving a smooth and damped trajectory, wherein the differences between muscle pulls around a joint determine the motion, and the sums determine the stiffness.

As physical pendula, the thigh, leg, and foot have natural periods. For example, in ordinary effortless walking, the period of the stride is almost precisely the period of the passive limb dangling from the hip. And this is as it should be for least energy expenditure. If an automatic spinal movement program is to be precise and smooth without unnecessary energy expense for a task, then, somehow or another, the program must have in it a representation of those moments. It is hard to imagine how the miraculous match is made between the physical limb and its representation, that is, what the rules are for laying down and modifying the program in the cord as the animal grows in size from infant to adult. A kind of spinal cord "learning" must be involved.

Stein presents some past and current speculation on the nature of such automata but is not happy with them. They are not explanations but nonce ideas; for, if challenged to make a working model, none, I think, could produce one starting with those ideas. Stein, at the end of Chapter 14, puts out a plea for the sort of mechanism that does not reflect, as I read him, outworn neo-Sherringtonian views imported from clinical neurology. He is more polite in his treatment of current notions than am I who know much less—and I apologize if I have misread him.

Bizzi and Abend are concerned with the conscious control of relatively unpracticed movement of the human arm under fairly restrictive conditions for a simple test. Their chapter, like Stein's, is written with admirable simplicity, and the experimental results stand forth clearly. I want to spend a little time on one of their points because of its potential for changing clinical understanding of stroke. To quote: "A concept that has emerged from these initial studies of multijoint movements is that there is a level in the CNS that specifies movement of the hand, rather than dealing with joint parameters (for the whole arm). The hand-trajectory plan may then be transformed into appropriate joint specifications."

This is a most powerful concept, the idea that the whole upper limb is servocontrolled to provide the proper trajectory for the hand in the space perceived.

For a long time we have known that the fibers that go from area 4 to the cord are not only a small fraction of the internal capsule or of the pyramids, but also that severance of the cortico-spinal tract high in the cord produces no paralysis. Quite the contrary, this severance alleviates Parkinsonism without producing paresis, as Putnam showed. (The operation was abandoned not so much because it entailed laminectomy, but because it violated some cherished superstitions.) Area 4 is not a motor center in the sense that it controls motoneurons directly. As far back as 1938, James Ward* reported that stimulation of a point in the foreleg region of cat cortex determined not a distinct muscle response but a distinct position of the forelimb in respect to the body. For whatever the initial position of the limb, stimulation of that point brought it to a specified final position.

I talked to Bizzi about this only to discover that he has repeated the same experiments in the monkey with substantially the same results. And I remember that Percival Bailey, in stimulating human cortex during surgery, had the same impression but not so clearly expressed and never published.

The notion is compelling that area 4 does not control muscles directly but rather the target spatial position to which muscular action brings the parts moved. It puts the organization of the movement as trajectory into the extrapyramidal system fed by capsular and pyramidal information, and even into the cord, while area 4 provides in executive fashion the spatially described goal or position sequence. In no trivial way, this idea makes area 4 more closely linked to the expectation of perceived limb position than to the effector mechanism whereby the position is attained. And that is in accord with the origins of the forebrain from the dorsal plate.

But the selfsame view poses the horrendous problem of the direct trajectory control which coordinates itself to the specifications laid down by cortex. It is interesting that monkeys, whose manual manipulative skill, good as it is, is much inferior to that of humans, do not show the same violent disruption of function as in humans, when area 4 is destroyed. I do not mean to extend speculation too much further and can only congratulate Bizzi and Abend on what seems to me a most remarkable new insight.

Finally, there is the anatomical elegance of Chapter 12 by Karten and

*Ward, J. (1938) The influence of posture on responses elicitable from the cortex cerebri of cats, *J. Neurophys.* **1**:463–475.

Kuljis. It is hard to pretend objectivity when someone else confirms by anatomical methods findings that one has made physiologically. In its fashion, the visual system of the frog is incredibly orderly, from the rods and cones to the central representations. The four functionally distinct laminae of visual process that are so clear to electrical recording have lacked anatomical grounds until now. But to find that there are also distinct transmitters for each function—the implication of this chapter—that is a great bonus.

To amplify or footnote the essay by Karten and Kuljis in terms of the physiological meaning, to expand on the interlamination in the colliculus of the projections from n. isthmi with those from the optic nerve, and to set forth the present physiological work so much in concord with the results they provide—that would occupy far more room than their marvelous contribution. It is enough to say that their methods have brought new hope to the analysis of frog vision, and I am profoundly grateful.

Taken together, these papers in honor of Ted Bullock made for a memorable session. If there was a problem, it was that the material was so rich that one began dreaming of related matters in the middle of a presentation. For this reason alone, we need to have the texts available.

It is good to see one's influence reflected in admirable works, and no more fitting tribute could have been offered to Ted than to display the fruits of his teaching developed in such novel ways.

Chapter Twelve

LAMINATION AND PEPTIDERGIC SYSTEMS IN THE FROG OPTIC TECTUM

HARVEY J. KARTEN and RODRIGO O. KULJIS

Departments of Neurobiology, Psychiatry, and Neurology
State University of New York at Stony Brook
Stony Brook, New York

Ted Bullock occupies a special place in the field of comparative neurobiology, not only for his many contributions to the field, but also even more for his personal qualities of warmth and generosity. His interests are so broad and his enthusiasm for biology so diverse and contagious that he has resisted any attempt to classify him "by" the animals he has studied. Ted has repeatedly demonstrated that the fundamental and continuing contribution of comparative neurobiology is in its value in the demonstration of universal patterns of organization of nervous tissue. He has epitomized the concept that comparative neurobiology is both a discipline and an experimental tool.

One of the favorite preparations of neurobiologists for the study of the visual system is the frog's optic tectum. Unquestionably, interest in this structure increased as a direct result of the work of the chairman of this afternoon's session, Jerry Lettvin. Lettvin and co-workers revealed an elegance of organization of both retinal ganglion cells and the optic tectum of the frog. In a sense, the subtleties were all the more striking, as the frog optic tectum initially seems to show little evidence of laminar differentiation when studied with more traditional histological methods.

Reexamination of the retino-tectal system with immunohistochemical methods for localization of transmitters has provided a novel picture of the frog optic tectum, yet one that corresponds closely to that provided by Lettvin et al. (1959). The presence of well-defined laminae and differential inputs from different categories of ganglion cells, revealed with histochemical methods, closely matches the results obtained on the basis of earlier studies using Golgi and electrophysiological methods.

PEPTIDERGIC SYSTEMS IN THE OPTIC TECTUM

The optic tectum of several classes of vertebrates demonstrates a particularly prominent pattern of lamination when examined in Nissl-stained material, with alternating layers of neurons and cell-free neuropil. The laminar patterns are particularly distinctive in birds and reptiles, but far less so in fish, amphibia, and mammals. The validity of the laminar patterns was reinforced by the demonstration that (1) specific afferent projections terminate in selected layers (Cajal, 1909; Cowan et al. 1961; Hunt et al., 1977), and that (2) efferent projections from the tectum to different remote targets arise from distinct and often different laminae (Cajal, 1909; Hunt et al., 1977; Benowitz and Karten, 1976; Brecha, 1978; Reiner and Karten, 1982).

Although tectal lamination is often fairly evident in Nissl-stained material in birds and reptiles, the laminar organization is less evident in the frog optic tectum. In the latter species, the superficial third of the tectum is relatively acellular and appears devoid of lamination in Nissl-stained material. Interestingly, P. Ramón Cajal (1894) described the presence of a subtle laminar pattern on the basis of vertical segregation of axon bundles and on the differential patterns of retinal axonal arbors at different depths. Ramón Cajal suggested a simplified schema to describe the optic tectum, with an external region dominated by retinal inputs, an intermediate zone with a mixed population of cell bodies, whose dendrites receive the retinal input, internuncial cells and some efferent projecting neurons, and a deep zone containing predominantly efferent neurons.

Lettvin et al. (1959) were able to take advantage of the differential axonal retinal arborizations within the superficial neuropil described by Ramón Cajal (1894, 1946). Lettvin et al. pointed out that since each different type of retinal terminal arbor presumably arose from a different type of ganglion cell, each type of ganglion cell presumably ended within a different sublamina of the superficial optic tectum. Recent experimental anatomical studies using degeneration, autoradiography, and, most recently, anterograde transport of horseradish peroxidase have supported the suggestion that unique categories of axons terminate within distinct sublaminae but have still not directly demonstrated that the retinal arbors within a single sublamina arise from a uniform population of ganglion cells. Using combined anatomical and physiological methods, Lettvin et al. (1959) concluded that a specific sublamina of the superficial optic tectum received input from a unique type of ganglion cell, distinguished by its physiological response characteristics. Lettvin et al. (1959) classified the ganglion cell response types with imaginative terms, suggesting a possible "neuroethological" role for each of the ganglion cell types (convexity, moving-edge, dimming, boundary, and contrast "detectors"). Thus, on the basis of afferent and efferent connections, distinct laminar patterns appear to have been reified.

A major advance in our attempts to further analyze laminar structures can be traced to the historically important landmark paper of Scharrer and Sinden (1949). These authors described a differential laminar distribution of acetylcholinesterase in the pigeon optic tectum. They suggested that "chemoarchitectonics" provided an additional major criterion for the morphological analysis of the nervous system. This approach has proven particularly fruitful with the recent methodological advances in our ability to identify and characterize different chemical transmitters and receptors using immunohistochemistry (IHC). The immunohistochemical approach not only provides information necessary for the analysis of the transmitters employed in individual components of the microcircuitry of a functional network but also has provided a powerful tool in defining subtle features of the architecture of the central nervous system.

IMMUNOHISTOCHEMICAL ANALYSIS OF THE LAMINAR ORGANIZATION OF THE FROG OPTIC TECTUM

Kuljis and Karten (1982) examined the distribution of various neuropeptides within the frog optic tectum. Peptide-like immunoreactivity within the retino-recipient portion of the anuran optic tectum displays an elaborate lamina-specific pattern of staining, with cell processes containing different peptides segregated within specific tectal sublaminae. Part of the peptide-containing cell processes originate from intrinsic tectal cells, whose somata are located in deep tectal layers. However, a substantial proportion of the peptide-containing processes within the tectum do not appear to originate from intrinsic tectal cells (Fig. 1). The latter processes presumably originate from somata located outside of the tectum. Many cell groups contribute processes to the tectum, including three pretectal nuclei, several tegmental fields, the nucleus isthmi, the cervical spinal cord dorsal gray columns, the suprapeduncular nucleus, a paramedian cell group rostral to the nucleus interpeduncularis, the deep layers of the contralateral tectum, and possibly the preoptic nucleus (Wilczynski and Northcutt, 1977). The most substantial single input to the optic tectum, however, is undoubtedly the retina, terminating precisely in that portion of the tectum displaying the most elaborate laminar pattern of peptide-like immunoreactivity (Kuljis and Karten, 1982; 1983a). Recent research, however, has failed

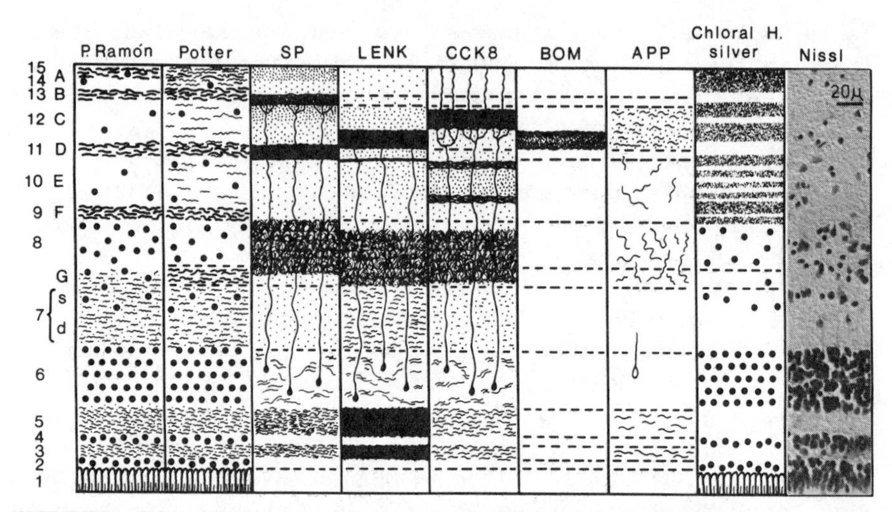

FIGURE 1. Schematic diagram of the organization of the optic tectum in *Rana pipens*. Horizontal correlation between the laminar organization in each individual column has been accurately depicted on the basis of double-labeling experiments. The relative distance of superficial tectal layers from the pial surface changes as layer A disappears toward caudal and medial levels. The relationships between the laminae depicted in the different columns are maintained despite these topographical variations. The width and depth relative to the pial surface of each of the layers of the superficial neuropil has not been depicted to accurate scale to emphasize the laminar distribution of peptides relative to each other. Lamination patterns attributed to Ramón Cajal (1894, 1946) and to Potter (1969) represent our interpretation of each author's description. Numbers designate tectal layers according to Ramón Cajal (1946) (1: ependymal cell layer; 15: immediately subpial layer). Letters designate tectal layers according to Potter. s: superficial (cell containing) portion of layer 7; d: deep (cell devoid) portion of layer 7. The cell contour depicted in the APP column represents occasional faintly labeled cell bodies. SP: substance P; LENK: leucine-enkephalin; CCK8: cholecystokinin octapeptide; BOM: bombesin; APP: avian pancreatic polypeptide. (Reproduced from Kuljis and Karten, 1982.)

to provide adequate evidence of the presence of neuropeptides within retinal ganglion cells in any class of vertebrate (Karten and Brecha, 1983; Brecha, 1983).

ARE GANGLION CELLS PEPTIDERGIC?

The identification of lamina-specific peptidergic systems in the optic tectum was a satisfying confirmation of Scharrer and Sinden's suggestion that histochemistry could be successfully applied to architectonics of the brain (1949). The IHC method revealed a pattern of lamination previously only hinted at with more traditional anatomical stains and provided a measure of objectivity in the classification of subtle lamination. However, equally intriguing was the question: What are the sources of the various peptidergic processes? Could we now use the IHC as a marker of specific cell types to identify various otherwise obscure afferent projections of the optic tectum? Furthermore, would it be possible to modify the peptidergic systems within the optic tectum as a result of either physiological or surgical manipulation of the animals?

The single most prominent input to the optic tectum is, of course, that from retinal ganglion cells. Despite the prominence of peptidergic staining in those tectal layers receiving retinal inputs, we initially thought that this was an unlikely source of the tectal peptides for two major reasons: (1) the optic nerve itself consistently failed to indicate the presence of peptidergic staining, and (2) only rare peptide-positive somata were seen in the retinal ganglion cell layer. These latter cells did not possess detectable axons and were believed to be displaced amacrine cells. The presence of numerous SP, LENK, and CCK8 somata in the deep layers of the tectum, with processes extending into the more superficial layers of the tectum, suggested that at least part of the superficial peptidergic staining was derived from these cells. One of the puzzles, however, was the lack of any BOM-positive somata within the tectum, or in any obvious tectal afferent neurons. In consideration of the possibility that the regulation of peptide synthesis of tectal neurons might be dependent on the presence of retinal inputs, we examined the tectum following retinal deafferentation (Kuljis and Karten, 1983a). We found that retinal ablation resulted in depletion of peptidergic staining in selected tectal laminae, as shown in Figure 2. Depletion occurred in those layers known to receive retinal afferents. Furthermore, partial lesions of the retina resulted in depletion in topographically corresponding portions of the optic tectum (Kuljis and Karten, unpublished observations).

Initially we proposed that this was consistent with trans-synaptic regulation of synthesis of tectal peptides by the retinal inputs. To test this still further, rather than removing the retina, we ligated, sectioned, or crushed the optic nerve but left the eye *in situ* (Kuljis, et al., 1984). Seven days later, we examined both the optic nerve stumps and the tectum. The tectum showed the now-anticipated pattern of lamina-specific depletion of SP, LENK, CCK8, and BOM. However, the pattern in the injured optic nerve was markedly altered. In contrast to the usual lack of evident peptidergic staining of the optic nerve, numerous SP, LENK, CCK8, and BOM axons were clearly evident within the distal (retinal) stump of the optic nerve, whereas the cerebral stump still failed to evidence any staining for these peptides (Fig. 3). Examination of the retina itself still failed to demonstrate any unequivocally positive peptidergic staining of ganglion cells.

We entertained the possibility that part of the immunoreactive processes in the optic nerve were sprouting fibers of the peptidergic amacrine cells in response to injury to the optic nerve. To test this possibility, seven days after the initial crush, horseradish peroxidase was placed on the freshly resectioned optic nerve stump. No evidence of retrograde transport to

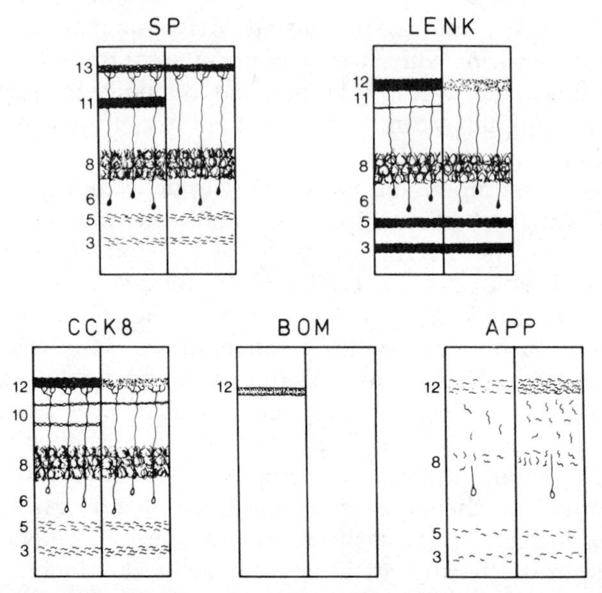

FIGURE 2. Schematic diagram of the effects of unilateral retinal deafferentation in the laminar organization of peptide-like immunoreactivity in the anuran optic tectum. The left column for each substance represents the pattern of peptide-like immunoreactivity in nondeafferented tecta contralateral to those deprived of retinal input. The right column for each substance shows the long-term effects of deafferentation. Postdeafferentation shrinkage in the superficial neuropil has not been depicted to facilitate band-to-band comparison between deprived tecta and their contralateral controls. Numbers represent the approximate location of tectal layers according to Ramón Cajal (1894, 1946). The relative distances between bands are out of scale to facilitate schematic comparative representation between both columns for each peptide. Empty cell contours depicted in the CCK8 and APP columns represent faintly labeled cell bodies. Compare with Figure 1. (Reproduced from Kuljis and Karten, 1983a.)

peptidergic amacrine cells was found. To further test the prospect that amacrine cells might contribute sprouting peptidergic axons to the retinal stump, we examined the retinal stump with progressively shorter survival times. Staining of processes within the retinal stump was clearly evident, even only one to two hours following the optic nerve crush. It is therefore unlikely that the immunoreactivity in the retinal stump was due to sprouting amacrine cells (Kuljis et al., 1984).

Simultaneous double staining for SP and LENK, or SP and CCK8, or SP and BOM, demonstrated that the peptidergic staining in the retinal stump was in different axons and did not appear to co-occur within single axons. The identity of the peptidergic processes in the retinal stump was demonstrated with electron microscopic immunohistochemistry. Numerous SP and LENK positive axons were found in the retinal stump. These axons were thin unmyelinated fibers (0.1–0.6 μm), with numerous large dense core vesicles and prominent growth cones in the vicinity of the crush (Fig. 4).

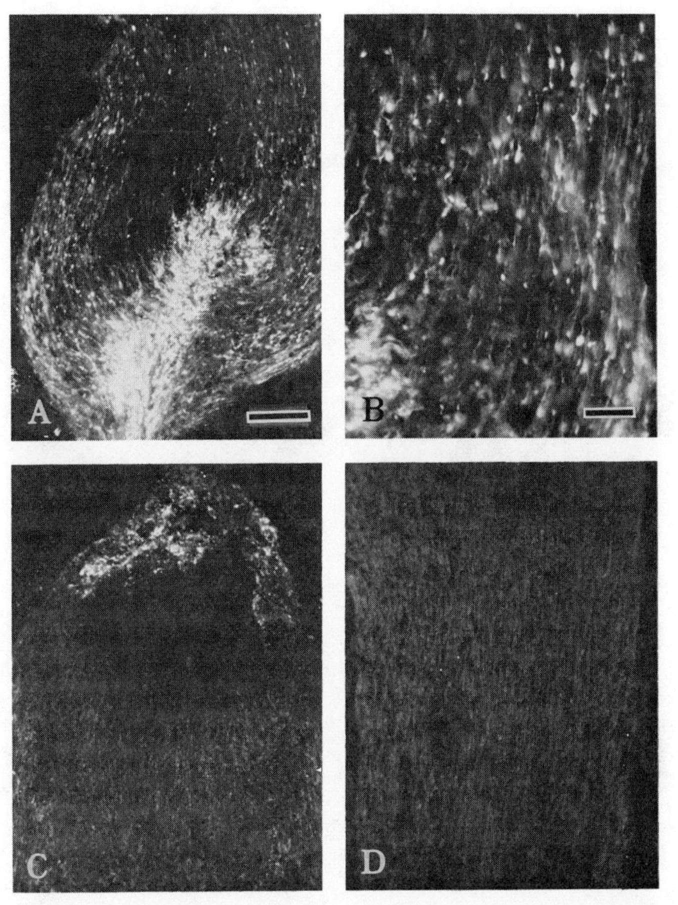

FIGURE 3. Substance P-like immunoreactivity in the optic nerve 7 days after ligation. Indirect fluorescence method. (*A*) Retinal stump, displaying abundant beadlike peptide-containing processes, some of them connected by thin filaments. Note that these peptide-positive axons increase in density toward the site of ligation (lower end of the photomicrograph), where especially high concentration of them occurs in the core of the nerve. Calibration bar: 100 μm. (*B*) Higher magnification of part of the field shown in (*A*). Note again the beadlike processes connected by thin filaments. Calibration bar: 30 μm. (*C*) Cerebral stump, displaying characteristic dust and clumplike accumulation of peptide-like immunoreactivity at the level immediately adjacent to the ligation (upper end of the photomicrograph). Note that the rest of the nerve is devoid of peptide-like immunoreactivity and that an area of markedly clear background is interposed between the peptide-containing area and the rest of the nerve. Magnification as in (*A*). (*D*) Contralateral optic nerve, without evident peptide-like immunoreactivity, as in normal optic nerves. Magnification as in (*A*). (Reproduced from Kuljis et al., 1984.)

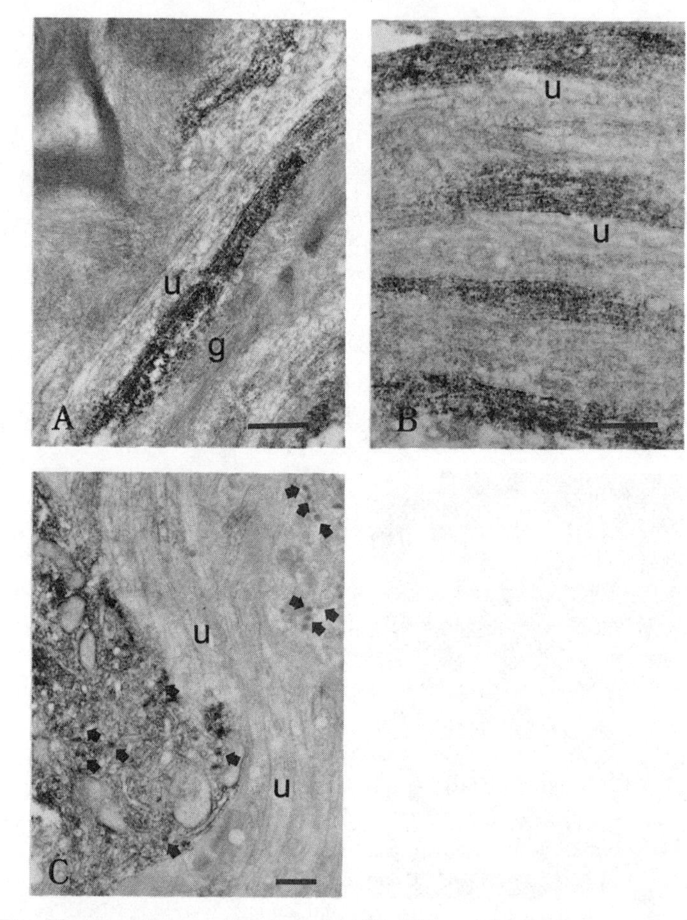

FIGURE 4. Electron microscopic demonstration of SP- and LENK-like immunoreactivities in the retinal stump of the optic nerve seven days after ligation. (*A*) SP-positive process longitudinally oriented along an unmyelinated fiber bundle. Note the presence of unlabeled axons and unlabeled glial cell processes (g) adjacent to the labeled axon. (*B*) SP-containing processes within a fascicle of unmyelinated axons. (C) LENK-positive dilated cell process within a bundle of unmyelinated axons. Note the abundance of cisternal and vesicular components, some of which are labeled, especially dense-core vesicles (arrows). Also note nearby unlabeled dilated cell process (upper right corner), containing organelles similar to those seen in the labeled process. These photomicrographs indicate the existence of SP- and LENK-like immunoreactivities within unmyelinated axons, which represent the majority of the fibers in the optic nerve. The fact that only a small proportion of axons and dilated processes (probably growth cones) were stained suggests that other subpopulations of axons may be present that contain different peptides and appear unstained in these specimens. Calibration bars: 0.5 μm; g: glial cell processes; u: unmyelinated axons. (Reproduced from Kuljis et al., 1984.)

These findings strongly indicated that retinal ganglion cells are capable of synthesizing and transporting peptides. Our previous failure to demonstrate peptides within ganglion cells is probably due to several factors. The most likely cause is the extremely thin diameter of the retinal ganglion cell axons, rendering IHC visualization difficult, combined with a relatively low concentration of neuropeptides within these fibers. Optic nerve crush resulted in a damming effect, with accumulation of peptides within the retinal stump. An additional contributing factor may be that the peptides are frequently transported in precursor form and only cleaved into their final product in the distal axon. In view of the rapid appearance of peptidergic staining within only two hours, we do not think that the appearance of peptides within the optic nerve stump is due to synthesis in response to injury.

DO CHEMICALLY SPECIFIC GANGLION CELLS INNERVATE SPECIFIC TECTAL LAMINAE?

The described findings strongly support the notion that retinal ganglion cells contain peptides. More striking, they indicate that different populations of peptidergic ganglion cells terminate in different sublaminae of the optic tectum. Though we have not yet been able to characterize the distinct somatic or dendritic morphologies of these ganglion cells, these findings provide interesting and direct support for at least one aspect of the original Lettvin hypothesis, that is, different types of ganglion cells terminate in different layers of the optic tectum. Furthermore, we may now add, they appear to utilize different transmitters/peptides and therefore have different presumed effects on postsynaptic structures. An exciting prospect for future studies is that their physiological responses may possibly be manipulated with pharmacological agents.

DO PEPTIDERGIC GANGLION CELLS REGENERATE AND DO THEY REINNERVATE THE APPROPRIATE TECTAL LAMINAE?

The ability of frog retinal ganglion cells to regenerate axons and regrow into the optic tectum is well established (Sperry, 1944; 1963). Maturana et al. (1959) reported that, following section of the optic nerve, the physiological responses of the reinnervated laminae of the optic tectum were similar to the normal adult animal. They suggested that the regenerating retinal ganglion cell axons were able to reinnervate appropriate and specific laminae. However, to date, no direct evidence of such lamina-specific reinnervation has been demonstrated with contemporary anatomical methods. The vast majority of studies of retinal reinnervation of the anuran optic tectum have been based on gross mapping studies using electrophysiological methods. Such methods are not well suited for the analysis of laminar specific reinnervation reported by Maturana et al. (1959).

Direct anatomical evidence showing that the specific category of retinal ganglion cell actually terminated in the appropriate lamina, however, was still lacking. In the event that the tectal laminae themselves might determine the morphology of the ingrowing axons, we still would not be able to state categorically that the specific ganglion cell reestablished connections with the appropriate lamina. Ideally, each different type of ganglion cell and its axon should be distinguishable on the basis of independent criteria, such as an assuredly type-specific antigen or transmitter.

Does lamina-specific reinnervation of the tectum by retinal ganglion cells occur? If so, how much time is required to reinnervate the tectum, and do the regenerating fibers occupy the appropriate tectal laminae? Following optic nerve crush, the optic tectum was examined with IHC methods after varying periods of time to determine whether any or all of the peptidergic retinal terminations are reestablished (Kuljis and Karten, 1983b; 1984). Animals were examined one to nine months following optic nerve crush. Clear evidence of reestablishment of SP was finally observed six to seven months postoperatively. The pattern of SP staining appeared identical to that seen in the normal adult animal. Unlike the normal animal, however, continuing evidence of SP staining of optic nerve axons was evident. Did this reflect regeneration of SP axons, or might it possibly be consequent to incomplete injury to the nerve? In support of the suggestion that this was due to lamina-specific regeneration, we found that not all categories of peptidergic axons were equally likely to regenerate. Thus, although the full pattern of SP peptidergic staining was restored, we found no indication of recovery of the normal pattern of LENK, CCK8, or BOM staining in the optic tectum (Kuljis and Karten, in preparation).

REGENERATIVE ABILITY OF DIFFERENT CLASSES OF GANGLION CELLS

Are all types of retinal ganglion cells equally capable of regenerating and reinnervating the appropriate laminae? With the exception of Maturana et al., (1959), studies of regeneration of the optic nerve in amphibia have almost exclusively been concerned with demonstrating the presence or absence of a simple retinotopic reinnervation. Since each tectal column subserving a given point in the visual field receives several distinct types of inputs (convexity, moving-edge, dimming, boundary, and contrast "detectors"), "appropriate" reinnervation also requires that all types of ganglion cells be capable of regeneration of their axons and terminals. The preceding observations suggest that either different categories of retinal ganglion cells have markedly different rates of regeneration despite similar axon diameters, or that not all categories of ganglion cells are equally capable of regeneration of their projections on the optic tectum. Further indication of the different responses of different types of ganglion cells was in our observations on the optic nerve itself. During the acute postoperative period, we were readily able to observe IHC staining for SP, LENK, CCK8, and BOM in the retinal stump distal to the crush. Following seven months survival, SP staining was still evident in the optic nerve and in the optic tectum. In contrast, there was no evident staining for LENK, CCK8, or BOM in the optic nerve at that time (Kuljis and Karten, in preparation).

These observations suggest that SP-containing cells may possess more regenerative capacity than LENK-, CCK8-, or BOM- positive cells. In addition, these findings indicate that SP-containing retinal ganglion cell axons are not only capable of successful regeneration but are also capable of reinnervating a specific tectal sublamina with remarkable precision. The latter finding provides strong experimental support to the chemoaffinity hypothesis of specific innervation (Sperry, 1963), and to electrophysiological studies indicating specific reinnervation of different tectal laminae by different functionally defined retinal ganglion cell classes (Maturana et al., 1959). The presence of different transmitters/modulators in different categories of ganglion cells also may provide powerful tools for pharmacological analysis of the differential actions of individual categories of ganglion cells.

REFERENCES

Benowitz, L. I. and H. J. Karten (1976) Organization of the tectofugal visual pathway in the pigeon: A retrograde transport study. *J. Comp. Neurol.*, **167**:503–520.

Brecha, N. C. (1978) Some observations on the organization of the avian optic tectum: Afferent nuclei and their tectal projections. Ph.D. thesis, State University of New York at Stony Brook.

Brecha, N. (1983) Retinal neurotransmitters: Histochemical and biochemical studies. In *Chemical Neuroanatomy*, P. C. Emson, ed. Raven, New York.

Cajal, S. R. (1909) *Histologie du Système Nerveux de L'Homme et des Vertébrés* (1972 edition). Maloine, Paris.

Cowan, W. M., L. Adamson, and T. P. S. Powell (1961) An experimental study of the avian visual system. *J. Anat. (Lond.)* **95**:545–563.

Hunt, S. P., P. Streit, H. Kunzle, and M. Cuenod (1977) Characterization of the pigeon isthmo-tectal pathway by selective uptake and retrograde movement of radioactive compounds and by golgi-like horseradish peroxidase labeling. *Brain Res.* **129**:197–212.

Karten, H. J. and N. C. Brecha (1983) Localization of neuroactive substances in the vertebrate retina: Evidence for lamination in the inner plexiform layer. *Vision Res.* **23**:1197–1205.

Kuljis, R. O. and H. J. Karten (1982) Laminar organization of peptide-like immunoreactivity in the anuran optic tectum. *J. Comp. Neurol.* **212**:188–201.

Kuljis, R. O. and H. J. Karten (1983a) Modifications in the laminar organization of peptide-like immunoreactivity in the anuran optic tectum following retinal deafferentation. *J. Comp. Neurol.* **217**:239–251.

Kuljis, R. O. and H. J. Karten (1983b) Peptide-like immunoreactivity in regenerating anuran optic nerve fibers. *Abstr. Soc. Neurosci.* **9**:760.

Kuljis, R. O. and H. J. Karten (1984) Immunocytochemical observations of reinnervation of the optic tectum by peptidergic ganglion cells. *Proc. XIIIth Ann. Meet. Am. Acad. Neurol. Neurology* **34** suppl.:99.

Kuljis, R. O., J. E. Krause, and H. J. Karten (1984) Peptide-like immunoreactivity in anuran optic nerve fibers. *J. Comp. Neurol.* **226**:222–237.

Lettvin, J. Y., H. R. Maturana, W. S. McCulloch, and W. H. Pitts (1959) What the frog's eye tells the frog's brain. *Proc. Inst. Radio Eng., N.Y.* **47**:1940–1951.

Maturana, H. R., J. Y. Lettvin, W. S. McCulloch, and W. H. Pitts (1959) Evidence that cut optic nerve fibers in a frog regenerate to their proper places in the tectum. *Science* **130**:1709–1710.

Potter, H. D. (1969) Structural characteristics of cell and fiber populations in the optic tectum of the frog (*Rana catesbiana*). *J. Comp. Neurol.* **136**:203–232.

Ramón Cajal, P. (1894) *Investigaciones Micrográficas en el Encéfalo de los Batracios y Reptiles, Cuerpos Geniculados y Tubérculos Cuadrigeminos de los Mamíferos.* La Dereche, Zaragoza.

Ramón Cajal, P. (1946) El cerebro de los batracios. *Trab. Inst. Cajal Invest. Biol.* **38**:4–111.

Reiner, A. and H. J. Karten (1982) Laminar distribution of the cells of origin of the descending tectofugal pathways in the bird. *J. Comp. Neurol.* **204**:165–187.

Scharrer, E. and J. Sinden (1949) A contribution to the "chemoarchitectonics" of the optic tectum of the brain of the pigeon. *J. Comp. Neurol.* **91**:331–336.

Sperry, R. W. (1944) Optic nerve regeneration with return of vision in anurans. *J. Neurophysiol.* **7**:57–69.

Sperry, R. W. (1963) Chemoaffinity in the orderly growth of nerve fiber patterns and connections. *Proc. Natl. Acad. Sci., U.S.A.* **50**:703–710.

Wilczynski, W. and R. G. Northcutt (1977) Afferents to the optic tectum of the leopard frog. An HRP study. *J. Comp. Neurol.* **173**:219–230.

Chapter Thirteen

NEURONAL CIRCUITS FOR PATTERNING MOTOR ACTIVITY IN INVERTEBRATES

K. G. PEARSON

Department of Physiology
University of Alberta
Edmonton, Canada

Views on how nerve cells function changed radically in the 1950s as a result of the application of intracellular recording techniques. These advances renewed confidence that complex behavior could be explained in terms of neurons and that the mechanisms generating temporally patterned impulse sequences could be determined (Bullock, 1959). However, knowledge of the mechanisms for generating patterned motor activity did not come rapidly. In fact, it has only been within the last decade that we have gained any real understanding of the cellular basis of complex behavior (Tables 1 and 2). Progress had to await the discovery of suitable preparations (allowing intracellular recording from more than one neuron during the generation of patterned motor activity) and methods for identifying neurons reliably. It is surprising to many that knowledge about connectivity between identified neurons and the functioning of neuronal circuits has come so recently. The false perception that the field of neuronal circuit analysis has a long tradition presumably stems from the early success of theoretical models in reproducing observed patterns of motoneuronal activity (Reiss, 1964; Wilson and Waldron, 1968), as well as the prominence received by numerous qualitative schemes, none of which was based on

TABLE 1. Initial Descriptions of Neuronal Circuits for Patterning Motor Activity in Invertebrates[a]

Animal	System	Author(s)	Year
Aplysia	Gill withdrawal	Kupfermann et al.	1974
	Heartbeat	Koester et al.	1974
	Ventilation	Byrne	1983
Tritonia	Swimming	Getting et al.	1980
Helisoma	Feeding	Kaneko et al.	1978
Lymnaea	Feeding	Rose & Benjamin	1981
Leech	Swimming	Friesen et al.	1978
	Heartbeat	Thompson and Stent	1976
Crab	Heartbeat	Tazaki and Cooke	1979a
Lobster	Stomatogastric	Maynard	1972
	Heartbeat	Watanabe	1958
Crayfish	Tailflip	Zucker	1972
Locust	Jumping	Pearson et al.	1980
	Flight rhythm	Robertson and Pearson	1983
	Respiration	Burrows	1982
	Leg movement	Burrows and Siegler	1978
Fly	Flight	Koenig and Ikeda	1983

[a]This table lists only those initial studies in which direct connections between neurons were demonstrated by intracellular recording. For many of these systems, earlier studies reported patterns of synaptic activity in motoneurons but gave no data on connectivity.

direct observations of the proposed neuronal connections (Stein, 1971; Davis, 1969; Wilson, 1976). The net effect of the impressive recent success in what appears to be a long-established field has been a perceptible decline in interest in circuit analysis. It seems to many that few fundamental mysteries about the organization and functioning of neuronal circuitry remain to be solved. How valid is this view?

TABLE 2. Recent Description of Neuronal Circuits for Patterning Motor Activity in Invertebrates[a]

Animal	System	Reference
Tritonia	Swimming	Getting (1983a, b)
Leech	Swimming	Stent et al. (1978)
	Heartbeat	Calabrese (1979)
		Peterson (1983)
Crab	Heartbeat	Tazaki and Cooke (1979a)
Lobster	Stomatogastric	Miller and Selverston (1982)
Crayfish	Tailflip	Wine and Krasne (1982)
Locust	Jump	Pearson (1983)

[a]This table lists only those studies in which circuitry has been determined in sufficient detail to explain major features of the motor output pattern.

This question is not new, for even in the mid-1970s, when only very few circuits were known in any detail, Bullock (1976) wrote "The time is foreseeable when a skeptical panel member . . . will say of a proposed network analysis, 'Isn't this going to be just another case? Will the findings be of interest beyond those concerned with this animal and this action?' " Bullock suggested that circuit breaking could be justified for two reasons. The first is that the analysis of circuits, neuron by neuron, will lead to the discovery of "new degrees of freedom, additional complexity, and new phenomenology to be explained at lower levels." The second justification was "that rules, regularities, tendencies, and generalizations will be found . . . [which] . . . will simplify our picture and permit inductions in a field that needs them." Since this was written, our knowledge of neuronal circuitry has increased significantly: note that more than two-thirds of the studies listed in Table 1 are dated after 1976. With this recent accumulation of data, we are in a position to examine whether the justifications offered by Bullock have been substantiated. Thus the purpose of this chapter is to describe some of the new phenomena discovered since 1977 from studies on invertebrate motor systems and to assess whether any new general principles of neuronal functioning and organization have in fact emerged within this time. Since it is common today to divide the processes contributing to the patterning of motor activity into three groups—synaptic, cellular, and network—I have followed this practice in organizing the following sections.

SYNAPTIC PROPERTIES

Two phenomena recently discovered in invertebrate motor systems are graded transmitter release and multicomponent postsynaptic potentials. Both are clearly important in patterning motor activity in the systems in which they occur.

Graded Transmitter Release

Ted Bullock was one of the first to predict that many integrative events in nervous systems would be found to occur in the absence of nerve impulses. In 1959 he wrote, "I will venture to suggest that in the near future we will gain significant new insight . . . with respect to . . . the normal functional significance of *inter*cellular reactions mediated by graded activity without the intervention of all-or-none impulses." At that time, there were no definite examples of nerve cells that did not generate action potentials, and the only known case of graded interaction was from Bullock's own work with Watanabe on the cardiac ganglion of the lobster. Subsequently, the graded interactions observed by Bullock and Watanabe were shown to occur by way of electrotonic junctions. Further evidence for nonimpulsive interactions in invertebrate motor systems came with the discovery of

TABLE 3. Graded Transmitter Release in Arthropod Motor Systems

Animal	System	Cell type	Spiking(S) Nonspiking(NS)	Reference
Lobster	Stomatogastric	Motoneuron	S	Graubard et al., 1983
		Interneuron	NS	Graubard, 1978
	Ventilation	Interneurons	NS	Mendelson, 1971
Crayfish	Swimmeret	Motoneurons	S	Heitler, 1978
		Interneurons	NS	Heitler and Pearson, 1980
		Receptors	NS	Heitler, 1982
	Uropod	Interneurons	NS	Reichert et al., 1982
Crab	Ventilation	Motoneurons	S	Simmers and Bush, 1983
		Interneurons	NS	Simmers and Bush, 1980
	Posture	Receptors	NS	Bush, 1981
	Uropod	Receptors	NS	Paul, 1976
Locust	Walking	Interneurons	NS	Burrows, 1979b
	Posture	Interneurons	NS	Siegler, 1981
Cockroach	Walking	Interneurons	NS	Pearson and Fourtner, 1975

nonspiking receptor cells in the legs of crustacea (Ripley et al., 1968) and nonspiking interneurons in crustacea and insects (Mendelson, 1971; Pearson and Fourtner, 1975). Although the absence of action potentials in these cells suggested that they might interact with other neurons by graded release of chemical transmitter, this was not conclusively shown until intracellular recordings were made from both the pre- and postsynaptic neurons (Burrows and Siegler, 1978; Graubard, 1978; Blight and Llinas, 1980). Thus direct evidence for graded transmitter release in invertebrate motor systems has come only recently. With the clear examples we now have of this phenomenon, it is generally thought that graded transmitter release is a property of all nonspiking neurons. Since these neurons are now known to occur in many arthropod motor systems (Table 3), it seems that graded transmitter release is a very common phenomenon in animals belonging to this phylum. It should also be noted from Table 3 that graded release is also a property of some spiking neurons, the clearest examples being in the stomatogastric ganglion of the lobster (Graubard et al., 1983). I describe some of the main features of graded transmitter release in arthropod motor systems, but no attempt is made to review this topic extensively. Excellent recent reviews can be found elsewhere (Burrows, 1981; Simmers, 1981; Bush, 1981).

Graded transmitter release from nonspiking interneurons has been

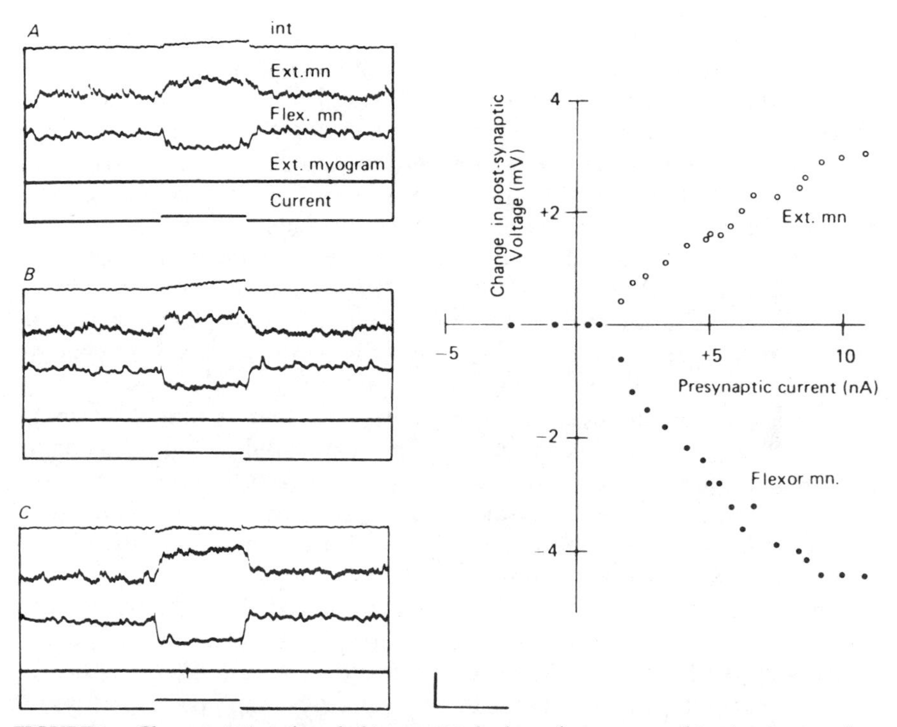

FIGURE 1. Characteristics of graded transmitted release from a nonspiking interneuron in a thoracic ganglion of the locust. (*A–C*) Depolarizing current pulses applied to the interneuron evoked a maintained shift in membrane potential in an extensor motoneuron and flexor motoneuron without the generation of action potentials in the interneuron (int). All three intracellular recordings were made simultaneously. The graph on the right shows the relationship between the magnitude of the injected current and the magnitudes of the postsynaptic responses. Calibrations: vertical, interneuron 40 mV, extensor motoneuron 4 mV, flexor motoneuron 8 mV; current 30 nA; horizontal 400 msec. (From Burrows, 1980.)

studied most extensively in locusts (reviewed by Burrows, 1981). Each thoracic ganglion in these animals contains about 3000 nerve cells, more than half of which are estimated to be confined entirely within the ganglion. Many, but not all, of these local neurons never produce action potentials under any conditions. Some of the characteristics of synaptic transmission from these nonspiking neurons are illustrated in Figure 1. The most obvious is that maintained depolarizations in the interneurons produce maintained shifts (depolarizing or hyperpolarizing) in the membrane potential of the postsynaptic neurons. The relation between the current injected into the interneuron and the shift in postsynaptic potential is fairly linear, and the magnitude of the postsynaptic potential change can be quite large. Each nonspiking interneuron usually has a widespread effect on many postsynaptic neurons, and these effects are distributed so as to produce a coordinated pattern of muscle contraction with clear functional significance.

The interactions between individual nonspiking interneurons have not yet been investigated extensively, but one interesting observation is that all direct interactions are inhibitory (Burrows, 1979b). This is similar to the situation in the swimming and heartbeat systems of the leech and the stomatogastric system of the lobster, where almost all interactions are inhibitory. Another interesting observation is that some nonspiking interneurons have a dual action: exciting some neurons and inhibiting others (Fig. 1). Whether these actions are both produced by direct monosynaptic connections has not been established.

A common phenomenon is that nonspiking interneurons in locusts release transmitter continuously under resting conditions. Thus movements of membrane potential in either direction from rest produce postsynaptic responses. A consequence of this phenomenon is that small unitary synaptic events in a nonspiking interneuron are expressed in the postsynaptic neurons (Burrows, 1979a). This ability of nonspiking interneurons to secrete transmitter continuously and have this release modulated by small fluctuations in membrane potential has led to the idea that much of the integration in systems of nonspiking interneurons may be localized in discrete regions of the neurons. At present, there is no direct evidence for this, but theoretical calculations show that passive spread of synaptic potentials might be limited. Moreover, the ultrastructure of thoracic interneurons showing input and output synapses localized close to each other (Watson and Burrows, 1983) is consistent with the notion of local integration.

One of the obvious functional advantages of graded transmitter release is that it does away with the need for encoding and decoding sequences of spike trains and allows a continuous relationship between input and output (Pearson, 1976). Thus it is not surprising that nonspiking interneurons have been found where precise tonic control of motoneuronal activity is necessary, such as in the regulation of posture (Siegler, 1981). This advantage is also realized in other animals, such as the cockroach, in which a single nonspiking interneuron is able to smoothly control the discharge rate of leg motoneurons over a wide range (Pearson and Fourtner, 1975), in decapod crustacea, where single, nonspiking proprioceptors continuously control the motor output for the regulation of posture (Bush, 1981), and in the crayfish, where single, nonspiking interneurons mediate graded reciprocal inhibitory responses in the integration of sensory information from the tail fan (Reichert et al., 1983).

Nonspiking interneurons are also involved in the generation of rhythmic motor activity in three arthropod motor systems: insect walking (Pearson and Fourtner, 1975), crayfish swimmeret beating (Heitler and Pearson, 1980), and crab ventilation (Simmers and Bush, 1980). In none of these systems is there evidence that single nonspiking interneurons have the capacity to generate endogenous rhythmic oscillations in membrane potential. Thus the generation of these motor rhythms seems to depend primar-

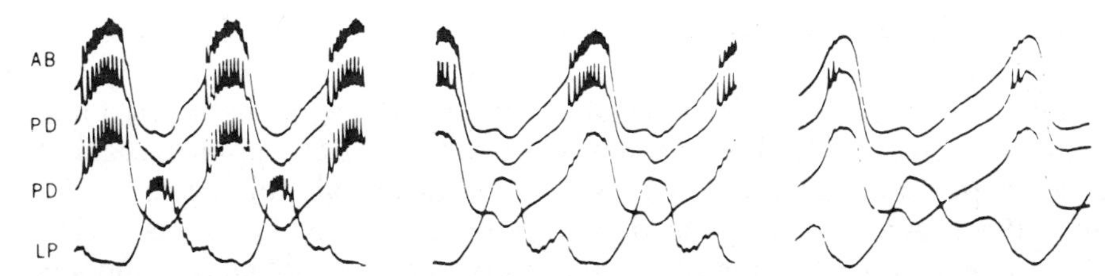

FIGURE 2. Rhythmicity and phasing of oscillations persist in stomatogastric neurons following abolition of spike activity. Left: Simultaneous intracellular recordings from neurons in the pyloric network under normal conditions. Middle: Seven minutes after perfusion of the posterior margin of the stomatogastric ganglion with $1.7 \times 10^{5}M$ tetrodotoxin. Right: Ten minutes after toxin perfusion almost all spike activity was blocked. Calibration: 10 mV, 0.5 sec. (From Raper, 1979.)

ily on interactions between nonspiking interneurons. The advantages of relying on graded interactions between nonspiking interneurons for the generation of motor rhythms in arthropods are not obvious, but one possibility may be that it provides a simple method of smoothly controlling the frequency of oscillations over a wide range.

Graded transmitter release is not confined to those neurons that do not generate action potentials. The best examples of spiking neurons capable of graded release have been reported in the stomatogastric ganglion of the lobster (Graubard et al., 1980, 1983). Almost all the neurons in this small ganglion (approximately 30 neurons), are motoneurons innervating muscles involved in the movement and digestion of food. These motoneurons must produce action potentials to transmit information to the muscles, but these action potentials are not necessary for the generation of the basic rhythm and for phasing the activity in the different neurons in the ganglion (Fig. 2; Raper, 1979). This ability for rhythmic patterned oscillations to persist in the absence of action potentials depends in part on the neurons being able to release transmitter in a graded fashion (Graubard et al., 1983). Many of the features of graded transmitter release from stomatogastric neurons are similar to those found for graded release from nonspiking interneurons in insects. In particular, many neurons are continuously releasing transmitter under resting conditions.

The finding of graded release from spiking neurons is important, for it demonstrates directly that transmitter can be released from the dendritelike processes of arthropod neurons. This possibility was first suggested from ultrastructural studies in which it was found that input and output synapses were intermingled on the distal segments of all neurites (King, 1976). A similar finding in insects (Watson and Burrows, 1983) has already been mentioned. Thus it seems probable that in many systems transmitter release from the dendrites of spiking neurons will prove to be an important process in integration. This is already accepted as a common mechanism is some vertebrate systems where dendrodendritic synapses are formed (Shepherd, 1979).

Multicomponent Postsynaptic Potentials

Another recently discovered synaptic phenomenon in an invertebrate motor system is the ability of some neurons to generate multicomponent postsynaptic potentials (PSPs) by way of monosynaptic connections (Getting, 1981). Activity in single presynaptic neurons evokes postsynaptic potentials with up to three separate components, but not all components are evoked in every postsynaptic neuron. Of the 20 synaptic connections among three classes of interneurons generating the swimming rhythm in *Tritonia*, seven show multicomponent PSPs. Within this group of seven, there are three types: excitatory-inhibitory, excitatory-inhibitory-excitatory, and inhibitory-excitatory. The durations of the different components vary over a wide range, the initial components lasting for less than a second and later components lasting up to 20 seconds. The various components of the PSPs can, however, be divided into three categories: those producing an effect during the time of the burst of the presynaptic neuron, those producing an effect over a period comparable to a single swim cycle, and those acting over many cycles.

The functional consequences of the occurrence of multicomponent PSPs in the *Tritonia* swimming system have been explored extensively by computer simulation (Getting, 1983a). Essentially, their occurrence leads to a dynamic reorganization of the synaptic interactions between neurons (Fig. 3), with these time-dependent changes being appropriate for the generation of the rhythmic reciprocal patterns of activity in swim interneurons. Early in the swim cycle, synergist neurons (C2 and DSIs, Fig. 3) mutually excite each other, and activity in C2 inhibits the VSIs. Later in the cycle, as a result of the expression of later components of the PSPs, C2 excites the VSIs and inhibits the DSIs, thus promoting a burst in the VSIs and terminating the C2-DSI burst. Removal of excitation from the VSIs terminates the VSI burst, thus removing inhibition from the DSIs and allowing the C2-DSI system to burst again.

Postsynaptic potentials with more than one component appear to be widespread in molluscan nervous systems. They were first described in the abdominal ganglion of *Aplysia* (Wachtel and Kandel, 1967), and since then they have been found in other *Aplysia* ganglia (Kandel, 1976) as well as in the brains of *Navanax* and *Tritonia*. The only example outside molluscs is in the sympathetic ganglia of lower vertebrates (Dodd and Horn, 1983).

CELLULAR PROPERTIES

Two of the most interesting cellular phenomena recently discovered in invertebrate motor systems are the ability of some neurons to generate long-lasting depolarizations (plateau or driver potentials) in response to a brief depolarization (Russell and Hartline, 1978; Tazaki and Cooke, 1979b)

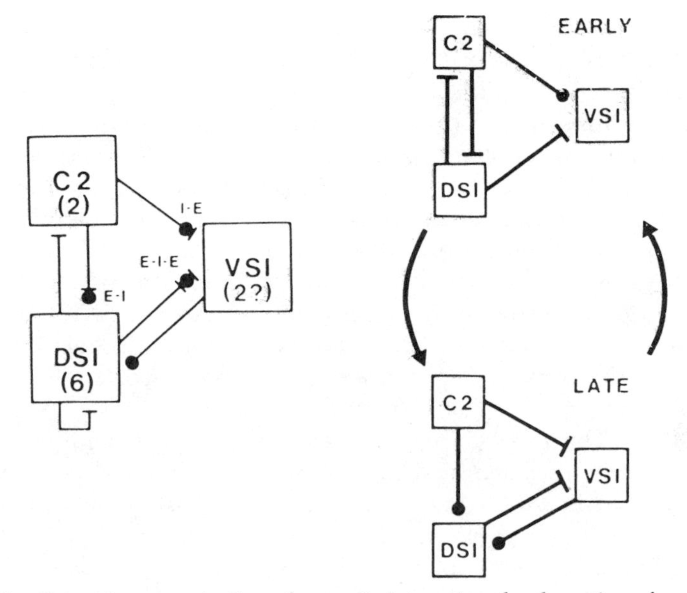

FIGURE 3. Dynamic reorganization of synaptic interactions by the action of multicomponent postsynaptic potentials in the swimming system of *Tritonia*. Left: Summary circuit diagram. Excitatory and inhibitory connections are shown as T bars and filled circles respectively. Mixed symbols indicate the multicomponent connections. DSI, dorsal swim interneurons; VSI, ventral swim interneurons. Right: The network on the left can be divided into two circuits depending on the initial and secondary action of each multicomponent connection. (From Getting, 1983a.)

and spiking between two thresholds in endogenously oscillating neurons (Robertson and Moulins, 1981). Because of the current widespread interest in the former and the unusual nature of the latter, these are the only cellular phenomena considered here. Of course, many other cellular properties are known to contribute to the patterning of motor activity, such as endogenous oscillations in membrane potential, postinhibitory rebound, adaptation, multiple spike initiating sites, and delayed excitation. The role of these processes has been discussed in a number of recent articles (Getting, 1983a,b; Miller and Selverston, 1982; Calabrese, 1979).

Plateau and Driver Potentials

Plateau potentials in nerve cells were first characterized in detail in neurons of the stomatogastric ganglion of the lobster (Russell and Hartline, 1978). When this ganglion is attached to the central nervous system, the majority of neurons can be induced to generate a long-lasting depolarization in response to a relatively brief depolarizing current pulse. There is a clear threshold for the initiation of the plateau potential and, once initiated, the membrane potential shifts in an all-or-none manner to the plateau level. To return the membrane potential to rest requires the injection of a suprathreshold hyperpolarizing current, and the induced repolarization is also an all-or-none phenomenon. Thus the transitions to the plateau level and back are threshold processes. In the absence of imposed hyperpolarizing currents, plateau potentials can last for more than 30 seconds (Dickinson and Nagy, 1983). The quasi-stable nature of plateau potentials means that under normal circumstances the return to resting potential requires inhibitory synaptic input from other neurons.

The obvious function of the plateau potentials in the stomatogastric ganglion is to produce large depolarizations and burst activity. In the generation of patterned motor activity, it appears that the depolarizations per se, and not the bursts of spikes, are of primary importance in ensuring rhythmicity and phasing of activity in different neurons (Anderson and Barker, 1981). The importance of plateau potentials in generating patterned motor activity is revealed by the weakening or total abolition of rhythmic motor activity when the capacity for neurons to generate plateau potentials is abolished. Cutting or blocking the main input nerve to the stomatogastric ganglion is one method for abolishing plateau potentials, since their generation depends on modulatory influences from neurons originating in other ganglia (Russell, 1979). Recently, individual neurons capable of inducing plateau potentials have been identified (Nagy and Dickenson, 1983; Russell and Hartline, 1981). A single neuron in the oesophageal ganglion, the anterior pyloric modulator (APM), can, when stimulated, induce plateau potentials in all neurons of the pyloric system (Fig. 4). Similarly, two identified neurons originating in the brain can induce plateau potentials (Russell and Hartline, 1981). The mechanism for inducing the capacity for plateau potentials is unknown, but it presumably depends on a neuromodulator altering the characteristics of voltage-dependent conductances in a manner analogous to that occurring in some snail neurons under the action of the peptide hormone vasopressin (Barker and Smith, 1976). The neuromodulator in the stomatogastric ganglion has not yet been identified, but one likely candidate is dopamine. Not only is dopamine located in the ganglion, but also exogenous application can induce bursting motor activity in isolated ganglia (Anderson and Barker, 1981).

Neurons in the cardiac ganglion of crabs and lobsters are also capable of generating long-lasting depolarizations resembling plateau potentials. These potentials have been termed "driver potentials" (Tazaki and Cooke, 1979a), since their obvious function is to generate bursts of activity independent of the process involved in producing rhythmicity. In fact, driver potentials occur in neurons incapable of endogenous rhythmicity. Driver potentials are regenerative, that is, elicited in an all-or-none manner by brief stimuli, and of a constant form. Sufficiently strong hyperpolarizing pulses applied near the peak of the driver potential can terminate the potential. The most obvious differences between stomatogastric plateau potentials and cardiac driver potentials is the shorter duration of the latter

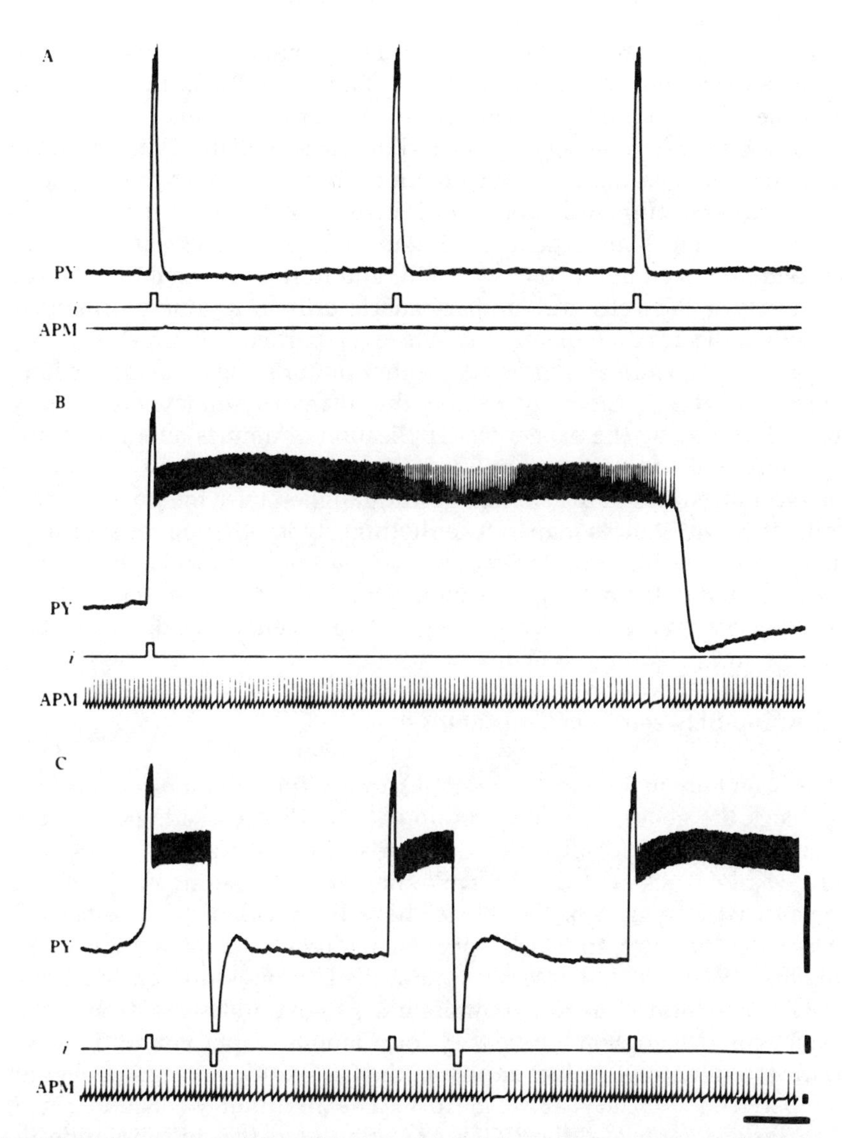

FIGURE 4. Plateau potentials in PY neurons of the lobster stomatogastric ganglion induced by stimulation of a single modulator neuron, the anterior pyloric modulator (APM). (*A*) In the absence of activity in the APM, brief depolarizing current pulses *i* produced a long plateau potential when APM was active. (*B*) The same current pulses produced a long plateau potential when APM was active. (*C*) Plateau potentials were terminated by the application of a brief hyperpolarizing current pulse. Calibrations: vertical 20 mV, horizontal 2 sec. (From Dickinson and Nagy, 1983.)

(approximately 250 msec). This difference is presumably because the termination of driver potentials must be by mechanisms intrinsic to the neurons, since there are no inhibitory neurons in the cardiac ganglion.

These recent findings of plateau and driver potentials in the neurons in crustacean ganglia raise the obvious question of whether these types of potentials occur in neurons of other invertebrate motor systems. The fact that similar types of regenerative depolarizations occur in molluscan neurons (involved in mucous secretion) and in nonneuronal tissues (such as salivary gland cells and cardiac muscle cells in a variety of animals) certainly indicates their ubiquitous nature. Of particular interest is whether plateau potentials can be induced in other motor systems by the action of neuromodulatory systems. We know that many rhythmic motor patterns can be induced by the exogenous application of amines and amino acids (walking, DOPA; leech swimming, 5HT; lamprey swimming, glutamate). The recent findings in the stomatogastric ganglion raise the possibility that modulatory substances may induce rhythmicity in other motor systems by changing the cellular properties of single neurons to enhance their "burstiness." Whether the natural activation of rhythmic motor systems relies on the induction of the capacity for long-lasting regenerative depolarizations remains an intriguing possibility.

Spiking between Two Thresholds

It has been known for a long time that strong depolarizations of nerve cells will block the generation of action potentials. This blockade has not been regarded as physiological since the levels of depolarization necessary to induce spike block are high. Recently, however, oscillating neurons in the stomatogastric system of the lobster have been found that normally depolarize sufficiently to block spike generation (Fig. 5; Robertson and Moulins, 1981, 1984). These neurons, called commissural gastric drivers (CDGs), are located in the commissural ganglia and send axons to the stomatogastric ganglion where they form monosynaptic connections with gastric motoneurons. Spikes appear in the CDGs only at membrane potentials between −60 mV and −30 mV. The rhythmic oscillations in the membrane potential of the CGDs are often of sufficient amplitude that spiking ceases near the peak depolarization. As a consequence, the CGDs can produce two bursts of spikes for one cycle of oscillation: one on the rising phase of the depolarization and the other on the falling phase. Only a single burst is produced if the depolarization fails to reach the upper threshold value.

The important consequence of this ability of the CGDs to cease firing when depolarized beyond −30 mV is that relatively small shifts in the mean membrane potential of the CGDs can produce major changes in the phasing of oscillations in postsynaptic neurons. This is illustrated in Figure 5 (right) where an upward shift in peak membrane potential caused the

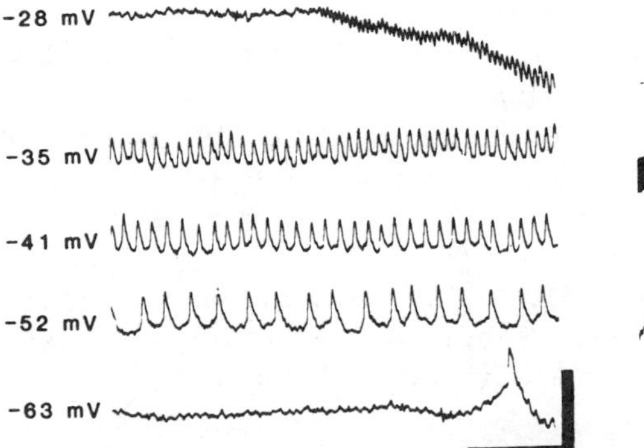

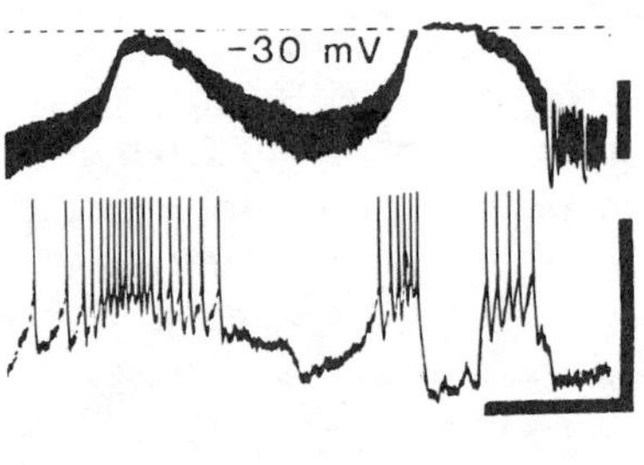

FIGURE 5. Neuron in the stomatogastric system of the lobster (the commissural gastric driver) that spikes between two thresholds. Left: Spikes generated at different membrane potentials. Note that continuous spiking only occurs when the membrane potential is in the range of −60 to −30 mV. Right: Change in phasing of postsynaptic response in a gastric mill motoneuron (bottom record) when the peak depolarization in the CGD neuron exceeded −30 mV (top record). The cessation of spiking above the upper threshold removed monosynaptic excitatory input from the gastric mill motoneuron and led to the appearance of two bursts of activity for one cycle of oscillation in CGD. Calibration: vertical 10 mV; horizontal 250 msec. (From Robertson and Moulins, 1981.)

postsynaptic neuron to generate two bursts per cycle, neither of which are in phase with the peak depolarization of the CGD neuron. The functional consequences of these dramatic alterations in phasing have not yet been established.

Since the CGDs are the only examples of neurons that show spike activity between two thresholds, it is impossible to assess the extent of this phenomenon. However, cases where such a mechanism would be appropriate can be easily imagined, as when two bursts of output are required for each cycle of movement. An example is in the walking system of the cockroach, where some foreleg motoneurons generate two bursts for each stepping cycle (Pearson and Iles, 1973). Another case is where a 180° phase shift is required without the use of inhibitory interneurons. This could be achieved if the presynaptic neuron oscillated with its peak depolarization going beyond the upper voltage threshold and with the minimum level of potential never going below the lower voltage threshold.

NETWORK PROPERTIES

Even though the neuronal circuitry has been reasonably well established in only a small number of invertebrate motor systems (Table 2), it is abundantly clear that there is enormous diversity in the circuitry in different systems, and, in general, the circuits are considerably more complex than anticipated from the relative simplicity of the motor output patterns. Interconnections are usually extensive, an extreme example being in the swimming system of *Tritonia* where each interneuron is connected to almost all the others (Getting, 1983a). Furthermore, numbers of neurons are not strongly related to the complexity of the motor output: 14 neurons are involved in the generation of the complex multiphase pattern in the stomatogastric pyloric system, whereas almost as many neurons (nine) are involved in the production of the simple single-phase output in the lobster and crab cardiac ganglia. With this degree of diversity and complexity, few generalizations can be made. There are, however, two features shared by many circuits: positive feedback and reciprocal inhibition.

Positive Feedback

In a variety of invertebrate motor systems small groups of neurons form mutually excitatory connections: *Tritonia* swimming (Getting, 1981), snail feeding (Kaneko et al., 1978), *Aplysia* ventilation (Byrne, 1983), stomatogastric pyloric system (Miller and Selverston, 1982), and crusacean heartbeat (Tazaki and Cooke, 1979a). In the majority of these systems, the mutual excitatory coupling occurs by means of electrical synaptic junctions. The one exception is in the swimming system of *Tritonia* where chemical synapses mediate excitatory coupling between interneurons active during dorsal flexion (the C2s and DSIs). The function of positive feedback in small numbers of functionally identical neurons appears to be to generate rapidly synchronized bursts in all the neurons of the set.

Mutual excitatory circuits for burst generation have not yet been found in annelids or insects. However, positive feedback pathways of a different kind exist in these animals. In the leech, for example, neurons capable of initiating the swimming rhythm receive excitatory input from the system generating the basic rhythm (Weeks, 1981). Presumably this positive feedback circuit helps to maintain the swimming rhythm, once initiated. Similarly, in the legs of insects positive feedback circuits function to maintain high levels of activity in leg motoneurons. During the preparatory phase of the jump, for example, the activity in leg extensor motoneurons is maintained by feedback from leg receptors excited by contraction of the extensor muscle (Heitler and Burrows, 1977).

Reciprocal Inhibition

Another common feature of network organization in invertebrate motor systems is reciprocal inhibition between neurons discharging in antiphase. This occurs in the swimming and heartbeat systems of the leech, the pyloric and gastric divisions of the stomatogastric system, the swimming system of *Tritonia*, and the ventilatory system of *Aplysia* (see Tables 1 and 2 for references). In all instances the reciprocally connected neurons are embedded within a larger network of neurons. This has led to difficulty in assessing the functional role of reciprocal inhibition and, in particular, whether it is important in the generation of the basic rhythmicity. Currently there is no clear example of mutual inhibitory interaction being solely responsible for rhythm generation. Nevertheless, in at least two systems reciproal inhibition has been shown to have a partial role in rhythm generation. In the heartbeat system of the leech, reciprocal inhibition between contralateral homologues functions to terminate endogenously initiated burst activity. Thus if one member of the pair is removed, the cycle duration increases dramatically (Peterson, 1983). Burst activity in the dorsal swim interneurons in the *Tritonia* swimming system is also terminated by inhibitory interneurons connected reciprocally with them (Getting, 1983a). It is less clear whether reciprocal inhibition is important for rhythm generation in other systems. What is clear, however, is that in all these systems, including leech heartbeat and *Tritonia* swimming, reciprocal inhibition is important for ensuring the correct phasing of activity. This has been shown particularly well in the pyloric system of the lobster stomatogastric ganglion (Miller and Selverston, 1982).

CONCLUSIONS

How well do the results from recent work on invertebrate motor systems support the contention made by Bullock (1976) that the analysis of neuronal circuitry is justified because it (1) leads to the discovery of new cellular and synaptic phenomena and (2) establishes the rules and constraints for the organization and functioning of neuronal systems? It is obvious that the first of these justifications has been fully substantiated. I have discussed just four recently discovered cellular and synaptic phenomena, but a number of other interesting processes have also been found: delayed excitation (Getting, 1983b), multiple spike-initiating sites (Calabrese, 1979), and different rates of recovery from inhibition (Miller and Selverston, 1982). With the now large number of known basic processes, it is natural to wonder how close we are to a complete catalog of cellular and synaptic phenomena. Of course, this is impossible to assess, but there are reasons for thinking that much more remains to be discovered in some areas. One obvious area is the modulation of cellular and synaptic properties by exogenous substances (either blood borne or released from modulator neurons). Another is the local integration within the dendritelike processes of neurons.

By comparison with the enormous advances in our knowledge of cellular and synaptic phenomena, there have been few insights into the general organization and functioning of neuronal networks. Part of the problem may be that we are attempting to generalize from a very limited set of data (note that we can regard the circuitry for only seven systems to be reasonably detailed, Table 2), and with more examples and a fuller knowledge of the systems already studied some important generalizations may emerge. However, the diversity and complexity of the now well-studied systems do not seem to support this view. In fact, as we have discovered more about each individual system we seem to move further away from general principles of organization. Just a few years ago, for example, Kristan (1980) made an excellent attempt to generalize about the mechanisms involved in the generation of rhythmic motor activity. His conclusions were quite reasonable, based on the data available at that time. However, in the space of just a few years, data have accumulated that are not consistent with some of the general ideas he suggested.

So, a fundamental question today is, "What are the reasons for the wide diversity in the organization and functioning of invertebrate motor systems?" Clearly, this is not related to the patterns of motor activity generated by these systems, for quite different mechanisms are used to generate fairly similar patterns of activity. Thus we must look beyond functional requirements. A clue for accounting for the diversity within different motor systems is a relationship between the existence of certain cellular and synaptic phenomena and the phyla in which these occur. For example, multicomponent PSPs have only been found in molluscs, and plateau and driver potentials found only in crustacea. Nonspiking interneurons and graded transmitter release are very common in insects and crustacea, but they have never been reported in molluscan motor systems and are not extensive in annelids. Finally, electrotonic interactions are extremely common in molluscs, crustacea, and annelids, but they have not yet been reported in insect motor systems. These obvious correlations between cellular and synaptic phenomena and the animals in which they occur suggest that much of the diversity in neuronal systems is related to evolutionary specializations within different phyla. Evolutionary processes may also explain much of the complexity we see in neuronal circuits. Certainly they all appear to be much more complex than is necessary for the generation of the motor patterns. One possibility is that the neuronal circuitry we observe is strongly related to the circuitry for the behaviors from which the existing behavior evolved. Some recent evidence to support this notion has come from studies on the organization of interneurons in the flight system of the locust (Robertson et al., 1982). The surprising observation in this system was that many important interneurons have their somata and numerous processes in abdominal ganglia. This was unexpected since wing movements are produced by activity in motoneurons contained largely within the second and third thoracic ganglia. The simplest explanation for the existence of interneurons in abdominal ganglia is that it is related to the evolutionary origin of the flight system. Wings probably evolved from thoracic appendages that were homologous to appendages distributed along all abdominal segments (Kukalova-Peck, 1978). The exact function of these appendages in the progenitors of flying insects is not known, but they may have been used for swimming or ventilation.

In any event, the occurrence of abdominally located interneurons in the locust flight system is presumably related to the fact that the flight system evolved from one concerned with controlling the movements of appendages on all body segments.

With these recent findings in the flight system of the locust, it will be of considerable interest to examine in other motor systems the relationship between the organization of the neuronal circuitry and the evolutionary origin of the behavior. The only way we can gain information on the evolution of nervous systems is by a comparative analysis of extant species. Thus a comparative approach toward gaining an understanding of the organization of invertebrate motor systems now seems more urgent than ever. This, of course, is an approach championed by Ted Bullock throughout his entire career.

Acknowledgments

I thank G. Boyan, D. Reye, and I. Gynther for their comments on early drafts of this manuscript.

REFERENCES

Anderson, W. W. and D. L. Barker (1981) Synaptic mechanisms that generate network oscillations in the absence of discrete postsynaptic potentials. *J. Exp. Zool.* **216:**187–191.

Barker, J. L. and T. G. Smith (1976) Peptide regulation of neuronal membrane properties. *Brain Res.* **103:**167–171.

Blight, A. R. and R. Llinas (1980) The non-impulsive stretch-receptor complex of the crab: a study of depolarisation release coupling at a tonic sensorimotor synapse. *Phil. Trans. R. Soc.* **B290:**219–276.

Bullock, T. H. (1959) Neuron doctrine and electrophysiology. *Science* **129:**997–1002.

Bullock, T. H. (1976) In search of principles in neural integration. In *Simpler Networks and Behavior,* J. C. Fentress, ed. Sinauer Associates, Sunderland, Mass., pp. 52–60.

Burrows, M. (1979a) Synaptic potentials effect the release of transmitter from locust nonspiking interneurons. *Science* **204:**81–83.

Burrows, M. (1979b) Graded synaptic interactions between local premotor interneurons of the locust. *J. Neurophysiol.* **42:**1108–1124.

Burrows, M. (1980) The control of sets of motor neurons by local interneurones in the locust. *J. Physiol.* **298:**213–233.

Burrows, M. (1981) Local interneurons in insects. In *Neurones without Impulses,* A. Roberts and B. M. H. Bush, eds. Cambridge University Press, Cambridge, pp. 199–221.

Burrows, M. (1982) Interneurones co-ordinating the ventilatory movement of the thoracic spiracles in the locust. *J. Exp. Biol.* **97:**385–400.

Burrows, M. and M. V. S. Siegler (1978) Graded synaptic transmission between local interneurones and motor neurones in the metathoracic ganglion of the locust. *J. Physiol.* **285:**231–255.

Bush, B. M. H. (1981) Non-impulsive stretch receptors in crustaceans. In *Neurones without*

Impulses, A. Roberts and B. M. H. Bush, eds. Cambridge University Press, Cambridge, pp. 147–176.

Byrne, J. H. (1983) Identification and initial characterization of a cluster of command and pattern-generating neurons underlying respiratory pumping in *Aplysia californica*. *J. Neurophysiol.* **49**:491–508.

Calabrese, R. L. (1979) Neural generation of the peristaltic and non-peristaltic heartbeat coordination modes in the leech *Hirudo medicinalis*. *Am. Zool.* **19**:87–102.

Davis, W. J. (1969) The neural control of swimmeret beating in the lobster. *J. Exp. Biol.* **50**:99–118.

Dickinson, P. S. and F. Nagy (1983) Control of a central pattern generator by an identified modulatory interneurone in crustacea. II. Induction and modulation of plateau properties in pyloric neurones. *J. Exp. Biol.* **105**:59–82.

Dodd, J. and J. P. Horn (1983) Muscarinic inhibition of sympathetic C neurones in the bullfrog. *J. Physiol.* **334**:271–291.

Friesen, W. O., M. Poon, and G. S. Stent (1978) Neuronal control of swimming in the medicinal leech. IV. Identification of a network of oscillatory interneurones. *J. Exp. Biol.* **75**:25–44.

Getting, P. A. (1981) Mechanisms of pattern generation underlying swimming in *Tritonia*. I. Neuronal network formed by monsynaptic connections. *J. Neurophysiol.* **46**:65–79.

Getting, P. A. (1983a) Mechanisms of pattern generation underlying swimming in *Tritonia*. II. Network reconstruction. *J. Neurophysiol.* **49**:1017–1035.

Getting, P. A. (1983b) Mechanisms of pattern generation underlying swimming in *Tritonia*. III. Intrinsic and synaptic mechanisms for delayed excitation. *J. Neurophysiol.* **49**:1036–1050.

Getting, P. A., P. R. Lennard, and R. I. Hume (1980) Central pattern generator mediating swimming in *Tritonia*. I. Identification and synaptic interactions. *J. Neurophysiol.* **44**:151–164.

Graubard, K. (1978) Synaptic transmission without action potentials: input-output properties of a nonspiking presynaptic neuron. *J. Neurophysiol.* **41**:1014–1025.

Graubard, K., J. A. Raper, and D. K. Harline (1980) Graded synaptic transmission between spiking neurons. *Proc. Natl. Acad. Sci. U.S.A.* **77**:3733–3735.

Graubard, K., J. A. Raper, and D. K. Hartline (1983) Graded synaptic transmission between identified spiking neurons. *J. Neurophysiol.* **50**:508–521.

Heitler, W. J. (1978) Coupled motorneurones are part of the crayfish swimmeret central oscillator. *Nature* **275**:231–234.

Heitler, W. J. (1982) Non-spiking stretch-receptors in the crayfish swimmeret system. *J. Exp. Biol.* **96**:355–366.

Heitler, W. J. and M. Burrows (1977) The locust jump. II. Neural circuits of the motor programme. *J. Exp. Biol.* **66**:221–242.

Heitler, W. J. and K. G. Pearson (1980) Non-spiking interactions and local interneurons in the central pattern generator of the crayfish swimmeret system. *Brain Res.* **187**:206–211.

Kandel, E. R. (1976) *Cellular Basis of Behavior*, W. H. Freeman and Co., San Francisco.

Kaneko, C. R. S., M. Merickel, and S. B. Kater (1978) Centrally programmed feeding in Helisoma: identification and characteristics of an electrically coupled pre-motor neuron network. *Brain Res.* **146**:1–22.

King, D. G. (1976) Organization of crustacean neuropil. II. Distribution of synaptic contacts on identified motor neurons in lobster stomatogastric ganglion. *J. Neurocytol.* **5**:239–266.

Koenig, J. H. and K. Ikeda (1983) Reciprocal excitation between identified flight motor neurons in *Drosophila* and its effect on pattern generation. *J. Comp. Physiol.* **150**:305–318.

Koester, J., E. Mayeri, G. Liebeswar, and E. R. Kandel (1974) Neural control of circulation in *Aplysia*. II. Interneurons. *J. Neurophysiol.* **37**:476–496.

Kristan, W. B. (1980) The generation of rhythmic motor patterns. In *Information Processing in the Nervous System*, H. Pinsker and W. D. Willis, eds. Raven, New York, pp. 241–261.

Kukalova-Peck, J. (1978) Origin and evolution of insect wings and their relation to metamorphosis, as documented in the fossil record. *J. Morphol.* **156**:53–126.

Kupfermann, I., T. J. Carew, and E. R. Kandel (1974) Local, reflex, and central commands controlling gill and siphon movements in *Aplysia*. *J. Neurophysiol.* **37**:996–1019.

Maynard, D. M. (1972) Simpler network. *Ann. N. Y. Acad. Sci.* **193**:59–72.

Mendelson, M. (1971) Oscillator neurons in crustacea ganglia. *Science* **171**:1170–1173.

Miller, J. P. and A. I. Selverston (1982) Mechanisms underlying pattern generation in lobster stomatogastric ganglion as determined by selective inactivation of identified neurons. IV. Network properties of the pyloric system. *J. Neurophysiol.* **48**:1416–1432.

Paul, D. H. (1976) Role of proprioceptive feedback from nonspiking mechanosensory cells in the sand crab, *Emerita analoga*. *J. Exp. Biol.* **65**:243–258.

Pearson, K. G. (1976) Nerve cells without action potentials. In *Simpler Networks and Behavior*, J. C. Fentress, ed. Sinaur Associates, Sunderland, Mass, pp. 99–110.

Pearson, K. G. (1983) Neural circuits for jumping in the locust. *J. Physiol. (Paris)* **78**:765–780.

Pearson, K. G. and C. R. Fourtner (1975) Nonspiking interneurons in the walking system of the cockroach. *J. Neurophysiol.* **38**:33–52.

Pearson, K. G., W. J. Heitler, and J. D. Steeves (1980) Triggering of the locust jump by multimodel inhibitory interneurons. *J. Neurophysiol.* **43**:257–278.

Pearson, K. G. and J. F. Iles (1973) Nervous mechanisms underlying intersegmental coordination of leg movements during walking in the cockroach. *J. Exp. Biol.* **58**:725–744.

Peterson, E. L. (1983) Generation and coordination of heartbeat timing oscillation in the medicinal leech. I. Oscillation in isolated ganglia. *J. Neurophysiol.* **49**:611–626.

Raper, J. A. (1979) Nonimpulse-mediated synaptic transmission during the generation of a cyclic motor program. *Science* **250**:304–306.

Reichert, H., M. R. Plummer, G. Hagiwara, R. L. Roth, and J. J. Wine (1982) Local interneurons in the terminal abdominal ganglion of the crayfish. *J. Comp. Physiol.* **149**:145–162.

Reiss, R. F. (1964) *Neural Theory and Modelling*. Stanford University Press, Stanford.

Ripley, S. H., B. M. H. Bush, A. Roberts (1968) Crab muscle receptor which responds without impulses. *Nature* **218**:1170–1171.

Robertson, R. M. and M. Moulins (1981) Firing between two spike thresholds: implications for oscillatiing lobster interneurons. *Science* **214**:941–944.

Robertson, R. M. and M. Moulins (1984) Oscillatory command input to the motor pattern generators of the crustacean somatogastric ganglion. II. Gastric rhythm. *J. Comp. Physiol.* **154**:473–492.

Robertson, R. M. and K. G. Pearson (1983) Interneurons in the flight system of the locust: distribution, connections and resetting properties. *J. Comp. Neurol.* **215**:33–50.

Robertson, R. M., K. G. Pearson, and H. Reichert (1982) Flight interneurons in the locust and the origin of insect wings. *Science* **217**:177–179.

Rose, R. M. and P. R. Benjamin (1981) Interneuronal control of feeding in the pond snail *Lymnaea stagnalis*. II. The interneuronal mechanism generating feeding cycles. *J. Exp. Biol.* **92**:203–228.

Russell, D. F. (1979) CNS control of pattern generators in the lobster stomatogastric ganglion. *Brain Res.* **177**:598–602.

Russell, D. F. and D. K. Hartline (1978) Bursting neural networks: a reexamination. *Science* **200**:453–456.

Russell, D. F. and D. K. Hartline (1981) A multiaction synapse evoking both EPSPs and enhancement of endogenous bursting. *Brain Res.* **223**:19–38.

Shepherd, G. M. (1979) *The Synaptic Organization of the Brain*. Oxford University Press, Oxford.

Siegler, M. V. S. (1981) Posture and history of movement determine membrane potential and synaptic events in non-spiking interneurons and motor neurons of the locust. *J. Neurophysiol.* **46**:296–309.

Simmers, A. J. (1981) Non-spiking interactions in crustacean rhythmic motor systems. In *Neurones without Impulses*, A. Roberts and B. M. H. Bush, eds. Cambridge University Press, Cambridge, pp. 177–198.

Simmers, A. J. and B. M. H. Bush (1980) Non-spiking neurones controlling ventilation in crabs. *Brain Res.* **197**:247–252.

Simmers, A. J. and B. M. H. Bush (1983) Central nervous mechanisms controlling rhythmic burst generation in the ventilatory motoneurones of *Carcinus maenas*. *J. Comp. Physiol.* **150**:1–22.

Stein, P. S. G. (1971) Intersegmental coordination of swimmeret motoneuron activity in crayfish. *J. Neurophysiol.* **34**:310–318.

Stent, G. S., W. B. Kristan, W. O. Friesen, C. A. Ort, M. Poon, and R. L. Calabrese (1978) Neuronal generation of the leech swimming movement. *Science* **200**:1348–1357.

Tazaki, K. and I. M. Cooke (1979a) Spontaneous electrical activity and interactions of large and small cells in cardiac ganglion of the crab, *Portunus sanguinolentus*. *J. Neurophysiol.* **42**:975–999.

Tazaki, K. and I. M. Cooke (1979b) Ionic basis of slow, depolarizing responses in cardiac ganglion neurons of the crab, *Portunus sanguinolentus*. *J. Neurophysiol.* **42**:1022–1047.

Thompson, W. J. and G. S. Stent (1976) Neuronal control of heartbeat in the medicinal leech. III. Synaptic relations of the heart interneurons. *J. Comp. Physiol.* **111**:309–333.

Wachtel, H. and E. R. Kandel (1967) A direct synaptic connection mediating both excitation and inhibition. *Science* **158**:1206–1208.

Watanabe, A. (1958) The interaction of electrical activity among neurons of lobster cardiac ganglion. *Jap. J. Physiol.* **8**:305–318.

Watson, A. H. D. and M. Burrows (1983) The morphology, ultrastructure, and distribution of synapses on an intersegmental interneurone of the locust. *J. Comp. Neurol.* **214**:154–169.

Weeks, J. C. (1981) Neuronal basis of leech swimming: separation of swim initiation, pattern generation and intersegmental coordination by selective lesions. *J. Neurophysiol.* **45**:698–723.

Wilson, D. M. (1967) An approach to the problem of control of rhythmic behavior. In *Invertebrate Nervous Systems*, C. A. G. Wiersma, ed. University of Chicago Press, Chicago, pp. 219–230.

Wilson, D. M. and I. Waldron (1968) Models for the generation of the motor output pattern in flying locusts. *Proc. Inst. Elec. Electron. Eng.* **56**:1058–1064.

Wine, J. J. and Krasne, F. B. (1982) The cellular organization of crayfish escape behavior. In *The Biology of Crustacea*, Vol. 4, *Neural Integration*, H. L. Atwood and D. L. Sandeman, eds. Academic Press, New York, pp. 241–292.

Zucker, R. S. (1972) Crayfish escape behavior and central synapses. I. Neural circuit exciting lateral giant fiber. *J. Neurophysiol.* **35**:599–620.

Chapter Fourteen

NEURAL CONTROL OF THE VERTEBRATE LIMB

Multipartite Pattern Generators in the Spinal Cord

PAUL S. G. STEIN

Department of Biology
Washington University
St. Louis, Missouri

CENTRAL PATTERN GENERATORS IN THE SPINAL CORD

Observations that the spinal cord can produce the motor programs underlying coordinated motor rhythms, such as stepping and scratching, date back to the beginning of this century (Brown, 1911; Sherrington, 1906, 1910a,b). These early data were utilized to support the "half-center" hypothesis, a bipartite organizational model of the spinal cord centers controlling the vertebrate limb (Brown, 1911). Such a model lacks the complexity to account for the multipartite features of vertebrate limb motor rhythms revealed by recent electrophysiological recordings during both stepping and scratching (Grillner, 1981; Stein, 1983, 1984). These recent data are consistent with the "unit burst generator" hypothesis, a multipartite organizational model that has the complexity to account for the temporal organization of each of several forms of both scratching and stepping (Grillner, 1981).

The generalization that the central nervous system contains central pattern generators (CPGs) that can produce the motor neuron activation patterns underlying rhythmic behavior in the absence of phasic sensory feedback is now an established concept of motor neurophysiology (Delcomyn, 1980; Grillner, 1981; Stein, 1978, 1983, 1984). Extensive examinations of the motor behavior of the vertebrate limb have been important in the development of this generalization. As long ago as 1910, deafferentation of a limb by means of dorsal rhizotomies was utilized to block phasic sensory feedback from the responding limb (Brown, 1911; Sherrington, 1910a,b). Both stepping movements (Brown, 1911; Grillner and Zangger, 1975; Sherrington, 1910a; Szekely et al., 1969) and scratching movements (Deliagina et al., 1975; Sherrington, 1910b; Stein and Grossman, 1980) can be observed in a deafferented vertebrate limb. In recent years, immobilization with a neuromuscular blocking agent has been utilized as an alternative method of blocking phasic sensory feedback. The motor neuron activation patterns characteristic of both stepping (Grillner and Zangger, 1979; Jacobson and Hollyday, 1982b; Perret and Cabelguen, 1980) and scratching (Berkinblit et al., 1978 a,b, 1980; Deliagina et al., 1975, 1981; Robertson et al., 1982, 1985; Stein and Grossman, 1980; Stein et al., 1982) can be observed in these preparations. Since the motor patterns observed in the immobilized preparation occur in the absence of "real" movements, they have been termed "fictive" motor programs. Most recently, an *in vitro* preparation of the turtle spinal cord, peripheral nerves, and a tactile receptive field has been utilized to produce a fictive scratch reflex (Keifer and Stein, 1983).

The preparations that can produce fictive motor programs are particularly attractive to motor neurophysiologists since they allow cellular recordings from spinal neurons during motor program generation (Berkinblit et al., 1978a,b, 1980; Grillner, 1981; Jankowska et al., 1967; Perret and Cabelguen, 1980; Robertson et al., 1982; Stein et al., 1982). It is the hope of experimentalists that such preparations can be utilized to reveal the neuronal mechanisms and the functional organization of spinal CPGs. Investigators have recently revealed important characteristics of the temporal patterns of limb motor rhythms and have revised long-held views of the functional organization of spinal CPGs. These recently revealed characteristics will be important because they assist in the formulation of new working hypotheses of spinal motor organization. These hypotheses can be tested by direct recordings from spinal neurons during the production of limb motor rhythms.

GRAHAM BROWN'S "HALF-CENTER" HYPOTHESIS: A BIPARTITE MODEL

A long-held view of spinal motor control is that each half of the limb enlargement contains a pair of "half-centers" that are reciprocally linked with each other by way of inhibitory synapses (Brown, 1911). This hypothesis was initially articulated by Graham Brown in 1911 to explain both his data and data gathered by Charles Sherrington (1910 a,b). This hypothesis asserts (1) the existence of a "flexor half center," whose activity drives *all* the flexors of the limb simultaneously, (2) the existence of an "extensor half-center," whose activity drives *all* the extensors of the limb simultaneously, and (3) the flexor half-center inhibits and is inhibited by the extensor half-center. If both half-centers are tonically excited, then the reciprocal inhibition between the half-centers, when combined with "fatigue" of each half-center, is sufficient to generate rhythmic alternation between flexors and extensors. Anders Lundberg has modernized this hypothesis and gathered data demonstrating that under some conditions there is a paired half-center organization in the spinal cord (Jankowska et al., 1967; Lundberg, 1981).

RECENT DATA: MULTIPARTITE CENTRAL PROGRAMS

Although under some conditions a paired half-center organization does exist in the spinal cord, the difficulty with this concept is that it lacks the complexity to explain recent electrophysiological recordings obtained in several vertebrates during stepping in newt (Szekely et al., 1969), chick (Jacobson and Hollyday, 1982a,b), and cat (Engberg and Lundberg, 1969; Forssberg et al., 1980; Grillner, 1981; Grillner and Zanger, 1975, 1979; Halbertsma, 1983; Perret and Cabelguen, 1980) and during scratching in turtle (Keifer and Stein, 1983; Robertson et al., 1982, 1985; Stein, 1983; Stein and Grossman, 1980; Stein et al., 1982) and cat (Berkinblit et al., 1978a,b, 1980; Deliagina et al., 1981). In each of these preparations, it is not possible to classify the activity patterns of all the muscles of the limb utilizing a bipartite classification scheme of flexors and extensors. Hip flexors can be activated at a different time of the motor cycle, when compared to the activation time of knee flexors (Grillner, 1981; Jacobson and Hollyday, 1982a,b; Lundberg, 1981; Perret and Cabelguen, 1980). In some motor programs knee extensors are active at the same time as hip flexors (Mortin et al., 1982; Robertson et al., 1982, 1985; Stein and Grossman, 1980; Stein et al., 1982). *In most preparations there are at least three distinct phases of the motor program cycle; in some preparations there are so many phases of the cycle that the best description of the system is to state that it is multipartite* [e.g., see Fig. 8 of Deliagina et al. (1981)].

There are many reasons for this multipartite output; they can be understood best by movement analyses of the multijointed vertebrate limb during different types of motor rhythms. During such movement analyses several factors become clear. First, some joints of the limb have more than one degree of freedom, for example, the hip can move in the flexion-extension dimension, in the abduction-adduction dimension, and in the dimension of rotation-counterrotation of the femur. Although the latter two dimensions are often ignored in analyses of forward stepping in the cat, they cannot be ignored during turning movements in the cat or during forward locomotion in amphibians and reptiles. Second, flexion at one joint can occur at several different phases of the flexion-extension movement cycle of another joint, according to the demands of the motor task. During *forward* stepping, *flexion* of the knee must occur during early swing phase when the hip is also *flexing*; during *backward* stepping, *flexion* of the knee must occur during early swing phase when the hip is *extending* (Miller et al., 1978). Third, for movement in certain dimensions, such as flexion-extension of the hip, there may be an alternating output of agonist and antagonist. For movement in other dimensions, such as abduction-adduction of the hip, there can be either cocontraction of agonist and antagonist when the amount of movement in that dimension is small, or alternation of agonist and antagonist when the amount of movement in that dimension is large. When all these factors are considered together, it is clear that the motor output pattern during any one vertebrate limb motor program will be multipartite. Since a vertebrate limb can perform multiple motor tasks, for example, forward stepping, backward stepping, rostral scratching, and caudal scratching, it is clear that there must be many different program generators and that each program generator must be capable of producing the multipartite motor output characteristic of its motor program.

THE SPINAL CORD CAN PRODUCE MULTIPLE FORMS OF A BEHAVIOR

For a long time it has been known that a given type of motor behavior can be produced in several forms, for example, walking can be either forward or backward. Most experiments with spinal preparations have dealt with only one form of a given motor behavior, for instance, usually only forward locomotion or rostral scratching. In our own work on the spinal turtle with swimming motor rhythms produced by electrical stimulation of the spinal cord, we mainly studied forward swimming (Lennard and Stein, 1977). In these experiments we also observed that stimulation of some sites in the spinal cord could lead to backward swimming (backpaddling). During both forms of the swim, the propulsive stroke occurs when the foot is held vertically and the toes of the foot are spread. The propulsive stroke of forward swimming is produced while the hip is extending (retracting); the propulsive stroke of backward swimming is produced while the hip is flexing (protracting). During the initial portion of the returnstroke of the swim, the knee flexes and the foot is held horizontally. The returnstroke of forward swimming is produced when the hip is flexing; the returnstroke of backward swimming is produced when the hip is extending. Therefore the spinal cord must produce one phase relationship of hip flexion-extension compared with the rest of the limb movement during forward swimming; another phase relationship must be produced during backward swimming. These experiments with swimming demonstrate that the different forms of a motor program can be produced in the absence of supraspinal input to the spinal cord.

The demonstration that multiple forms of a motor behavior can be produced in a spinal preparation in the absence of phasic sensory feedback has been made recently in our studies of several forms of the scratch reflex in the spinal turtle (Mortin et al., 1982, 1985; Robertson et al., 1982, 1985; Stein, 1983). Earlier descriptions of the motor program during the scratch reflex in the turtle were confined to descriptions of the rostral scratch reflex, previously termed "scratch reflex" (Stein and Grossman, 1980; Stein et al., 1982). In each form of the scratch reflex, the motor program produced by the spinal turtle is site specific, that is, tactile stimulation of a site on the body surface will elicit a limb movement during which the limb reaches toward and rubs against the stimulated site. Stimulation of a site on the shell bridge located rostral to the limb will elicit a rostral scratch reflex in which the foot reaches toward and rubs against the stimulated site. During the rostral scratch reflex, the knee extends while the hip is flexed (protracted). Stimulation of a site at the base of the tail located caudal to the hindlimb will elicit a caudal scratch reflex in which the foot reaches toward and rubs against the stimulated site. During the caudal scratch reflex the knee extends after the hip is extended (retracted). A third form of scratch reflex can be elicited by stimulation within the pocket region. The pocket is located between the hip and the shell bridge. Stimulation of a site in the pocket will elicit a pocket scratch reflex in which the side of the thigh, knee, and calf rubs against the stimulated site. During the pocket scratch reflex, the knee extends while the hip is extending. Thus during each of these three forms of the scratch reflex there is a characteristic phase relationship of the knee extension movement within the hip flexion-extension cycle (Mortin et al., 1985). The characteristic phase relationship of each form can also be observed when muscle activation patterns are recorded (Mortin et al., 1982; Robertson et al., 1985). In addition, in the immobilized turtle, the characteristic phase relationship of each form can also be observed in the relationship of knee extensor motor neuron activity within the hip motor neuron activity cycle (Robertson et al., 1982, 1985). Each of the three forms of the scratch reflex is therefore cyclic, multipartite, and centrally programmed within the spinal cord. These data imply that the control center responsible for the generation of the motor rhythms of the vertebrate limb must be quite complex. Any working hypothesis of the functional organization of this control center must take into account the full motor capacity of the spinal cord. This means that the half-center hypothesis must be replaced with a more general hypothesis that can account for these complexities.

GRILLNER'S "UNIT BURST GENERATOR" HYPOTHESIS: A MULTIPARTITE MODEL

Recently Sten Grillner has formulated a hypothesis that deals with the motor complexity of the spinal cord (Grillner, 1981). In this hypothesis he notes that certain components of spinal cord CPGs can display half-center organization, for example, there is reciprocal inhibition between the portion of the control center (unit generator) activating hip flexion and the portion of the control center activating hip extension. In addition, he recognizes that different phase relationships can exist when one movement, for example, hip flexion, is compared with another movement, say, knee flexion, and therefore a separate control center must be postulated for each direction of each dimension of movement. Moreover, the Grillner model recognizes that under some conditions there can be repeated movements in one direction in the absence of any neural activity causing movements in the alternate direction. An example of this is the "B phase deletion" exhibited by the spinal turtle during some examples of the rostral scratch reflex (Stein and Grossman, 1980; Stein et al., 1982). In this behavior the limb performs multiple rhythmic hip flexions, combined with a properly phased knee extension, but there is no activation of hip extensor motor activity. Thus although a reciprocal relationship may exist between the hip flexor center and the hip extensor center, the expression of that rhythmicity is not necessary for motor rhythm production. The Grillner hypothesis therefore postulates that the unit generator controlling each direction of each dimension of movement has intrinsic rhythmicity.

The Grillner model can also deal with the several forms of a motor behavior. Grillner recognizes that the type of synaptic connections between the unit generators may change according to the demands of the motor task. For forward stepping, he postulates that there is reciprocal *excitation* between the hip flexor generator and the knee flexor generator; for backward stepping, there is reciprocal *inhibition* between the hip flexor generator and the knee flexor generator. The Grillner model is not the only one that has been proposed to replace the Brown model, for example, see Berkinblit et al. (1978b). The advantage of the Grillner model is that it is sufficiently general to deal with the recent results on vertebrate limb motor rhythms.

MULTIACTION MUSCLES AND MULTIPLE MUSCLES FOR AN ACTION

There are still other complexities that we must consider. The unit burst generator hypothesis of Grillner assumes that there is a control center for each direction of each dimension of movement. Some muscles of the vertebrate limb, for example, femorotibialis (vastus), act to produce only one direction of one dimension of movement, for example, knee extension. Most muscles of the vertebrate limb act across several dimensions of movement, for example, the iliotibialis muscle of the turtle serves to flex the hip, to abduct the hip, and to extend the knee. The Grillner model gives the testable prediction that an iliotibialis motor neuron should receive excitatory synaptic drive from the hip flexion center, the hip abduction center, and the knee extension center. In addition, intracellular recordings from motor neurons during fictive motor programs show that each motor neuron receives inhibitory synaptic drive during the portion of the motor cycle when the motor neuron is quiescent (Berkinblit et al., 1980; Grillner, 1981; Perret and Cabelguen, 1980; Robertson et al., 1982; Stein et al., 1982). We therefore have the additional testable hypothesis that an iliotibialis motor neuron can receive inhibition from the hip extensor center, the hip adductor center, and the knee flexor center. Which of these excitatory and inhibitory potentials will dominate the behavior of the motor neuron will of course depend upon the particular motor program that is generated. Grillner gives us the advantage of a testable hypotheses that can be examined in several preparations.

Although, for the muscles of the hip, knee, and ankle, it is possible that a single control center exists for each direction of each dimension of movement, a more complex situation exists in the foot and toes. Recently Burke and his co-workers (O'Donovan et al., 1982) completed a thorough examination of two anatomical synergists in the cat, the flexor digitorum longus and the flexor hallucis longus. Each of these muscles works in parallel to plantarflex the distal phalanges of the toes. The muscles have different histochemical compositions and are physiologically active at different times of the motor cycle. Whether histochemically different synergic motor units are active at different times of the movement cycle for other muscles has yet to be determined.

CONCLUSIONS

The vertebrate spinal cord has considerable motor organization, even when deprived of phasic sensory feedback and supraspinal neural information. One of the challenges for experimentalists who study motor control will be to elucidate the functional organization of this motor network and to determine what neuronal mechanisms are responsible for the generation of motor rhythms.

Acknowledgments

The author's research is supported by N.I.H. Grant NS-15049. I thank Gail Robertson and Lawrence Mortin for many helpful discussions and editorial assistance.

REFERENCES

Berkinblit, M. B., T. G. Deliagina, A. G. Feldman, I. M. Gelfand, and G. N. Orlovsky (1978a) Generation of scratching. I. Activity of spinal interneurons during scratching. *J. Neurophysiol* **41:**1040–1057.

Berkinblit, M. B., T. G. Deliagina, A. G. Feldman, I. M. Gelfand, and G. N. Orlovsky (1978b) Generation of scratching. II. Nonregular regimes of generation. *J. Neurophysiol.* **41:**1058–1069.

Berkinblit, M. B., T. G. Deliagina, G. N. Orlovsky, and A. G. Feldman (1980) Activity of motoneurons during fictitious scratch reflex in the cat. *Brain Res.* **193:**427–438.

Brown, T. G. (1911) The intrinsic factors in the act of progression in the mammal. *Proc. R. Soc. Lond. B* **84:**308–319.

Delcomyn, G. (1980) Neural basis of rhythmic behavior in animals. *Science* **210:**492–498.

Deliagina, T. G., A. G. Feldman, I. M. Gelfand, and G. N. Orlovsky (1975) On the role of central program and afferent inflow in the control of scratching movements in the cat. *Brain Res.* **100:**297–313.

Deliagina, T. G., G. N. Orlovsky, and C. Perret (1981) Efferent activity during fictitious scratch reflex in the cat. *J. Neurophysiol.* **45:**595–604.

Engberg, I. and A. Lundberg (1969) An electromyographic analysis of muscular activity in the hindlimb of the cat during unrestrained locomotion. *Acta Physiol. Scand.* **75:**614–630.

Forssberg, H., S. Grillner, and J. Halbertsma (1980) The locomotion of the low spinal cat. I. Coordination within a hindlimb. *Acta Physiol. Scand.* **108:**269–281.

Grillner, S. (1981) Control of locomotion in bipeds, tetrapods, and fish. In *Handbook of Physiology, Section 1: The Nervous System, Volume 2: Motor Control,* V. B. Brooks, ed. American Physiological Society, Bethesda, Md., pp. 1179–1236.

Grillner, S. and P. Zangger (1975) How detailed is the central pattern generator for locomotion? *Brain Res.* **88:**367–371.

Grillner, S. and P. Zangger (1979) On the central generation of locomotion in the low spinal cat. *Exp. Brain Res.* **34:**241–261.

Halbertsma, J. (1983) The stride cycle of the cat: the modelling of locomotion by computerized analysis of automatic recordings. *Acta Physiol. Scand. Suppl.* **251:**1–75.

Jacobson, R. D. and M. Hollyday (1982a) A behavioral and electromyographic study of walking in the chick. *J. Neurophysiol.* **48:**238–256.

Jacobson, R. D. and M. Hollyday (1982b) Electrically evoked walking and fictive locomotion in the chick. *J. Neurophysiol.* **48:**257–270.

Jankowska, E., M. G. M. Jukes, S. Lund, and A. Lundberg (1967) The effect of DOPA on the spinal cord. 5. Reciprocal organization of pathways transmitting excitatory action to alpha motoneurons of flexors and extensors. *Acta. Physiol. Scand.* **70:**369–388.

Keifer, J. and P. S. G. Stein (1983) *In vitro* motor program for the rostral scratch reflex generated by the turtle spinal cord. *Brain Res.* **266:**148–151.

Lennard, P. R. and P. S. G. Stein (1977) Swimming movements elicited by electrical stimulation of turtle spinal cord. I. Low-spinal and intact preparations. *J. Neurophysiol.* **40:**768–778.

Lundberg, A. (1981) Half-centres revisited. *Adv. Physiol. Sci.* **1:**155–167.

Miller, S., D. Mitchelson, and P. D. Scott (1978) Coupling of hip and knee movement during forwards and backwards stepping in man. *J. Physiol. (Lond.)* **277:**45P–46P.

Mortin, L. I., J. Keifer, and P. S. G. Stein (1982) Three forms of the turtle scratch reflex. *Abstr. Soc. Neurosci.* **8:**159.

Mortin, L. I., J. Keifer, and P. S. G. Stein (1985) Three forms of the scratch reflex in the spinal turtle: movement analyses. *J. Neurophysiol.* **53**:1501–1516.

O'Donovan, M. J., M. J. Pinter, R. P. Dum, and R. E. Burke (1982) Actions of FDL and FHL muscles in intact cats: functional dissociation between anatomical synergists. *J. Neurophysiol.* **47**:1126–1143.

Perret, C. and J. M. Cabelguen (1980) Main characteristics of the hindlimb locomotor cycle in the decorticate cat with special reference to bifunctional muscles. *Brain Res.* **187**:333–352.

Robertson, G. A., J. Keifer, and P. S. G. Stein (1982) Central programs for three forms of the turtle scratch reflex. *Abstr. Soc. Neurosci.* **8**:159.

Robertson, G. A., L. I. Mortin, J. Keifer, and P. S. G. Stein (1985) Three forms of the scratch reflex in the spinal turtle: central generation of motor patterns. *J. Neurophysiol.* **53**:1517–1534.

Sherrington, C. S. (1906) Observations on the scratch-reflex in the spinal dog. *J. Physiol. (Lond.)* **34**:1–50.

Sherrington, C. S. (1910a). Flexion-reflex of the limb, crossed extension reflex, and reflex stepping and standing. *J. Physiol. (Lond.)* **40**:28–121.

Sherrington, C. S. (1910b) Notes on the scratch-reflex of the cat. *Q. J. Exp. Physiol.* **3**:213–220.

Stein, P. S. G. (1978) Motor systems, with specific reference to the control of locomotion. *Annu. Rev. Neurosci.* **1**:61–81.

Stein, P. S. G. (1983) The vertebrate scratch reflex. *Symp. Soc. Exp. Biol.* **37**:383–403.

Stein, P. S. G. (1984) Central pattern generators in the spinal cord. In *Handbook of the Spinal Cord, Vols. 2 and 3: Anatomy and Physiology,* R. A. Davidoff, ed. Marcel Dekker, New York, pp. 647–672.

Stein, P. S. G. and M. L. Grossman (1980) Central program for scratch reflex in turtle. *J. Comp. Physiol.* **140**:287–294.

Stein, P. S. G., G. A. Robertson, J. Keifer, M. L. Grossman, J. A. Berenbeim, and P. R. Lennard (1982) Motor neuron synaptic potentials during fictive scratch reflex in turtle. *J. Comp. Physiol.* **146**:401–409.

Szekely, G., G. Czeh, and G. Voros (1969) The activity pattern of limb muscles in freely moving and deafferented newts. *Exp. Brain Res.* **9**:53–62.

Chapter Fifteen

CONTROL OF MULTIJOINT MOVEMENT

EMILIO BIZZI and WILLIAM K. ABEND

Department of Psychology
Massachusetts Institute of Technology
Cambridge, Massachusetts

An important goal in the study of arm movements is to understand what mechanical variables are controlled by the motor system. Studies of the mechanical properties of the musculoskeletal apparatus may be useful in gaining insights into the "rules" of the neural controller. This approach is based on the assumption that the neural control system has developed not only to control but also to take advantage of the musculoskeletal apparatus. The hope is that the information gained by an analysis of muscle mechanical properties will lead to a deeper understanding of the neuronal mechanisms subserving motor control.

In the following sections we first describe some studies of a single joint, the results of which have led to the notions of *final position control* and *reference trajectory*. We then discuss the control of multiple-degree-of-freedom movements. In considering multijoint movements, we take into consideration how single-degree-of-freedom movements are controlled. However, it seems unlikely that the control of polyarticular movements involves simply a concatenation of several monoarticular control modules since, as discussed, multiple-degree-of-freedom movements involve new control problems that do not arise in the single-degree case. Because of these new problems the multiple-degree case is richer experimentally than the single-degree situation. Introducing the full complexity of movement into the experimental paradigm has led to the formulation of a new concept of posture and movement. This concept subsumes what is known about single-degree movement characteristics, and single-degree control is handled as a special case for the multijoint controller.

OPEN-LOOP CONTROL OF FINAL POSITION

This section summarizes experiments directed at understanding some of the mechanisms subserving single-degree-of-freedom movements. As discussed later, the results of initial experiments on head movements were replicated in the context of elbow movements (Bizzi et al., 1976, 1978; Polit and Bizzi, 1979). The studies indicate that an animal can execute a simple pointing movement and maintain a new position in the absence of proprioceptive feedback. These findings have important implications regarding the functional relationship between descending commands and posture.

Monkeys were trained to make coordinated, horizontal eye-head movements directed at visual targets. In the intact animal, the unexpected application of a constant load was followed by an increase in electromyographic (EMG) activity of the neck muscles, presumably due to an increase in muscle spindle and tendon organ activity. As shown in Figure 1, despite these changes in the flow of proprioceptive activity, the head reached its "intended" final position *after* the constant load was removed, a fact suggesting that the program for final position was maintained during load application and was not readjusted by proprioceptive signals acting at segmental or suprasegmental levels. On the basis of this result, Bizzi et al. (1976) concluded that the central program establishing final position is not dependent on a readout of proprioceptive afferents generated during the movement but is preprogrammed. It should be stressed that the load disturbances were totally unexpected and that the monkeys were *not* trained to move their head to a certain position, but chose to program a head movement together with an eye movement to perform a visual discrimination task (Bizzi et al., 1976).

In a second set of experiments Bizzi et al. (1976) examined the effect of stimulating proprioceptors only during head movements. To this end, an inertial load was used to modify the trajectories without causing a steady-state disturbance (Fig. 2). As a result of the sudden and unexpected increase in inertia during centrally initiated head movements a number of changes in head trajectory, relative to unloaded movements, were observed. An initial decrease in head speed, due to the inertial load, was followed by a relative increase in speed (which resulted from the kinematic energy acquired by the load being transmitted to the decelerating head). The increased speed led to an overshoot in head position, and this was followed by a return to the intended position (Fig. 2).

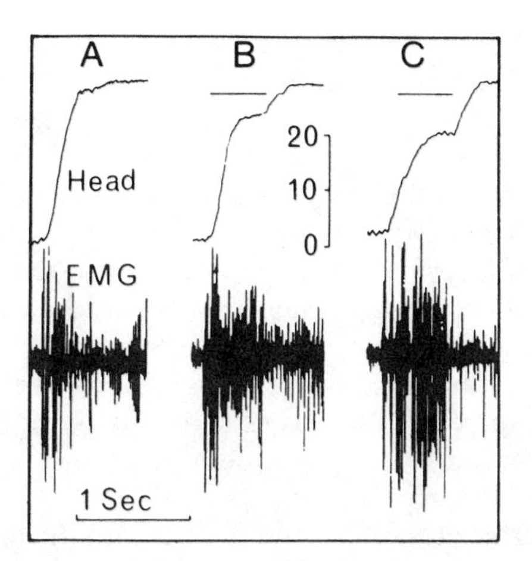

FIGURE 1. Typical visually triggered horizontal head movements in chronically vestibulectomized monkey to appearance of target at 40° but performed in total darkness. (*A*) An unloaded movement. (*B*) A constant-force load (315 gm-cm) was applied at the start of the movement, resulting in an undershoot of final position relative to (*A*) despite increase in EMG activity. (*C*) A constant-force load (726 gm-cm) was applied. Note head returns to same final position after removal of the load. Vertical calibration in degrees; time marker is 1 sec; EMG recorded from left splenius capitis. [From Bizzi, Polit, and Morasso (1976) *J. Neurophysiol.* **39**:435–444.]

The changes in head trajectory brought about by the sudden and unexpected increase in head inertia induced corresponding modifications in the length and tension of the neck muscles. The agonist muscles were, in fact, first shortened, and then the shortening of the same muscles was facilitated during the overshoot phase of the head movement induced by the kinetic energy of the load. Such loading and unloading did, of course,

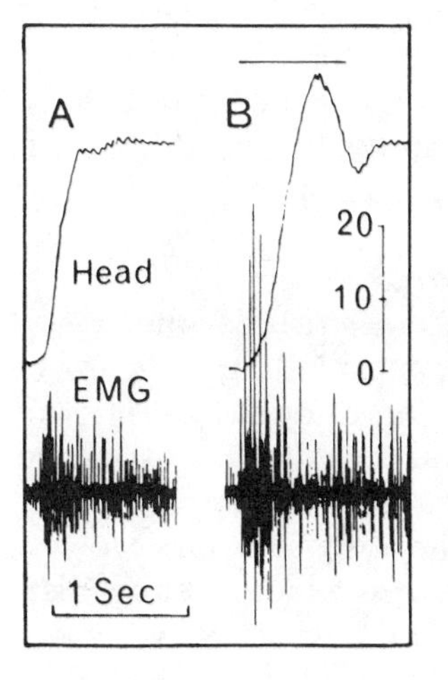

FIGURE 2. Typical head responses of a chronically vestibulectomized monkey to sudden appearance of target at 40°. (*A*) An unloaded movement; (*B*) a load of approximately six times the inertia of the head applied at the start of the movement, as indicated by the force record. Both movements were performed in total darkness, the light having been turned off by the increase in EMG (splenius capitis). Peak force exerted by the monkey is approximately 750 gm-cm; head calibration is in degrees; time marker is 1 sec. [From Bizzi, Polit, and Morasso (1976) *J. Neurophysiol.* **39**:435–444.]

provoke the classical muscle spindle response mediated by group Ia and group II afferent fibers which, in turn, affected the agonist EMG activity. Figure 2*B* shows that there was first a greater increase in motor unit discharge during muscle stretch than would have occurred if no load were applied, followed by a sudden decrease in activity at the beginning of the overshoot phase. Therefore, during a head movement, an unexpected inertial load induced a series of waxing and waning proprioceptive signals from muscle spindles, tendons, and joints, but the *intended head position* was eventually reached, even in the complete absence of other sensory cues (visual and vestibular). This observation, together with those on the effect of constant-torque loads, suggests that the central program establishing final head position is not dependent on a readout of proprioceptive afferents generated during the movement but, instead, is preprogrammed.

To provide a further test of the hypothesis that the final head position is preprogrammed, Bizzi et al. (1976) investigated the attainment of final head position when monkeys were deprived of neck proprioceptive feedback. To ensure "open loop" conditions, the animals were vestibulectomized two to three months previous to the deafferentation [vestibulectomized monkeys recover eye-head coordination (Dichgans et al., 1973)]. After deafferentation, the animals were still able to make accurate visually evoked responses (Bizzi et al., 1976). The experiment dealt with a constant torque applied during centrally initiated movements (Fig. 3). Just as with intact animals, when the load was applied unexpectedly at the beginning of a visually triggered movement, the posture attained by the head fell short of the intended final position. After removal of the constant torque, the head attained a position that was not significantly different from the one reached by the head in the no-load case (Fig. 3). These results can be explained by modeling the neck muscles as opposing springs. The head will lie in a position at which the forces exerted by the agonist and antagonist springs are equal and opposite. Of course, if the head is moved to a new position by an external force, but the length-tension properties of the springs are not changed, then, when the external force is removed, the head will move back to the original position. This resting position could be more permanently modified by changing the stiffness (obtained by dividing the distance a spring is stretched into the magnitude of the resulting force increment) or the rest length of one or both of the springs, as described by Hooke's law. Just as with springs, the force exerted by a muscle increases as the muscle is stretched; the spring characteristics of the muscle are adjusted by changing the neural input (Rack and Westbury, 1969). We therefore postulate that the motor program specifies, through a selection of a set of length-tension properties of agonist and antagonist muscles, an equilibrium point between these two sets of muscles that correctly positions the head in relation to a visual target. In the absence of other forces, the final head position will be determined by that interaction of agonists and antagonists which results in a position where the tensions on the two sets of muscles are equal and opposite (Asatryan and Feldman, 1965; Feld-

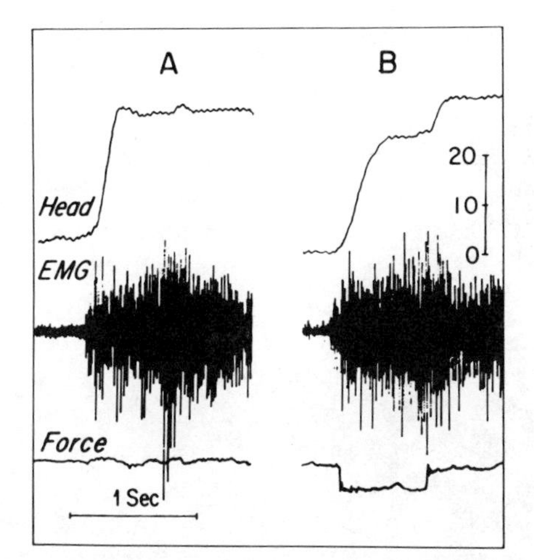

FIGURE 3. Typical movements of a chronically vestibulectomized monkey, with sectioned dorsal roots (C1-T3) made open-loop (in total darkness). (*B*) constant-force load (315 gm-cm) was applied at the start of movement, resulting in an undershoot while the load was on. Similarity of EMG pattern in (*A*) and (*B*) shows lack of a stretch reflex. Peak force in (*B*) approximately 315 gm-cm. Vertical calibration in degrees. [From Bizzi, Polit, and Morasso (1979) *J. Neurophysiol.*, **39**:435–444.]

man, 1974a, b; Rack and Westbury, 1969). Given this model, it is not surprising that the head overshoot during inertial loading is corrected with a return movement to the intended position, because an increase in antagonist tension and, hence, a return head movement (Fig. 2*B*). By the same token, because head position is related to muscle length and the load, an undershoot is observed when a constant external opposing torque is applied (Figs. 1 and 3). The same hypothesis explains why the head moves to the intended final position when the constant torque is removed.

Thus it seems that final head position in both intact and deafferented preparations should be viewed as an equilibrium point dependent on the firing rate of the alpha motoneurons (MN) innervating agonists and antagonists; the length-tension properties of the muscles involved in maintaining the posture; the passive, elastic properties of the musculoskeletal apparatus; and the external load. In the intact animal, however, in parallel with this basic process, the proprioceptive system participates in the process of reaching final position by increasing muscle stiffness when a load disturbance is applied (Rack and Westbury, 1969). In fact, any stimulation of the proprioceptive apparatus, by virtue of its reflex connections, will modify the firing rate and the recruitment of alpha MNs and therefore force the selection of a new length-tension curve (Rack and Westbury, 1969).

In a complementary set of experiments involving forearm movements performed by rhesus monkeys, Polit and Bizzi (1979) extended the previously described findings on the final position control of the head. The monkey's forearm was fastened to an apparatus that permitted flexion and

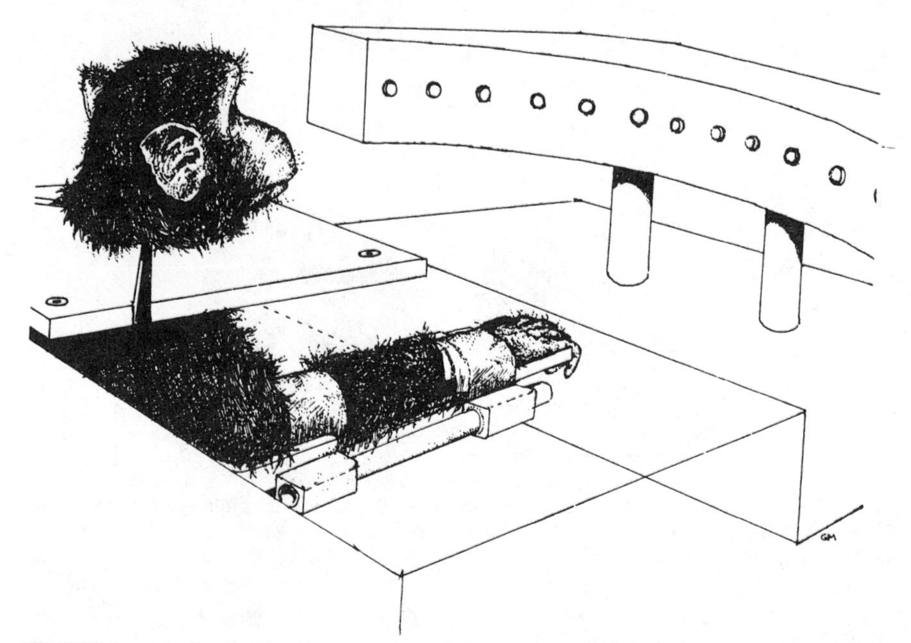

FIGURE 4. Monkey set up in arm apparatus. Arm is strapped to splint, which pivots at elbow. Target lights are mounted in perimeter arc at 5° intervals. During the experimental session the monkey was not permitted to see its arm, and the room was darkened. [From Polit and Bizzi (1979) *J. Neurophysiol.* **42:**183–194.]

extension about the elbow in the horizontal plane (Fig. 4), and the monkey was trained to point at target lights with the forearm. Several target lights were spaced at 10° intervals along an arch, which was centered on the axis of rotation of the elbow. The monkey was trained to move the forearm into a 12–15° wide target zone centered on the illuminated target light and hold that position for roughly one second. Experiments were conducted in a darkened room so that the monkey saw only the target lights. A torque motor in series with the elbow pivot was used to apply positional disturbances to the arm. On random pointing trials, the initial position of the forearm was displaced. Despite these changes, the intended final arm position was always attained; this was true whether the torque motor had displaced the forearm further away from, closer to, or even beyond the intended final position. There were no significant differences among the final positions achieved in these three conditions.

Naturally, the attainment of the intended arm position of this experiment could be explained by assuming that afferent proprioceptive information modified the original motor command. However, the results of previous work on final head position suggest an alternative hypothesis: that the motor program underlying arm movement specifies, through the selection of a new set of length-tension curves, an equilibrium point between agonists and antagonists that correctly positions the arm in relation to the visual target. To investigate this hypothesis, Polit and Bizzi (1979), retested the monkeys' pointing performance after they had undergone a bilateral C1-T3 dorsal rhizotomy (Taub et al., 1965, 1966, 1975). The animals were again required to produce pointing movements in an "open-loop" mode, as no proprioceptive activity could reach the spinal cord, and visual feedback of the arm position was not present. Under these conditions, the animals could still produce pointing responses very soon after surgery (within two days in one case). The forearm was again displaced (at random times) immediately after the appearance of the target light and released just prior to the activation of motor units in the agonist muscles. For each target position, t-tests were performed for differences between the average final position of movements with undisturbed and disturbed initial positions. No significant differences were found. These observations suggest that, as in the case of head position control, the final forearm position is directly programmed through alpha MN activity, which selects the appropriate length-tension relationship for each of the muscles involved in the movement. The final limb position is reached when the tension generated by the agonists is equal and opposite to that generated by the antagonists. This view implies that, for each limb position, particular levels of alpha MN activity to the agonist muscles correspond to particular levels of activity to the antagonists.

A recent study of the EMG of agonist and antagonist muscles acting on the elbow and wrist joints in humans showed that a change from one posture to another involved a modulation of EMG activity in both flexors and extensors. Although the tonic EMG activity of flexors and extensors at each posture was variable, the *ratio* between the alpha neuronal inputs to the muscles was significantly less variable, and no effect of the direction, amplitude, or velocity of the movement was detected (Lestienne et al., 1981). Once the final position was reached, a dynamic characteristic of the program underlying this agonist-antagonist innervation was noted. There was usually a progressive attenuation in the agonist and antagonist EMG activity, without any change in final arm position (Polit and Bizzi, 1979; Bizzi, Prablanc, and Hogan, in preparation). This finding indicates that the central programmer might gradually select a series of length-tension curves for agonists and antagonists that may perhaps differ in slope but all specify the same final position.

It is tempting to speculate that this representation of posture as an equilibrium point between agonist and antagonist length-tension curves (see schematic representation of length-tension curves in Fig. 5) also had implications for movement (Feldman, 1974a, b; Cooke, 1979). If the central nervous system (CNS), were abruptly to specify new length-tension relationships (through a change in alpha MN activity) for the muscles, movement would occur until a new equilibrium point was reached. Clearly, the suggestion that the CNS may control simple movements by specifying *only* final position is attractive because, in this way, a single process would

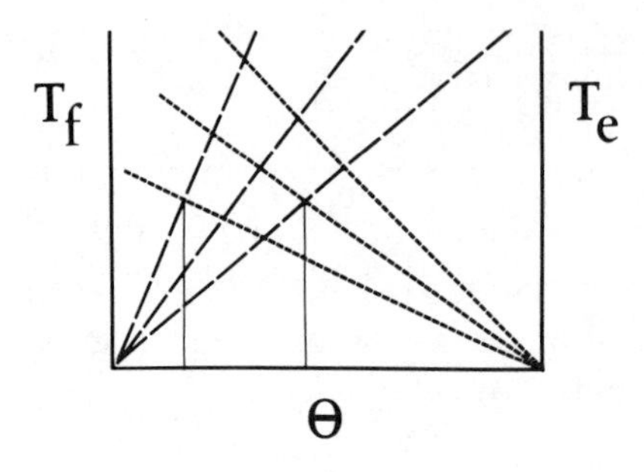

FIGURE 5. Schematic representation of flexor T_f and extensor T_e length-tension curves. Θ represents joint angle. [From Bizzi, Accornero, Chapple, and Hogan (1982) *Exp. Brain Res.* **46**:139–143.]

subserve both posture and movement (Kelso and Holt, 1980; Sakitt, 1980). The details of the trajectory would, in fact, be determined only by the inertial and visco-elastic properties of the limbs and muscles. However, recent experimental findings reviewed in the next section, indicate that the CNS actively controls the trajectory in addition to the final position.

TRAJECTORY FORMATION IN SINGLE-DEGREE-OF-FREEDOM ARM MOVEMENTS

The goal of this series of experiments was the determination of whether the final position control model is sufficient to account for all of the characteristics of elbow movements. This has been investigated by determining the time course of the neural signals executing the transition from an initial to a final position. If the transition to the final alpha MN levels is achieved briskly (compared to the time course of the development of a motor unit twitch in response to a single action potential), then the speed of the movement could not be centrally controlled but would be determined only by the inertial and viscoelastic properties of the musculoskeletal apparatus. Alternatively, if the change to the final alpha MN signal occurs slowly, this would indicate that there is active central control of the trajectory of the movement in addition to control of the final position.

Monkeys performing a pointing task similar to that described in the previous sections were studied (Fig. 4). Again, the main experimental procedure involved the use of force and positional disturbances which were applied with a torque motor coupled to the shaft of the pivot arm on which the elbow rested (Fig. 4). In some experiments the animal was prevented from detecting disturbances of forearm position by surgically interrupting sensory roots conveying afferent activity from the arm, neck, and upper torso (C1 to T3). After deafferentation, the forearm pointing responses, which had been learned in the preoperative state, could be easily evoked by presenting the targets (Bizzi et al., 1981, 1982, 1984). The movements were similar to those observed before deafferentation. As in the preoperative recordings, the EMG activity usually appeared as a moderate burst that gradually blended with the tonic activity characteristic of the holding phase. Although it is entirely possible that deafferentation, like any other CNS lesion, might have induced modifications in motor programming, the fact that we obtained similar results in intact and deafferented animals suggests that the same basic mechanism continued to control these simple movements.

Three experimental paradigms were used to determine the time course of the alpha MN signals. The first two experimental manipulations relied on quickly forcing the forearm into the upcoming new final position. As the intact animal began its movement toward a new target position, a brief torque pulse (150 msec), whose onset was triggered by the initial increase in EMG in the agonist muscles, drove the elbow quickly to the intended final position (Fig. 6*B* and *C*). However, instead of remaining in this position, the forearm returned to an intermediate point between the initial and

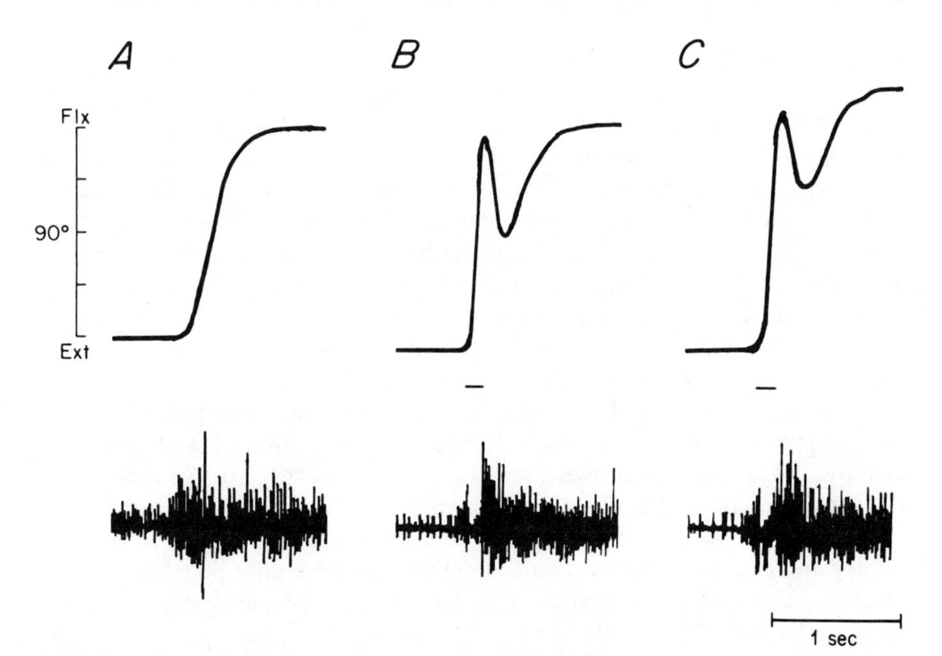

FIGURE 6. Torque pulse assisting the natural movement of the arm toward the target position. Upper trace: arm position with an elbow angle of 90° at the scale midpoint flexion upward; lower trace: flexor (biceps) EMG. Bar beneath the position trace indicates duration of the torque pulse. (*A*) Control movement in the absence of a torque pulse. (*B*) and (*C*) Two movements with assisting torque pulses. The arm reaches the intended target position early in the movement, transiently returns to an intermediate position, and then proceeds to the final equilibrium position. Note the unloading reflex in the EMG trace in (*B*) and (*C*). Flx, flexion; Ext, extension. [From Bizzi and Abend (1983) *Motor Control Mechanisms in Health and Disease*, J. E. Desmedt, ed. Raven, New York.]

final positions before reversing direction again and then continuing its trajectory toward the position specified by the target. Note that the return movement to the intermediate position was in the direction of extension while the EMG activity was present in flexor muscles. This experiment suggests that these simple forearm movements do not result from rapid shifts in the equilibrium point. According to the hypothesis of "final position control," we would expect the steady-state equilibrium position to be achieved after a delay due only to the dynamics of muscle contraction. Because individual motor units, recruited at low levels of force, reach their peak force in 60–80 msec after the onset of the action potential in the muscle fiber (cf. Milner-Brown et al., 1973; Collatos et al., 1977; Desmedt, 1981), we would expect that the net muscle force rises to nearly its final value in roughly 200 msec (Hogan, 1984). However, our results indicate that for a movement of at least 600 msec duration (Fig. 6), the mechanical expression of the alpha MN activity does not reach steady state until at least 450 msec have elapsed following the onset of action potentials in the muscle.

Similar conclusions can be drawn from a second experiment that was performed in deafferented animals. The torque motor was used to suddenly displace and maintain the arm to what would be the location of the next target (Fig. 7*A*). The animals could not have expected a reward, as no new target was illuminated. In fact, because of the absence of any proprioceptive or visual information regarding arm position, the animals were unaware of the displacement. Now, with the arm still constrained by the servomotor, the target light corresponding to the new arm position was illuminated. To the trained monkey, the appearance of this target light represented a signal to start the neural process involved in pointing to the target. This process became manifest through the appearance of EMG activity in the proper set of muscles, after the usual reaction time (Fig. 7*A*). After a predetermined time had elapsed following the onset of the EMG activity, the torque motor was turned off, releasing the arm. At this point, the arm was in exactly the correct position for receiving a reward. It is therefore remarkable that the forearm did not remain stationary. Instead, it moved toward the position from which it had originally been displaced, and then changed direction and returned to the position specified by the target light (Fig. 7*A*). While the to-and-fro movement took place, the agonist muscle developed an EMG pattern comparable to that observed during normal, undisturbed movements. Thus, in the presence of flexor muscle activity, we observed movement in an extensor direction. This remarkable finding cannot be explained if muscles are regarded purely as force generators, but it is readily explained if the length dependence of muscle force is taken into account. It should be pointed out that if alpha MN activity evoked by the target light had rapidly achieved levels appropriate for the new final position, then no return movement should have taken place (see schematic Fig. 5). The fact that a return movement *did* occur indicates that

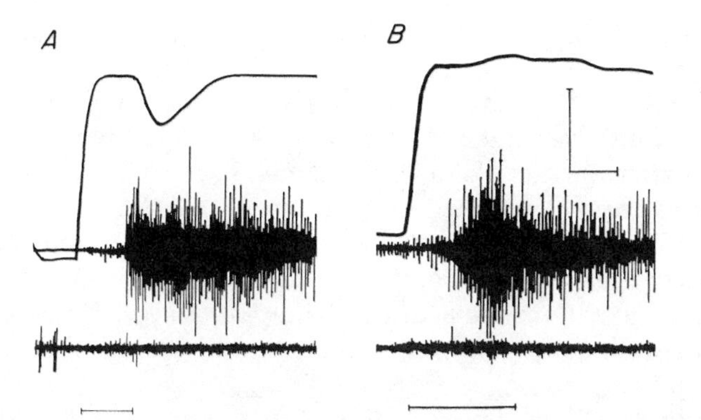

FIGURE 7. (*A*) Forearm movements of a deafferented animal in the absence of visual feedback. Displacement of the forearm to a flexion position at which a target light was displayed. At the termination of the servomotor action (indicated by horizontal bar), note the movement of the forearm toward extension and subsequent return to the position specified by the target. Flexor activity evoked by the target light is similar to that observed during undisturbed movements of same amplitude. Lower EMG corresponds to extensor muscles. Two-hundred instances of this behavior have been observed in two animals. (*B*) Same animal. Displacement of the forearm to a flexion position at which a target light was displayed. No movement of the forearm at the termination of servomotor action. EMG activity from flexors is triggered by the appearance of the light and is similar to that observed during undisturbed movements. Lower EMG corresponds to extensor muscles. There have been 220 instances like the one displayed here observed in two monkeys. Time calibration, 500 msec. Vertical bar represents joint angle position, 20°. [From Bizzi, Accornero, Chapple, and Hogan (1982) *Exp. Brain Res.*, **46**:139–143.]

the control signal shifted slowly toward the final position. This conclusion is consistent with the observation that the amplitude of the movement toward the starting position decreased as the period of servo restraint of the arm was prolonged. Finally, when the servo action was maintained after the appearance of the evoked EMG activity for a period corresponding to the normal movement duration, then no significant movement of the arm was observed after it had been released by the servo (Fig. 7*B*). These findings suggest the existence of a gradually changing control signal during movement of the forearm from one position to the next and are not consistent with a view postulating a steplike shift to a final equilibrium point. Beyond this, these experiments indicate that the neural input to the muscles can be interpreted as specifying an equilibrium position, plus a stiffness about the equilibrium position. Thus, in the transition from the initial to the final position, the alpha MN activity is defining intermediate equilibrium positions, which constitute a reference trajectory whose end point is the desired final position. As a result, following the cessation of the assist pulse (see Fig. 6), or at the cessation of torque motor action (as in Fig. 7), the limb heads for an intermediate position before reaching its final termination.

In a third set of experiments further confirmation was obtained of the view indicating a gradual change in the control signal establishing the final equilibrium point. Both intact and deafferented animals were used and the following procedures adopted. Before the onset of visually triggered movements, the limb was clamped in its initial position; it was released at various times after the onset of evoked agonist EMG activity. The duration of the holding phase was varied randomly from 100–600 msec. The acceleration of the limb immediately after the release was measured and plotted as a function of the holding time, that is, the time elapsed since the beginning of the EMG activity in the agonists. The plot of acceleration in intact and deafferented monkeys showed a gradual increase for holding times up to 400–600 msec.

These findings are consistent with the notion of a reference trajectory that specifies a gradual shift in the equilibrium point. As the equilibrium point moves farther away from the position at which the limb had been restrained, progressively larger torques are generated, resulting in progressively increasing values of acceleration following release and progressively faster movement trajectories. These findings are not in accord with the hypothesis postulating that arm trajectory is controlled by a simple rapid shift to a final equilibrium position. According to this hypothesis, the trajectory of the released arm should have been simply delayed, but its shape should not have been affected.

Physiologically, we do not know how the gradual shift in equilibrium point is programmed. Some of the factors responsible for the progressive increase in tension are the twitch contraction time of the muscle fibers (about 80 msec, see Collatos et al., 1977), the recruitment order, and the firing rate of arm muscle MN (Desmedt and Godaux, 1978; Freund and Budingen, 1978; Henneman, 1965; Tanji and Kato, 1973).

It should be emphasized that the results described here have been obtained by analyzing forearm movements performed at moderate speeds. In very fast movements, the shift in equilibrium point must be more abrupt and may even transiently involve a shift to a position beyond the intended equilibrium point. This would amount to a pulse step command of the type known to control eye movements and fast limb movements (Desmedt and Godaux, 1978; Freund and Budingen, 1978; Ghez and Vicario, 1978a,b; Robinston, 1964). Thus it is conceivable that, for simple one-degree-of-freedom movements such as those discussed in the first and second sections, one of the roles of the "reference trajectory" is to specify movement speed. The notion of reference trajectory is taken up again in the context of multijoint arm movements.

TRAJECTORY FORMATION IN MULTIPLE-DEGREE-OF-FREEDOM ARM MOVEMENTS

We now wish to apply the ideas derived from studies of simple one-degree-of-freedom movements to the multijoint context. Over a long time, there were only a few studies of how the CNS coordinates the large number of degrees of freedom of movement of the human arm (Marey, 1901; Muybridge, 1901; Bernstein, 1967), but since 1966 there has been an increasing interest in this area (Abend, et al., 1982; Amassian and Eberle, 1982; Amassian, et al., 1982; Delatizky, 1982; Flash and Hogan, 1982; Georgopoulos et al., 1981, 1982, 1983; Gilman et al., 1976; Hocherman et al., 1983; Hofsten, 1979; Hollerbach, 1982; Hollerbach and Atkeson, 1983; Hollerbach and Flash, 1982; Kalaska et al., 1983; Kots and Syrovegin, 1966; Lacquaniti and Soechting, 1982; Morasso, 1981, 1983; Morasso and Mussa-Ivaldi, 1982; Soechting and Lacquaniti, 1981; Soechting and Ross, 1983; Viviani and Terzuolo, 1980). One reason for this interest is that most natural movements involve simultaneous rotations about multiple joints. In addition, multijoint movements involve new control issues not present in the case of one-joint movements. The experimental situation is therefore richer, and the possibility of new insights is created. We begin by discussing these new problems and then discuss how the single-joint control concepts can be brought to bear on the problems of multijoint control.

Multijoint arm movements involve kinematic and dynamic issues that do not arise in the case of one-degree-of-freedom movements (Hollerbach, 1982; Horn and Raibert, 1977; Raibert, 1976). The kinematic problem is that of trajectory formation. *Trajectory* refers to the path taken by the hand as it moves from one location in movement space to another and the speed with which the hand moves along the path. For a one-joint movement about the elbow with the wrist fixed, the path of the hand is mechanically constrained to an arch with a radius equal to the length of the forearm. Consequently the speed and stiffness profiles of the movement must be controlled, but only the final position of the path need be specified. If, in addition, rotation about the shoulder joint is allowed, then the hand could approach a target along any of a vast number of paths. Now the controller must plan the entire path in addition to the speed and stiffness characteristics, and the path, speed, and stiffness of the hand will be the result of a complex interplay of the effects of the rotation of each of the involved joints.

Three differences between the mechanical properties of multijoint movements and those of one-degree-of-freedom movements are of particular interest. *First,* in multijoint movements the torque required to move one joint is dependent on the position of the other joints. *Second,* there are joint interactional effects that result from muscles that span more than one joint (e.g., biceps brachii). *Third,* the dynamic formulations of multijoint movements contain interaction terms that are of purely mechanical origin. As a result of these cross-coupling effects, the rotation of one joint due to shortening of a muscle anatomically associated with that joint will effect movements of all of the other joints in the linkage. One interaction torque results from reactional forces, which are proportional to joint acceleration. As illustrated in Figure 8*A*, if joint 1 is caused to rotate in the indicated direction, a reaction torque will act on joint 2. Similarly, joint 1 will rotate if a

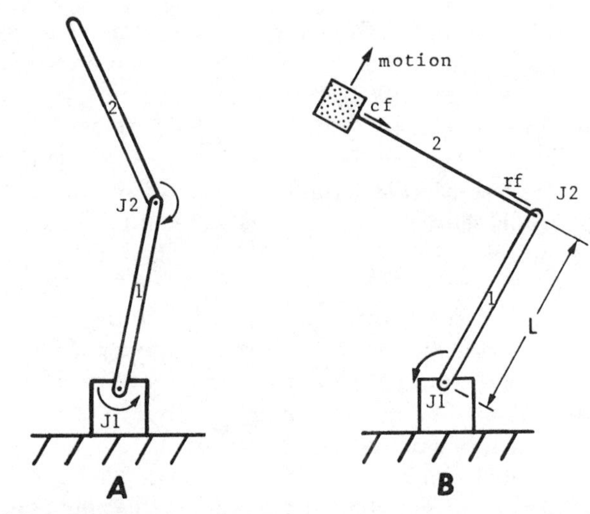

FIGURE 8. Schematic illustration of interaction torques. (*A*) Horizontal two-link arm is fixed at one end to a wall by way of joint 1 (J1). Second degree of freedom is provided by joint 2 (J2). If J1 is caused to rotate in indicated direction, a reaction torque will result in J2 motion. (*B*) Motion at J1 caused by "centripetal" interaction torque (see text); rf, reaction force; cf, centripetal force. [From Bizzi and Abend (1983) *Motor Control Mechanisms in Health and Disease*, J. E. Desmedt, ed., Raven, New York.]

torque is produced at joint 2. Figure 8*B* illustrates the "centripetal" interaction torque. Here, link 2 is assumed to have been caused to rotate about joint 2 and is represented as a mass at the end of a rotating cable; the centripetal force acting on the mass establishes a reaction force (proportional to the square of the joint-2 angular velocity), which acts on joint 1 through lever arm L. A third interaction torque, referred to as the Coriolis torque, is proportional to the product of the angular velocities of the two joints and acts about joint 1. The magnitudes of all these interactional forces are affected by the particular trajectory of the arm and can be quite large (Hollerbach and Flash, 1982). If there is no provision for these coupling effects, motion about one joint would cause other joints to flail, so that errors in joint motion and hand motion would occur. No such behavior is observed in normal subjects. Even when the speed of movement to a target is varied, so that the magnitude of the interactional torques varies, no clear change in trajectory characteristics is seen (Hollerbach and Atkeson, 1983). It appears that the interactional torques are taken into consideration in the planning of movement.

These issues in multijoint control raise many questions. Does the motor system first develop a trajectory plan and then determine those muscle stiffnesses that will generate appropriate joint movements? If so, what are the criteria used to select a particular trajectory? Is the trajectory plan specified in joint coordinates, or is the plan Cartesian, specifying the path of the hand and, therefore, requiring a transformation into joint coordinates? How are the necessary joint torques determined? Are they computed on a real-time basis, or is some memory approach used? What consideration is given to the joint interaction torques? If compensation is provided, are specific compensating torques injected, or is the stiffness of the linkage raised to overwhelm their effects? Finally, what is the role of final position control and of reference trajectories in the production of multijoint movements? Are there characteristics of these movements that, as in single-degree-of-freedom movements, indicate that factors other than final control are operative? Does a reference trajectory specify speed, the path, and stiffness?

To gain some insight into these questions, two-degree-of-freedom arm movements performed by normal adult humans were recorded. The wrist are braced, so that only movements about the shoulder and elbow joints were allowed. The subject, maintaining his arm in a horizontal plane passing through the shoulder (so that gravitational effects were constant), grasped the handle of a light-weight, two-joint mechanical linkage and moved the handle to each of a series of visual targets which were disturbed in the movement space (Fig. 9*A*). The geometry of the experiment arrangement was such that the signals from potentiometers of the two mechanical joints could be used to compute the hand and joint trajectories.

In an initial study using this apparatus, Morasso (1981) instructed subjects simply to move their hand from one target to another; there was no instruction regarding speed or accuracy. Data similar to Morasso's are presented in Figure 9, where several movements are illustrated. Two findings are of interest. First, the path taken by the hand from one target to another was straight or only gently curved (Fig. 9*B*). A similar result has been obtained in the monkey (Gilman, et al., 1976; Georgopoulos et al., 1981) and in other studies of humans (Abend, et al., 1982; Soechting and Lacquaniti, 1981). This finding may be expected on the basis of casual observation of human behavior. It is of interest because the straight hand movements result from the combined effects of rotation of both joints; convoluted movements would result if the two degrees of freedom were not perfectly coordinated. Therefore the tendency to produce straight hand paths suggests that path planning occurs in addition to final position control.

A second finding reported by Morasso (1981) is illustrated in Figure 9*C*, *D*, and *E*, where kinematic data are presented for three movements which were performed in different parts of the work space. The data show that the joint position and the joint velocity traces vary widely from movement to movement, whereas the hand speed profile is always roughly bell shaped, even when the joint angular velocities are complex. The work space independence of the hand speed is consistent with, but of course does not prove, the notion that the CNS plans a movement in terms of the hand kinematics and then transforms the plan into joint coordinates. This

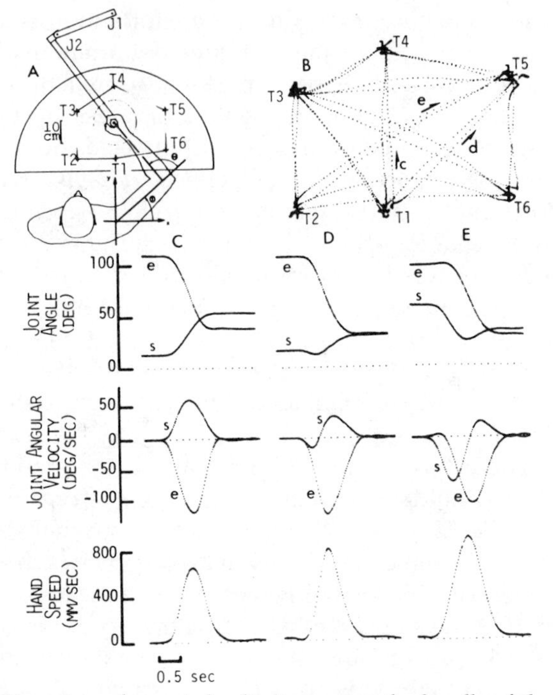

FIGURE 9. (*A*) Plan view of a seated subject grasping the handle of the two-joint hand-position transducer (designed by N. Hogan). The right arm was elevated to shoulder level and moved in a horizontal work space. Movement of the handle was measured with potentiometers located at the two mechanical joints of the apparatus (J1, J2). A horizontal semicircular plate located just above the handle carried the visual targets. Six visual target locations (T1 through T6) are illustrated as crosses. The digitized paths between targets and the curved path were obtained by moving the handle along a straight edge from one target to the next and then along a circular path; movement paths were reliably reproduced. (*B*) A series of digitized handle paths (sampling rate, 100 Hz) performed by one subject in different parts of the movement space. The subject moved his hand to the illuminated target and then waited for the appearance of a new target. Targets presented in random order. Arrows show direction of some of the hand movements. (*C, D,* and *E*) Kinematic data for three of the movements, the paths of which are shown in (*B*). Letters show correspondence, for example, data under (C) are for path c in (*B*); e, elbow joint; s, shoulder, angles measured as indicated in (*A*). [From Bizzi and Abend (1983) *Motor Control Mechanisms in Health and Disease,* J. E. Desmedt, ed., Raven, New York.]

is in agreement with Bernstein's statement: "The hypothesis that there exist in the higher levels of the CNS projections of space, and not projections of joints and muscles, seems to me to be at present more probable than any other" (1967, p. 50). In Bernstein's view, such a scheme reduces the number of variables controlled by the CNS.

Since subjects tend to produce roughly straight hand paths when given no instruction other than to move the hand to a target, a group of subjects was instructed to approach targets by way of curved paths (Abend et al.,

1982). There was no instruction regarding accuracy or speed. The subjects first performed a series of movements in the absence of visual feedback of arm position and then repeated the experiment with feedback present. Typical results are illustrated in Figure 10 where, for each of five movements, the hand path and the time course of the hand speed and path curvature are presented. When the subject was told only to move his hand to the target, the path was almost straight and the speed profile for the hand was bell shaped (movement 1). The curved movements (2–5), however, had remarkable curvature and speed profiles. Consider movement 4. Although the path is smooth, there is a midcourse peak in curvature. The initial portion of the hand speed curve is quite similar to that for the straight movement, but the rest of the curve shows a midcourse speed valley followed by a second peak. The speed valley corresponds in time to the curvature peak, whereas the speed peaks occurred during relatively straight parts of the path. The characteristics of movement 5 are quite different from those of movement 4 although the instruction was the same. The path appears to be composed of a series of three gently curved segments that meet at more highly curved regions. The two peaks in the curvature profile correspond temporally with the speed valleys; speed

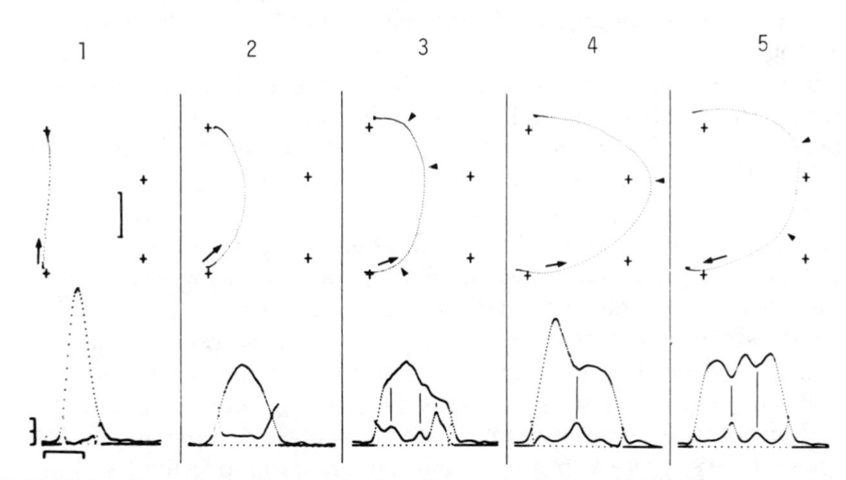

FIGURE 10. Five movements recorded from the same subject. In 1 the subject was told only to move her hand to the target; in 2 to 5, to use a curved path to reach the target. For each movement, the hand path is illustrated by plotting the hand location every 10 msec. Crosses represent the locations of targets T1, T4, T5, T6 (see Fig. 9*A*). Arrows indicate direction of movement along the trajectory. Bar under 1 represents 10 cm and applies to all five movements. The hand speed and trajectory curvature profiles below each trajectory begin when the target was illuminated and are shown superimposed; in each case, the curvature profile is the more shallow of the two. Ordinate: full bar represents a speed of 200 mm/sec; half bar represents a curvature of 1/150 mm (inverse of radius of curvature). Abscissa scale: 500 msec. In 3, 4, and 5 arrowheads indicate points of local curvature maxima along the trajectory; these curvature maxima are also denoted by vertical lines over the curvature profiles. [From Abend, Bizzi, and Morasso (1982) *Brain* **105**:331–348.]

peaks occur at the midcourse region of each path segment. The speed valleys and/or curvature peaks for shallow curved movements were sometimes less distinct than those for more highly curved movements. For example, trajectory 3 appears to be composed of four segments, and the curvature trace contains three distinct peaks. However, the speed trace contains no valleys, although it is quite irregular, compared to the trace for the straight movement. This subject occasionally produced a trajectory such as movement 2, with a relatively constant curvature and a bell-shaped speed profile. The presence of curvature peaks and associated hand speed irregularities are typical features of curved trajectories.

The path and speed discontinuities, rather than being properties of a special kinematic pattern, characterize a wide variety of movements. This has been investigated in three ways (Abend et al., 1982). First, it was shown that the characteristics are independent of the part of the horizontal work space employed. Second, the results were unchanged when the subject attempted to mimic constant-curvature, nonconstraining guide paths (Fig. 11). Third, the wrist splint was removed, and the horizontal movement plane of the hand was lowered from shoulder to waist level. In this way the work space of the arm was altered, a small vertical movement component was required, and the number of degrees of freedom was increased. Despite these changes, the character of the hand kinematics was unchanged.

There is no a priori reason why the hand should slow when the curvature is high if the hand speed is an expression of the rotation of each of the active joints. Therefore the question arises as to whether the speed valleys might reflect inertial effects of one of the two active joints. Ordinarily, during the course of a curved movement, a joint comes to zero speed and reverses direction. However, two findings argue against the speed valleys being a reflection of a joint reversal. First, joint reversals were required for some straight movements but did not impart a hand speed valley (Fig. 9*E*). Second, subjects occasionally produced curved movement with bell-shaped speed profiles, even though a joint reversal was required. These points suggest that the speed valley is the result of factors imposed by the multijoint controller rather than events involving individual joints.

In summary, there appears to be a tendency to move the hand along roughly straight paths, and this tendency persists even when the subject is in the midst of producing curved trajectories. Usually, curved paths had a segmented appearance, as if the multijoint controller were approximating a curve with a series of low-curvature segments. In addition, there are hand speed characteristics that appear to be related to the character of the path. Discontinuities in movement characteristics have been noted previously (Brooks et al., 1973; Morasso and Mussa-Ivaldi, 1982; Navas and Stark, 1968; Viviani and Terzuolo, 1980; von Hofsten, 1979). These findings may serve as clues in understanding the organization of the central motor controller.

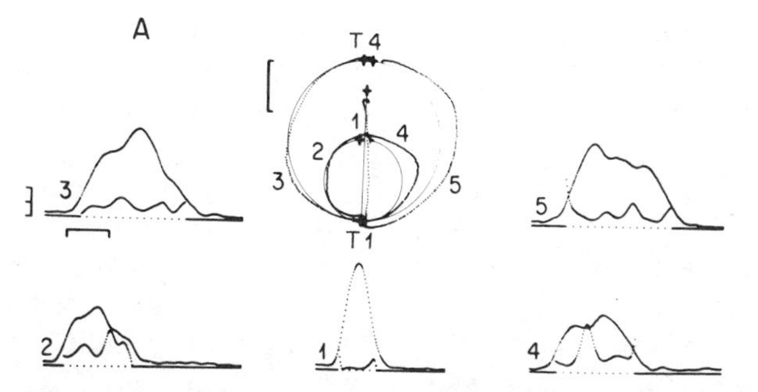

FIGURE 11. Movements made by one subject who was told to mimic guide paths. Thin continuous lines superimposed on the path plots denote constant curvature guide paths that were drawn on the target panel. Crosses represent the targets at the ends of the guides. For each curved trajectory, arrows indicate the direction of the movement, and arrowheads indicate points of local maxima in trajectory curvature. Numbers in small type show the correspondence between trajectories and associated velocity and curvature profiles; numbers in large type (T1, T4) denote targets located in the same orientation at T1 and T4 in Figure 9. [Modified figure from Abend, Bizzi, and Morasso (1982) *Brain* **105:**331–348.]

DISCUSSION

We wish to emphasize two points regarding posture control. First, the important multijoint control problems requiring explanation concern trajectory formation, not posture control. A particular arm posture is a function of the equilibrium position of each joint. There is no reason to expect that posture maintenance of a multijoint limb is managed by the CNS in any fundamentally different manner than that of a single-degree-of-freedom situation. Second, there are experimental situations in which the final posture may be affected by peripheral feedback. In the studies we outlined, the movements under investigation were large elbow movements. Experimental modifications of peripheral feedback may cause final position errors when movements are small (Sanes and Evarts, 1983) or are carried out at joints that are more distal than the elbow (Day and Marsden, 1982).

As in the case of elbow movements, several features of multijoint movements indicate that the trajectory of the movement is controlled, in addition to the final position. These features include the tendency for the hand to move along straight paths, the segmented appearance of the curved movements, and the temporal correspondence between path curvature peaks and speed discontinuities. It is unlikely that such movement characteristics can be accounted for by assuming that movements are simply passive transitions to new joint postures and governed only by the inertial and viscoelastic properties of the arm (Delatizky, 1982). Instead, they probably indicate the presence of specific trajectory strategies that are determined by a central reference signal that optimizes some variable, such as the distance moved, the energy dissipated, or mechanical stress and wear.

In a study that directly addresses this issue, it was found that a model based on minimizing the rate of change of hand acceleration adequately describes the characteristics of human arm movement (Flash and Hogan, 1982).

A concept has emerged from these initial studies of multijoint movements that there is a level in the CNS that specifies movement of the hand, rather than dealing with joint parameters. The hand-trajectory plan may then be transformed into appropriate joint specifications. Hand control could be a convenient organizing principle for multijoint control, since in many natural situations the goal of an arm movement is to position the hand appropriately and to control the response of the hand to an arbitrary external force input.

The notions of final position control and reference trajectory can be generalized to serve as a basis for hand control. For single-degree-of-freedom movements, a reference trajectory determines a dynamic equilibrium in the interactions of the muscles active about a moving joint. The equilibrium point that is chosen determines the joint stiffness. Of course, in the one-degree case, joint-position control cannot be distinguished from hand-position control. It may be hand position that is controlled in the special situation of one-degree-of-freedom movement, just as it seems to be the hand that is referenced in multiple-degree movements. A corollary of this idea is that the trajectory of multijoint movements is also determined as a dynamic equilibrium state. Now, however, to control the conformation of the entire linkage, the reference signal would have to specify the interaction of all the arm muscles. We speculate that what might actually be under control is hand posture, which is a resultant of the activity levels in all of the arm muscles. The interaction of the arm muscles can be described by considering the net stiffness of the hand. Hand stiffness can be characterized by observing the static force required to maintain the hand away from equilibrium. Static forces can be determined for each of a series of directions of hand displacement. In this way a hand-stiffness field is determined. Preliminary results indicate that the stiffness field for each work-space location can be characterized by its shape, size, and orientation (Hocherman et al., 1983; Mussa-Ivaldi, Bizzi, and Hogan, 1985, in press). This result is significant in a number of ways. First, it provides a new and rich description of posture and allows an evaluation of the way in which the spinal and supraspinal structures control the characteristic parameters of the stiffness field. In addition, on-going experimental studies directed at ascertaining the range of adaptive changes of the stiffness field in a variety of conditions are providing new insights into the complex interactions and constraints between alpha-motoneurons and muscle geometry. Beyond this, the identification of the shape and orientation of the elastic force field in each part of the work space will allow testing of the possible relationship between the static field at particular work-space locations and multijoint arm movement trajectories that pass through these locations.

Acknowledgments

This research was supported by National Institute of Neurological and Communicative Disorders and Stroke Research Grants NS09343, NS06416, NS00747; National Institute of Arthritis, Metabolism, and Digestive Diseases Grant AM26710; National Aeronautics and Space Administration Grant NAG-126; and National Eye Institute Grant EY02621.

REFERENCES

Abend, W. K., E. Bizzi, and P. Morasso (1982) Human arm trajectory formation. *Brain* **105**:331–348.

Amassian, V. E. and L. Eberle (1982) Quantitative relations between forelimb joint angles during vertical trajectories. *J. Physiol. (Lond.)* **326**:53–54P.

Amassian, V. E., L. Eberle, and D. Batson (1982) Some factors underlying specifiic forelimb trajectories. *Abstr. 12th Annu. Meet., Soc. Neurosci.*, San Francisco.

Asatryan, D. G. and A. G. Feldman (1965) Biophysics of complex systems and mathematical models. Functional tuning of nervous system with control of movement or maintenance of a steady posture. I. Mechanographic analysis of the work of the joint on execution of a postural task. *Biophysics* **10**:925–935.

Bernstein, M. (1967) *The Co-ordination and Regulation of Movements*. Pergamon Press, Oxford.

Bizzi, E., N. Accornero, W. Chapple, and N. Hogan (1981) Central and peripheral mechanisms in motor control. In *New Perspectives in Cerebral Localization*, R. A. Thompson, ed. Raven, New York, pp. 23–34.

Bizzi, E., N. Accornero, W. Chapple, and N. Hogan (1982) Arm trajectory formation in monkeys. *Exp. Brain Res.* **46**:139–143.

Bizzi, E., N. Accornero, W. Chapple, and N. Hogan (1984) Posture control and trajectory formation during arm movement. *J. Neurosci.* **4**:2738–2744.

Bizzi, E., P. Dev, P. Morasso, and A. Polit (1978) Effect of load disturbances during centrally initiated movements. *J. Neurophysiol.* **41**:542–556.

Bizzi, E., A. Polit, and P. Morasso (1976) Mechanisms underlying achievement of final head position. *J. Neurophysiol.* **39**:435–444.

Brooks, V. B., J. D. Cooke, and J. S. Thomas (1973) The continuity of movements. In *Control of Posture and Locomotion*, R. B. Stein, K. B. Pearson, R. S. Smith, and J. B. Redford, eds. Plenum, New York, pp. 257–272.

Collatos, T. C., V. R. Edgerton, J. L. Smith, and B. R. Botterman (1977) Contractile properties and fiber type compositions of flexors and extensors of elbow joint in cat: implications for motor control. *J. Neurophysiol.* **40**:1292–1300.

Cooke, J. D. (1979) Dependence of human arm movements on limb properties. *Brain Res.* **165**:366–369.

Day, B. L. and C. D. Marsden (1982) Accurate repositioning of the human thumb against unpredictable dynamic loads is dependent upon peripheral feed-back. *J. Physiol. (Lond.)* **327**:393–407.

Delatizky, J. (1982) Final position control in simulated planar arm movements. *Abstr. 12th Annu. Meet., Soc. Neurosci.*, San Francisco.

Desmedt, J. E. (1981) The size principle of motoneuron recruitment in ballistic or ramp voluntary contractions in man. In *Motor Unit Types, Recruitment and Plasticity in Health and Disease. Prog. Clin. Neurophysiol.* Vol. 9, J. E. Desmedt, ed. Karger, Basel, pp. 97–136.

Desmedt, J. E., and E. Godaux (1978) Ballistic skilled movements: load compensation and patterning of the motor commands. In *Cerebral Motor Control in Man: Long Loop Mechanisms. Prog. Clin. Neurophysiol.* Vol. 4, J. E. Desmedt, ed. Karger, Basel, pp. 21–25.

Dichgans, J., E. Bizzi, P. Morasso, and V. Tagliasco (1973) Mechanisms underlying recovery of eye-head coordination following bilateral labyrinthectomy in monkeys. *Exp. Brain Res.* **18**:548–562.

Feldman, A. G. (1974a) Change of muscle length due to shift of the equilibrium point of the muscle-load system. *Biofizika* **19**:534–538.

Feldman, A. G. (1974b) Control of muscle length. *Biofizika* **19**:749–951.

Flash, T. and N. Hogan (1982) Evidence for an optimization strategy in arm trajectory formation. *Abstr. 12th Annu. Meet., Soc. Neurosci.*, San Francisco.

Freund, H.-J. and H. J. Budingen (1978) The relationship between speed and amplitude of the fastest voluntary contractions of human arm muscles. *Exp. Brain Res.* **31**:1–12.

Ghez, C. and D. Vicario (1978a) The control of rapid limb movement in the cat. I. Response latency. *Exp. Brain Res.* **33**:173–189.

Ghez, C. and D. Vicario (1978b) The control of rapid limb movement in the cat. II. Scaling of isometric force adjustments. *Exp. Brain Res.* **33**:191–202.

Georgopoulos, A. P., J. F. Kalaska, R. Caminiti, and J. T. Massey (1982) On the relations between the direction of two-dimensional arm movements and cell discharge in primitive motor cortex. *J. Neurosci.* **2**:1527–1537.

Georgopoulis, A. P., J. R. Kalaska, and J. T. Massey (1981) Spatial trajectories and reaction times of aimed movements: effects of practice, uncertainty, and change in target location. *J. Neurophysiol.* **46**:725–743.

Georgopoulos, A. P., J. F. Kalaska, J. T. Massey, and R. Caminiti (1983) Cortical mechanisms of two-dimensional aimed arm movements. IX. Static (positional) factors cannot account fully for movement-related cell discharge in motor cortex and area 5. *Abstr. 13th Annu. Meet., Soc. Neurosci.*, Boston.

Gilman, S., D. Carr, and J. Hollanberg (1976) Kinematic effects of deafferentation and cerebellar ablation. *Brain* **99**:311–330.

Henneman, E. (1965) Functional significance of cell size in spinal motoneurons. *J. Neurophysiol.* **28**:560–580.

Hocherman, S., F. A. Mussa-Ivaldi, E. Bizzi, and N. Hogan (1983) Control of multi-joint arm posture. *Abstr. 13th Annu. Meet., Soc. Neurosci.* **9**:179.

Hofsten, C. von (1979) Development of visually directed reaching: the approach phase. *J. Hum. Movement Stud.* **5**:160–178.

Hogan, N. (1980) Mechanical impedance control in assistive devices and manipulators. Proceedings of the Joint Automatic Controls Conference 1, TA10-B.

Hogan, N. (1984) An organising principle for voluntary movements. *J. Neurosci.* **4**:2745–2754.

Hollerbach, J. M. (1982) Computers, brains and the control of movement. *Trends NeuroSci.* **5**:189–192.

Hollerbach, J. M. and C. Atkeson (1983) Kinematic features of unrestrained vertical arm movements. *Abstr. 13th Ann. Meet., Soc. Neurosci.*, Boston.

Hollerbach, J. M. and T. Flash (1982) Dynamic interactions between limb segments during planar arm movement. *Biol. Cybern.* **44**:67–77.

Horn, B. K. P. and M. H. Raibert (1977) Configuration space control. Massachusetts Institute of Technology Artificial Intelligence Lab. Memo No. 458, Cambridge, Massachusetts.

Kalaska, J. F., A. P. Georgopoulos, and R. Caminiti (1983) Cortical mechanisms of two-dimensional aimed arm movements. VIII. Cell discharge in motor cortex and area 5 varies with movement direction, not with final position of the arm. *Abstr. 13th Annu. Meet., Soc. Neurosci.*, Boston.

Kelso, J. A. S. and K. G. Holt (1980) Exploring a vibratory system analysis of human movement production. *J. Neurophysiol.* **43:**1183–1196.

Kots, Y. M. and A. M. Syrovegin (1966) Fixed sets of invariants of interactions of the muscles of two joints used in the execution of single voluntary movements. *Biofizika* **11:**1061–1066.

Lacquaniti, F. and J. F. Soechting (1982) Coordination of arm and wrist motion during a reaching task. *J. Neurosci.* **2:**399–408.

Lestienne, F., A. Polit, and E. Bizzi (1981) Functional organization of the motor process underlying the transition from movement to posture. *Brain Res.* **230:**121–131.

Marey, E. J. (1901) La locomotion animale. *Traité de Physique Biologique,* J. Arsonval, ed. Masson, Paris, vol. 1, pp. 229–287.

Milner-Brown, H. S., R. B. Stein, and R. Yemm (1973) The contractile properties of hand motor units during voluntary isometric contractions. *J. Physiol. (Lond.)* **228:**285–306.

Morasso, P. (1981) Spatial control of arm movements. *Exp. Brain Res.* **4:**223–227.

Morasso, P. (1983) Three-dimensional arm trajectories. *Biol. Cybern.* **48:**1–8.

Morasso, P. and F. A. Mussa-Ivaldi (1982) Trajectory formation and hand writing: a computational model. *Biol. Cybern.* **45:**131–142.

Mussa-Ivaldi, F-A., Hogan, N., anad Bizzi, E. Neural, mechanical, and geometric factors subserving arm posture in humans. *J. Neurosci.,* 1985, in press.

Muybridge, E. (1901) *The Human Figure in Motion.* Chapman and Hall, London.

Navas, F. and L. Stark (1968) Sampling on intermittency in hand control system dynamics. *Biophys. J.* **8:**252–302.

Polit, A. and E. Bizzi (1979) Characteristics of motor programs underlying arm movements in monkeys. *J. Neurophysiol.* **42:**183–194.

Rack, P. M. H. and D. R. Westbury (1969) The effects of length and stimulus rate on tension in the isometric cat soleus muscle. *J. Physiol. (Lond.)* **204:**443–460.

Raibert, M. (1976) A state space model for sensorimotor control and learning. Massachusetts Institute of Technology Artificial Intelligence Lab. AIM 351, Cambridge, MA.

Robinson, D. A. (1964) The mechanics of human saccadic eye movement. *J. Physiol. (Lond.)* **174:**245–264.

Sakitt, B. (1980) A spring model and equivalent neural network for arm posture control. *Biol. Cybern.* **37:**227–234.

Saltzman, E. (1979) Levels of sensorimotor representation. *J. Path. Psychol.* **20:91**–163.

Sanes, J. N. and E. V. Evarts (1983) Effects of perturbations on accuracy of arm movements. *J. Neurosci.* **3:**977–986.

Soechting, J. F. and F. Lacquaniti (1981) Invariant characteristics of a pointing movement in man. *J. Neurosci.* **1:**710–720.

Soechting, J. F. and B. Ross (1983) Psychophysical identification of coordinate representation of human arm. *Abstr. Soc. Neurosci.* **9:**1033.

Tanji, J. and M. Kato (1973) Firing rate of individual motor units in voluntary contraction of abductor digiti minimi muscle in man. *Exp. Neurol.* **40:**771–783.

Taub, E., R. C. Bacon, and A. J. Berman (1965) Acquisition of a trace-conditioned avoidance response after deafferentation of the responding limb. *J. Comp. Physiol. Psychol.* **59:**275–279.

Taub, E., S. J. Ellman, and A. J. Berman (1966) Deafferentation in monkeys: effect on conditioned grasp response. *Science* **151:**593–594.

Taub, E., I. A. Goldberg, and P. Taub (1975) Deafferentation in monkeys: pointing at a target without visual feedback. *Exp. Neurol.* **46:**178–186.

Viviani, P. and C. Terzuolo (1980) Space-time invariance in motor skills. In *Tutorials in Motor Behavior,* G. E. Stelmach and J. Requin, eds. North-Holland, Amsterdam, pp. 525–533.

Chapter Sixteen

NEURONAL OSCILLATORS IN MAMMALIAN BRAIN

RODOLFO R. LLINÁS

Department of Physiology and Biophysics
New York University Medical Center
New York, New York

The issue of oscillations in central nervous systems (CNS) has been a topic of extensive study in both vertebrate and invertebrate forms throughout most of this century (cf. Carpenter, 1982). This topic is one of the many to which T. H. Bullock has given much attention, both at the single-cell level as well as at the system point of view (Bullock, 1979). The oscillatory properties of neuronal tissue are clearly among the most fundamental aspects of the organization of both sensation and motor performance. The question to be tackled relates to whether these oscillatory mechanisms, so conspicuously present at the neuronal level of lower vertebrates as well as invertebrates, are present in the mammalian brain only as circuit properties or are also in some measure reflections of the intrinsic electrical properties of single neurons. The thesis developed in this chapter relates to neuronal oscillation as viewed with the newly acquired perspective provided by the electrophysiological properties of mammalian neurons *in vitro*.

NEURONAL OSCILLATION IN MAMMALIAN CNS

One of the truly remarkable findings at the beginning of this century relating to the electrical activity in mammalian brain was the discovery of the electroencephalograhic field potentials by Hans Berger (1929). With rather primitive tools he demonstrated the presence of rhythmic electrical activities on the surface of the cranium and related the different rhythms to given states of consciousness. The development of this initial discovery over the ensuing five decades has yielded much important information relating to the physiology and pathology of the CNS. However, some of the points that have remained obscure relate to the nature of and the mechanism by which these surface rhythmic potentials are generated.

For the most part, in the 1960s and 1970s, the accepted site of origin for the most prominent oscillatory rhythms was the thalamus. In fact, the accepted view was that the oscillatory behavior was a property of neuronal circuits, where the neuronal system was organized to produce, following an early excitatory postsynaptic potential (EPSP), a rapid synaptic inhibition (IPSP). These EPSP-IPSP sequences, when organized so that they occurred in a cyclic manner, were considered to be the normal basis for the previously mentioned rhythm. The minimum neuronal machinery required for such a system to exist was agreed to be a two-neuronal chain organized in a loop such that, as in the thalamus, an activated excitatory neuron projecting to the cortex would excite, by way of axon collaterals, an inhibitory interneuron. This interneuron would return axons to the cell of origin, implementing the oscillatory behavior (by way of a negative feedback) with a delay (Anderson and Sears, 1964).

In more recent years, the *in vitro* study of the electrical activity in brain slices, in particular, the electrophysiology of the inferior olive and thalamus, has provided us with further information regarding possible mechanisms for such oscillations. Indeed, today, the most likely mechanism to be considered when reviewing the oscillatory properties of the CNS would not necessarily lie in the circuitry but rather in the intrinsic ionic properties of individual neurons (Llinás and Yarom, 1981b; Jahnsen and Llinás, 1984b). There is, therefore, a clear shift of emphasis from the properties of circuits to the properties of single neurons. Undoubtedly, inhibitory feedback will always play a role in the generation of neuronal oscillations, if only by aiding the synchronization of single oscillators into sets of coupled oscillators, which in turn give the macroscopic field potentials observable at the surface of the cranium.

IONIC MECHANISMS INVOLVED IN OSCILLATION OF INFERIOR OLIVE: THE ISSUE OF REBOUND EXCITATION

In recent *in vitro* experiments using brainstem slices, Dr. Yosef Yarom and I (Llinás and Yarom, 1981a,b) demonstrated that inferior olive (I.O.) neurons have a set of ionic conductances that are activated in such a manner as to convey to these cells intrinsic oscillatory properties. Thus, as shown in Figure 1*A*, the firing of these cells is characterized by an initial

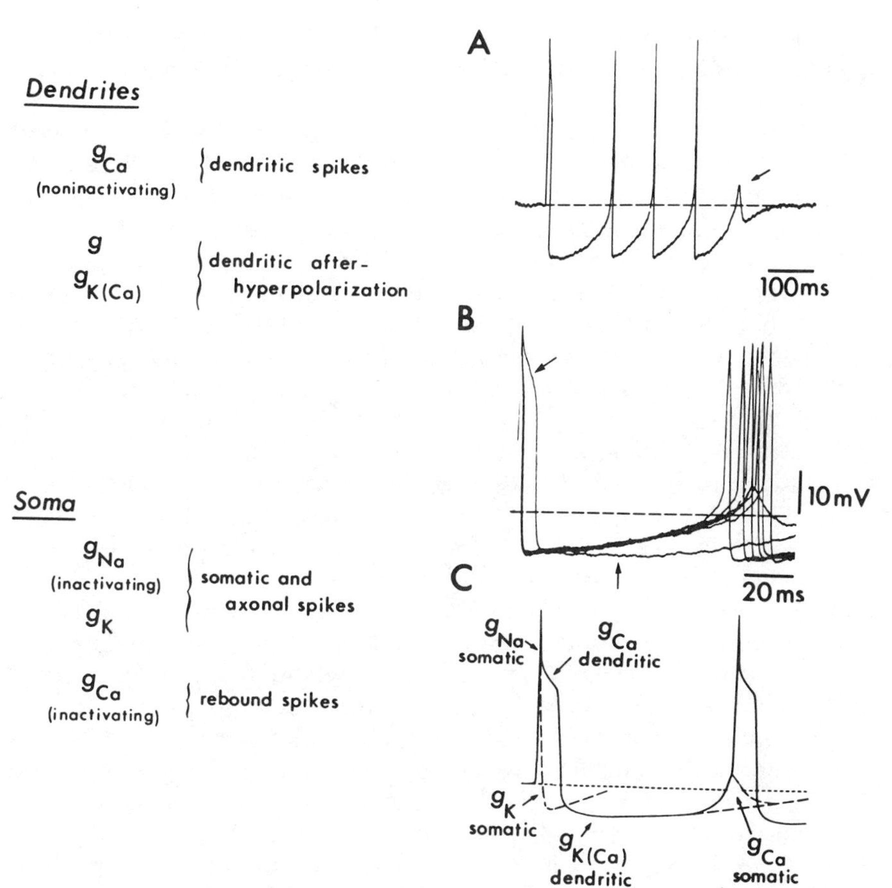

FIGURE 1. Distribution of ionic conductances and oscillatory properties of inferior olivary (I.O.) neurons. The distribution of ionic conductances is described on the left of the figure for dendrites and soma. On the right, oscillation of I.O. cell is shown after the addition of harmaline to the bath. The initial spike is followed by three action potentials and a subthreshold rebound indicated by an arrow. Note that the threshold for the rebound response is negative to the resting membrane potential. (*B*) Similar oscillation consisting of 15 spikes displayed such that the oscilloscope is triggered by an action potential in the oscillatory train. Note that, as in (*A*), the first spike of the sequence is longer lasting (arrow) and has a more prolonged after-hyperpolarization (arrow). The action potentials that followed are shorter lasting and are generalized after an interval of approximately 100 msec. (*C*) The ionic conductances responsible for each one of the components of the oscillatory response are illustrated. (Unpublished observations by R. Llinás and Y. Yarom.)

fast-rising action potential (a somatic sodium spike), which is prolonged to 10–15 msec by an after-depolarization. The plateau after depolarization is followed by an abrupt, long-lasting after-hyperpolarization, which totally silences the spike-generating activity. The hyperpolarization is typically terminated by a sharp, active rebound response, which arises at the membrane potential negative to resting level, overshoots the resting membrane potential, and can once again activate the cell. The oscillatory tendency of the I.O. neuron shown in Figure 1 was enhanced by addition of harmaline to the bath (see later discussion).

The results of experiments in which a calcium-free Ringer solution or calcium blocker was used, or in which barium was substituted for calcium, demonstrate that the after-depolarization-hyperpolarization sequence following the initial spike is due to the activation of an initial calcium conductance (Llinás and Yarom, 1981a) followed by a potassium conductance. The analysis of extracellular field potentials has further demonstrated that the initial after-depolarization is produced by activation of a voltage-dependent calcium conductance located in the dendrites (Llinás and Yarom, 1981b). This conductance is very similar to that first demonstrated in Purkinje cells (Llinás and Hess, 1976; Llinás and Sugimori, 1980a,b) and is present in the dendrites of other central neurons (Schwartzkroin and Slawsky, 1977; Wong, Prince and Basbaum, 1979).

In I.O. cells the initial broad calcium spike elicits a strong increased permeability to potassium, which is most prominent in the dendrites (as determined from the amplitude and polarity of the extracellular field potential). The very high membrane conductance obtained during this period of hyperpolarization is sufficient to render the cell inexcitable to the extent that even powerful synaptic inputs are totally shunted. In effect, during this period there is a virtual "clamping" of the membrane at the potassium equilibrium potential. This potassium conductance is calcium dependent and similar to that initially described in invertebrate neurons (Meech and Standen, 1975). Indeed, when barium is substituted for calcium, the after-hyperpolarization is completely abolished (Llinás and Yarom, 1981b). This is not unexpected, as barium does not activate the calcium-dependent potassium conductance (Eckert and Lux, 1976).

The duration of the after-hyperpolarization is thus modulated by the amount of calcium entering the dendrites during the after-depolarization. Indeed, if the dendritic calcium action potential is smaller, the duration of the after-hyperpolarization is decreased, although its amplitude does reach the potassium equilibrium potential (see Fig. 1*A*). Prolonged dendritic spikes generate after-hyperpolarizations as long as 200–250 msec. This can be seen in Fig. 1*B* where the sweep was triggered with the rising phase of the spike. Note that the broadest after-depolarization generated by a calcium-dependent dendritic spike is followed by a prolonged after-hyperpolarization (arrows). This point is central to the oscillatory properties of I.O. cells, as it indicates that calcium entry determines the cycle time of the oscillator.

Probably the most significant factor uncovered by the *in vitro* studies with respect to neuronal oscillation, however, was the finding that the membrane potential returned to the baseline in a rather abrupt manner at the end of the after-hyperpolarization, overshooting the initial resting potential. This rebound response is due to the activation of a somatic calcium-dependent action potential and results from a second voltage-dependent calcium conductance, which is unusual in that it is inactive at resting membrane potential (−65 mV). The membrane after hyperpolarization deinactivates this conductance, and thus, as the membrane potential returns to baseline, a "low threshold" calcium spike is generated (Llinás and Yarom, 1981a,b). The rebound potential can be increased sufficiently by small changes in the resting membrane potential to produce a full sodium spike activation, which in turn can set forth the whole sequence of events once again, and thus will fire in an oscillatory manner with a frequency of 5–10 Hz. The dependence of deinactivation of this low threshold spike on membrane potential is shown in Figure 2. After blocking g_{Na} with tetrodotoxin, calcium-dependent spikes are generated; as the membrane is hyperpolarized, the amplitude and rate of rise increases (Fig. 2*A*). The second derivative of the spike (lower trace in the upper panel) gives the rate of rise of the spike. This value is plotted in Figure 2*B* for two cells (open and

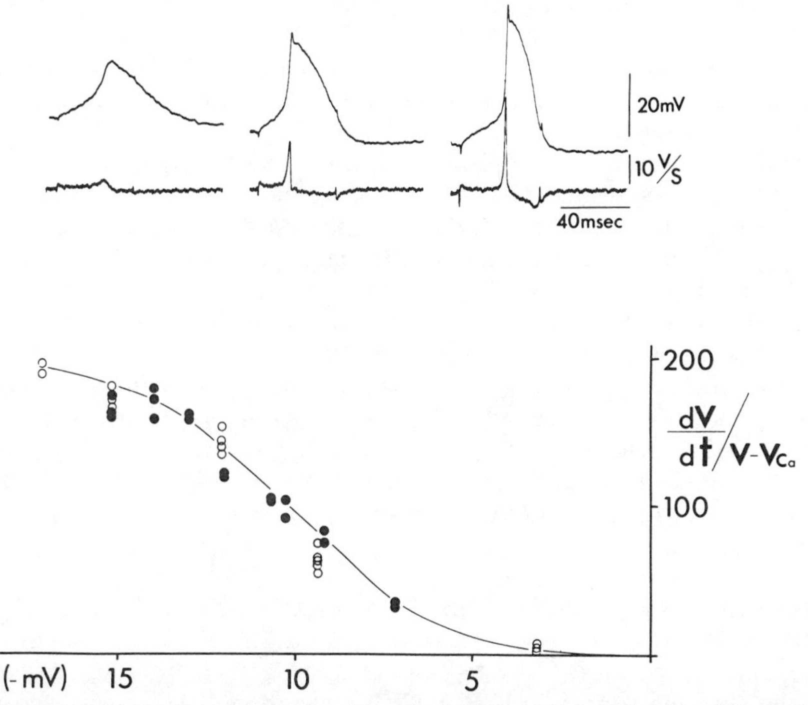

FIGURE 2. Calcium-dependent rebound spike. In the upper panels rebound potentials produced by square current pulse injection at three different levels of membrane potential after addition of TTX to the bath. In the record to the left the action potential was observed following hyperpolarization of 3 mV from rest, a second one 8 mV, and the third, 12 mV. Differences in the rate of rise in the action potential are seen in differentiated records immediately below the action potentials. In the lower panel, the amplitude of the first derivative of the action potentials illustrated above is plotted against membrane hyperpolarization from rest. (Modified from Llinás and Yarom, 1981b.)

closed circles) as a function of membrane potential (Llinás and Yarom, 1981b).

PHARMACOLOGICAL ASPECTS OF THE REBOUND EXCITATION

Among the most spectacular pharmacological effects of drugs capable of increasing neuronal oscillation in higher vertebrates are the tremors that follow the administration of harmaline, an alkaloid of *Pegamus harmala.* This tremor has been known since the turn of the century (Neuner and Tappeiner, 1894). It can produce a 10–12 Hz tremor in intact and decerebrated cats (Villablanca and Riobo, 1970; Lamarre et al., 1971; de Montigny and Lamarre, 1973), as well as in other mammals, and it seems to operate almost specifically at the inferior olive level. Indeed, intracellular recording from Purkinje cells demonstrated years ago that harmaline tremor was accompanied by all-or-none Purkinje cell EPSPs (de Montigny and Lamarre, 1973; Llinás and Volkind, 1973). The reversal of these large synaptic potentials (Llinás and Volkind, 1973) was the direct demonstration that such activation of Purkinje cells was indeed due to the activation of the inferior olivary cell, which terminates on the Purkinje cell as climbing fibers (Eccles, Llinás, and Sasaki, 1966).

Since this oscillatory firing reflects intrinsic conductance changes in each inferior olivary neuron, its basic frequency cannot be easily modified. That is, individual I.O. cells are limited cycle oscillators. Indeed, normally, I.O. cells cannot be made to discharge with frequencies much higher than 15 cycles per second. Inferior olive axons generate a short burst of repetitive firing during each cycle. The interval between these bursts is determined by the powerful after-hyperpolarizations separating the calcium plateaus. Activation of the I.O. during harmaline intoxication is seen in the cerebellar cortex as a burst of climbing fiber inputs with a basic 1–2 Hz frequency (cf. Llinas, 1981). Recent *in vitro* experiments using brainstem slices have shown that the application of harmaline hyperpolarizes I.O. neurons and produces an exaggerated rebound response (Yarom and Llinás, 1981).

OSCILLATORY NEURONAL INTERACTIONS WITHOUT ACTION POTENTIALS

In the slice preparation treated with harmaline, and occasionally in the absence of harmaline, the membrane potential of I.O. neurons tends to oscillate, proceeding in an almost perfect sinusoidal waveform. An example of such spontaneous oscillation is shown in Figure 3. In (*A*) three traces are superimposed; the consistency of the oscillation is further demonstrated in (*B*). Here, a Lissajous figure illustrates the constant periodicity of the harmonic oscillation of the membrane potential. This sinusoidal modulation of membrane potential, *which is not related to the generation of*

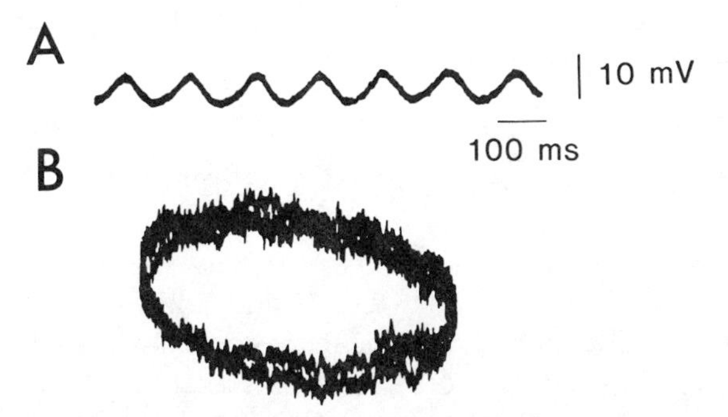

FIGURE 3. Spontaneous oscillatory potentials in inferior olivary cells in the presence of tetrodotoxin. (*A*) Oscillatory potentials occurring at close to resting potential, having a peak-to-peak average of approximately 8 mV and a peak-to-peak frequency of approximately 5 Hz. (*B*) A Lissajous figure generated by utilizing a record similar to that in (*A*) in the vertical plane of the CRT while the horizontal was fed by a sinusoidal oscillator. (Unpublished observations by R. Llinás and Y. Yarom.)

action potentials in I.O. neurons, is observed throughout the I.O. nucleus. The frequency (5 cycles per second) is the same for neighboring cells, and there is total coherence in the oscillation of different neurons. The frequency of oscillations has been observed to vary in different preparations and under different pharmacological conditions. This different frequency in the oscillatory rhythm probably reflects the metabolic state of the given preparation. Regardless of this variability, the synchrony of the ensemble emphasizes the importance of the electrotonic coupling between cells and suggests the presence of an underlying chemical oscillatory mechanism (cf. Neu, 1980) which modulates the membrane conductance.

The issue of interest here is that, as proposed in a closing remark of one of the most outstanding papers to be published over the past 25 years in neuronal integration (Bullock, 1959), communication between cells can take place in the central nervous system even at the mammalian level *without the generation of action potentials.* There is no question that the inferior olive represents a unique nucleus where several unusual events come together to give this particular neuronal circuit rather special properties. Indeed, we describe here mechanisms by which both depolarization and hyperpolarization increase membrane excitability. Second, the presence of electrical coupling between neurons allows oscillatory properties that would be otherwise far more difficult to implement. Third, oscillatory phenomena such as shown in Figure 3 do happen in *in vitro* conditions even in the absence of sodium spiking (i.e., after application of TTX to the bath or after removal of extracellular sodium). Although this may be a unique example of the intrinsic properties of the neurons being displayed at the multineuronal level, it seems clear that this type of oscillation is also present *in vivo.* Indeed, it has been recently demonstrated by multiple extracellular Purkinje cell recordings from cerebellar hemispheres in the rat, where up to 16 simultaneous recordings were obtained, that the type of oscillatory behavior observed *in vitro* actually takes place *in vivo* (Bower and Llinás, 1982, 1983).

THE THALAMUS: ITS OSCILLATORY PROPERTIES

A second example of an intrinsic oscillatory neuron has been described *in vitro* in the guinea pig thalamus (Llinás and Jahnsen, 1982; Jahnsen and Llinás, 1984a,b). Thalamic neurons demonstrate ionic mechanisms similar to those encountered in I.O. cells. This oscillatory behavior was found in all parts of the thalamus, including both geniculate nuclei. The ionic conductances present in thalamic neurons and the manner in which they may be combined to generate oscillatory firing in thalamic cells are summarized in Figure 4. As in the I.O., a strong TTX-insensitive rebound calcium spike can be obtained at membrane potential levels negative to -60 mV. Combined with the hyperpolarizing potassium conductance and an A current, it enables thalamic cells to generate oscillatory responses at frequencies near 6 Hz. When they are depolarized, the distribution of the electrical activity in these cells differs from that of the I.O. neurons. Although there is a dendritic calcium conductance, it is not as powerful as that in the I.O., and somatic firing may not activate the after-depolarization/depolarization sequence. Thus tonically depolarized thalamic cells can fire at high frequencies. Thalamic cells also display, as stated before, an early potassium conductance (A current) similar to that described in invertebrate neurons (Hagiwara et al., 1961; Connor and Stevens, 1971) and a noninactivating sodium conductance similar to that seen in Purkinje cells (Llinás and Sugimori, 1980a). The point to be emphasized here is not the differences between these two groups of cells but rather that they are both capable of distinct oscillatory behavior in the absence of recurrent inhibition; the capacity to generate oscillatory behavior is inherent to individual cells.

As in the I.O. the spatial distribution of specific ionic conductances is such that thalamic neurons can behave as intrinsic oscillators. In contrast to the I.O. neurons, thalamic cells have two distinct functional states. When depolarized, they fire tonically and can transmit information in a continuous manner, but, when hyperpolarized, they fire in an oscillatory manner within a period of about 10 Hz (Llinás and Jahnsen, 1982). Although only thalamic neurons have displayed this versatility (switching between tonic and phasic responses), it is quite probable that such properties may be present in other CNS neurons.

ROLE OF THE CYCLIC ACTIVITY IN THE CNS

For the I.O. to exert an influence on motoneuronal pools along the neuraxis, the oscillatory behavior of single neurons must be synchronized to yield oscillations in an ensemble of cells. Such a mechanism may be

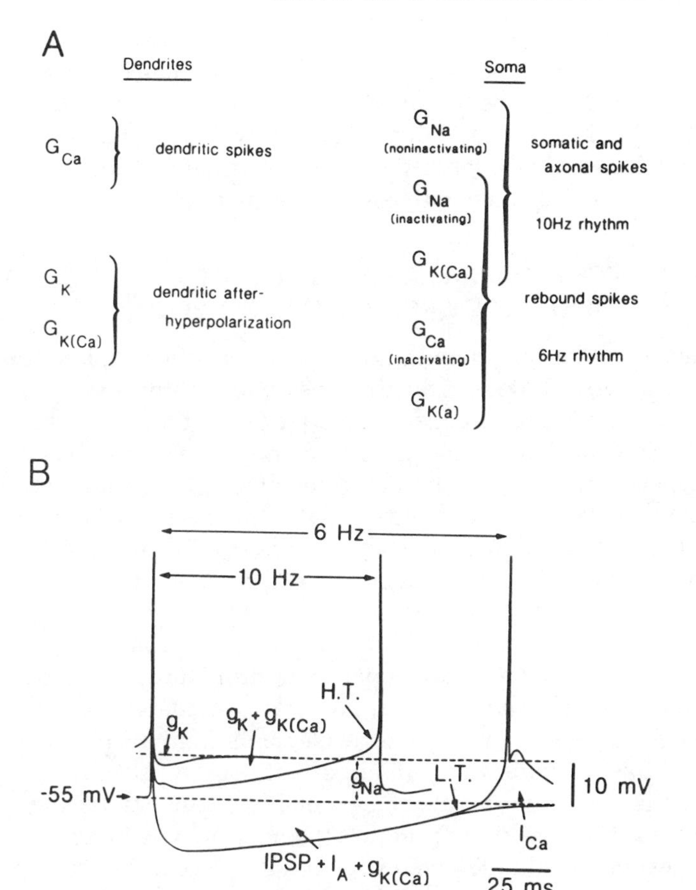

FIGURE 4. Distribution of ionic conductances in thalamic neurons and ionic mechanism responsible for two frequencies of oscillation. (*A*) Description of ionic conductances encountered in different parts of a thalamic neuron. (*B*) Two distinct frequencies, 6 and 10 Hz, were observed *in vitro*. The frequencies could be modulated by the level of resting potential by way of the expression of different conductances at these resting levels. (Modified from Jahnsen and Llinás, 1984b.)

subserved by the electrical coupling between I.O. neurons (Llinás et al., 1974; Llinás and Yarom, 1981a). This coupling is most probably related to the presence of gap junctions (cf. Bennett and Goodenough, 1978) at the olivary glomeruli as well as directly between dendrites (Sotelo et al., 1974; King, 1976; Gwyn et al., 1977). The oscillatory behavior of single cells would become synchronized through coupling such that the I.O. may generate a phasic modulation of the motoneurons in brainstem and spinal cord by way of vestibulo- and reticulospinal pathways (de Montigny and Lamarre, 1973; Llinás and Volkind, 1973). The main function of this oscillatory output would be to synchronize the activation of muscle groups throughout the body to generate organized motor responses. The ability of brain regions to recruit groups of motoneuronal pools in this way is essential to the generation of even the simplest coordinated movement, since sets of muscles must be activated in a specific temporal sequence. With respect to the thalamic oscillation, it seems evident that oscillatory activity at that nucleus is closely related to changes in states of awareness. Indeed oscillatory spindling is an early sign of incipient sleep.

SIGNIFICANCE OF OSCILLATORY PROPERTIES OF NEURONS AND THE QUESTION OF REBOUND EXCITATION

The question to be considered next is that of distribution. It seems evident that at least two cell types in the CNS are capable of intrinsically generating oscillatory activity with frequencies very close to those observed clinically and experimentally in some motor tremors. Further, a link may exist between specific nuclei and specific forms of tremor. In fact, the I.O. and associated nuclei may be directly related to the 8–10 Hz physiological tremor observed in higher vertebrates (Llinás, 1984), whereas phenomena such as the alpha rhythm seem clearly related to the thalamus and to the state of consciousness. Both cases have the common characteristic of cell activation and reactivation by way of a rebound excitatory phenomenon.

It is important to note that anodal break firing (postanodal exaltation) is a rebound response and a general property of excitable tissues observable to varying degrees in most excitable elements from axons to central dendrites. At this juncture, then, an important point must be made: Rebound excitation is a special example of a general phenomenon usually produced by deinactivation of sodium channels following a hyperpolarization. At the normal resting membrane potential, a certain percentage of these channels are in the inactivated state, and hyperpolarizing the membrane can reincorporate them into the active channel pool. I.O. and thalamic cells are examples of a special case where a new ionic mechanism (deinactivation of a g_{Ca}) greatly exaggerates this tendency to rebound depolarization. The importance of the enhanced rebound depolarization producing excitation in these cases is clear; it is the most reliable way to activate a neuron since activation of a g_{Ca} is preceded by a hyperpolarization that insures a maximal level of sodium-dependent electroresponsive "readiness" during the rebound. Hyperpolarization of the membrane may be viewed metaphorically as the stretching of a bow, the rebound as the release of the arrow. Synchronization is then attained by the simultaneous release of arrows, and the interval is the time necessary to stretch the bows once again.

Acknowledgment

Research was supported by United States Public Health Service grant NS-13742 from the National Institute of Neurological and Communicative Disorders and Stroke.

REFERENCES

Anderson, P. and T. A. Sears (1964) The role of inhibition in the phasing of spontaneous thalamo-cortical discharge. *J. Physiol. (Lond.)* **173:**459–480.

Bennett, M. V. L. and D. A. Goodenough (1978) Gap junctions, electrotonic coupling and intercellular communication. *Neurosci. Res. Prog. Bull.* **16**(3):377–463.

Berger, H. (1929) Ueber das Elektrenkephalogramm des Menschen. *Arch. Psychiat.* **87:**527.

Bower, J. and R. Llinás (1982) Simultaneous sampling and analysis of the activity of multiple, closely adjacent, cerebellar Purkinje cells. *Abstr. Soc. Neurosci.* **8:**830.

Bower, J. and R. Llinás (1983) Simultaneous sampling of the responses of multiple, closely adjacent, Purkinje cells responding to climbing fiber activation. *Abst. Soc. Neurosci.* **9:**607.

Bullock, T. H. (1959) Neuron doctine and electrophysiology. *Science* **129:**997–1002.

Bullock, T. H. (1979) Evolving concepts of local integrative operations in neurons. In *The Neurosciences: Fourth Study Program,* F. O. Schmitt and F. G. Worden, eds. M.I.T. Press, Cambridge, pp. 43–49.

Carpenter, D. O., ed. (1982) *Cellular Pacemakers,* Vol. 1: *Mechanisms of Pacemaker Generation.* Wiley, New York.

Connor, J. A. and C. Stevens (1971) Prediction of repetitive firing behavior from voltage clamp data on an isolated neuron soma. *J. Physiol. (Lond.)* **213:**31.

Eccles, J. C., R. Llinás, and K. Sasaki (1966) The excitatory synaptic action of climbing fibres on Purkinje cells of the cerebellum. *J. Physiol. (Lond.)* **182:**268–296.

Eckert, R. and H. D. Lux (1976) A voltage-sensitive persistent calcium conductance in neuronal somata of Helix. *J. Physiol. (Lond.)* **254:**129–151.

Gwyn, D. G., G. P. Nicholson, and B. A. Flumerfelt (1977) The inferior olivary nucleus in the rat: A light and electron microscopic study. *J. Comp. Neurol.* **174:**489–520.

Hagiwara, S., K. Kusano, and N. Saito (1961) Membrane changes of Onchidium nerve cell in potassium-rich media. *J. Physiol. (Lond.)* **155:**470–489.

Jahnsen, H. and R. Llinás (1984a) Electrophysiological properties of guinea-pig thalamic neurones: An *in vitro* study. *J. Physiol. (Lond.)* **349:**205–226.

Jahnsen, H. and R. Llinás (1984b) Ionic basis for the electroresponsiveness and oscillatory properties of guinea-pig thalamic neurones *in vitro. J. Physiol. (Lond.)* **349:**227–247.

King, J. W. (1976) The synaptic cluster (glomerulus) in the inferior olivary nucleus. *J. comp. Neurol.* **165:**387–400.

Lamarre, Y., C. de Montigny, M. Dumont, and M. Weiss (1971) Harmaline-induced rhythmic activity of cerebellar and lower brain stem neurons. *Brain Res.* **32:**246–250.

Llinás, R. (1981) Electrophysiology of the cerebellar networks. In *Handbook of Physiology,* Vol. 2: *The Nervous System,* Part 2, V. B. Brooks, ed., American Physiological Society, Bethesda, Md., pp. 831–976.

Llinás, R. (1984) Rebound excitation as the physiological basis for tremor: A biophysical study of the oscillatory properties of mammalian central neurons in vitro. In *Movement Disorders: Tremor,* L. J. Findley and R. Capildeo, eds. Macmillan, London, pp. 165–182.

Llinás, R., R. Baker, and C. Sotelo (1974) Electrotonic coupling between neurons in cat inferior olive. *J. Neurophysiol.* **37:**560–571.

Llinás, R. and R. Hess (1976) Tetrodotoxin-resistant dendritic spikes in avian Purkinje cells. *Proc. Natl. Acad. Sci. U.S.A.* **73:**2520–2523.

Llinás, R. and H. Jahnsen (1982) Electrophysiology of mammalian thalamic neurons *in vitro. Nature* **297:**406–408.

Llinás, R. and M. Sugimori (1980a) Electrophysiological properties of *in vitro* Purkinje cell somata in mammalian cerebellar slices. *J. Physiol. (Lond.)* **305:**171–195.

Llinás, R. and M. Sugimori (1980b) Electrophysiological properties of in vitro Purkinje cell dendrites in mammalian cerebellar slices. *J. Physiol. (Lond.)* **305**:197–213.

Llinás, R. and R. Volkind (1973) The olivo-cerebellar system: Functional properties as revealed by harmaline-induced tremor. *Exp. Brain Res.* **18**:69–87.

Llinás, R. and Y. Yarom (1981a) Electrophysiology of mammalian inferior olivary neurons *in vitro*. Different types of voltage-dependent ionic conductances. *J. Physiol. (Lond.)* **315**:549–567.

Llinás, R. and Y. Yarom (1981b) Properties and distribution of ionic conductances generating electroresponsiveness of inferior olivary neurons *in vitro*. *J. Physiol. (Lond.)* **315**:569–584.

Meech, R. W. and N. B. Standen (1975) Potassium activation in Helix aspersa neurons under voltage clamp: A component mediated by calcium influx. *J. Physiol. (Lond.)* **249**:211–239.

Montigny, C. de and Y. Lamarre (1973) Rhythmic activity induced by harmaline in the olivo-cerebellar-bulbar system of the cat. *Brain Res.* **53**:81–95.

Neu, J. C. (1980) Large populations of coupled chemical oscillators. *SIAM J. Appl. Math* **38(2)**:305–316.

Neuner, A. and H. Tappeiner (1894): Ueber bie Wirkungen der Alkaloide von Peganum harmala, insbesonders des Harmalins. *Arch. Exp. Pathol. Pharmakol.* **36** I: 69.

Rutherford, J. G. and D. G. Gwyn (1977) Gap junctions in the inferior olivary nucleus of the squirrel monkey, *Saimiri sciureus*. *Brain Res.* **128**:374–378.

Schwartzkroin, P. A. and M. Slawsky (1977) Probable calcium spikes in hippocampal neurons. *Brain Res.* **135**:17–161.

Sotelo, C., R. Llinás, and R. Baker (1974) Structural study of the inferior olivary nucleus of the cat: morphological correlates of electrotonic coupling. *J. Neurophysiol.* **37**:541–559.

Villablanca, J. and F. Riobo (1970) Electroencephalographic and behavioral effects of harmaline in intact cats and in cats with chronic mesencephalic transection. *Psychopharmacologia* **17**:302–313.

Wong, R. K. S., D. A. Prince, and A. I. Basbaum (1979) Intradendritic recordings from hippocampal neurons. *Proc. Natl. Acad. Sci. U.S.A.* **76**:986–990.

Yarom, Y. and R. Llinás (1981) Oscillatory properties of inferior olive cells. A study of guinea pig brain stem slices *in vitro*. *Abstr. Soc. Neurosci.* **7**:864.

Part Four

INTEGRATIVE PROCESSES: BEHAVIOR

Commentary

WALTER HEILIGENBERG

Scripps Institution of Oceanography
University of California, San Diego
La Jolla, California

Our search for the neuronal bases of animal behavior has succeeded increasingly in recent years. This is due to a vast improvement in neurophysiological and neuroanatomical techniques, as well as to a growing influence of ethology. Where the former have provided more powerful tools, the latter has significantly guided the strategy of our approach by reminding us that organisms have evolved in adaptation to particular ecological conditions and that the study of their physiology should therefore be guided by evolutionary considerations.

Whereas, for example, earlier investigators of sensory systems were primarily concerned with the physical exactness of their stimulus regimes, they often failed to ask whether these forms of stimulation approached any of those stimulus situations to which the animal had evolved to respond. By studying, for example, the responses of higher-order acoustical neurons to white noise bursts, clicks, and pure tones of various amplitudes and durations, we may find ourselves in a jungle of classes of neurons with no apparent order in sight. To the extent, however, that we recruit stimulus regimes that mimic those to which the animal is adapted to respond in its natural behavior, we discover marvels of neuronal specialization, and meaningful classes of neurons can soon be defined. Moreover, sensory, motor, and central-nervous specializations of a given species should reveal their significance, as comparative studies of related species will identify behavioral specializations obviously linked to these neuronal specializations. The comparison of ecological and behavioral adaptations in Cyclostomes, for example, should ultimately help to understand why the telencephalic hemispheres of hagfish are two to three times larger than those of lampreys, to name just one such case mentioned in Chapter 20 by Northcutt.

In studying an information-processing system of considerable complexity, it is imperative to know the natural forms of inputs to this system and the conditions under which it normally functions. In the case of a man-made system this means that we know the purpose for which it has been designed. In the case of the animal this means that we know the set of natural stimuli as well as the ecological constraints under which it has evolved. Without this knowledge, we might waste time and effort in studying epiphenomena, and we may only by accident stumble upon the relevant properties of the system. In view of the staggering complexity of brains, our limitations of time and resources demand this approach all the more.

The exploration of the animal's natural behavioral repertoire and the comparative study of species with different behavioral and ecological specializations are indispensable guides for the conception of meaningful physiological experiments on integrative central-nervous mechanisms. This becomes particularly obvious in the investigation of sensory information processing. After behavioral analysis has helped us to determine the nature of a stimulus complexity that controls a biologically adaptive behavioral response, we can, by stepwise modification and simplification of this stimulus complexity, single out those physical variables and their particular mutual relations that are relevant to the control of this behavior. This analysis then enables us to postulate sensory and central-nervous specializations that are logically required for the perception and computation of those relevant stimulus variables. Starting out from the sensory level and advancing through levels of higher-order neurons, we can then successively determine the particular neuronal implementation of abstract computational algorithms postulated on the basis of behavioral experiments, as demonstrated in Konishi's Chapter 19 on directional hearing in owls.

Physiological experiments that disregard the behavioral significance of brain structures are countless, and in spite of their often impressive technical accomplishments have told us little abut the functional significance of these structures. By gross electrical stimulation of fiber tracts, for example, we may certainly learn about the existence of pathways. But unless such stimuli sufficiently mimic the orchestration of activities in assemblies of nerves that characterize responses to natural stimulus patterns, we may never understand the integrative properties of these structures. We know so little about the integrative function of the cerebellum, for example, because we have not been very successful in identifying behavioral responses that could be used as assays to determine relevant input patterns to this structure. By contrast, we have learned much, for example, about visual control of motor responses in flies, auditory orientation in frogs, owls and bats, or multimodal representation of space in snakes and owls because behavioral assays have told us what to look for specifically in the nervous systems of these animals (see Chapter 5 by Knudsen, Chapter 18 by Hartline, and Chapter 19 by Konishi).

One may argue, however, that structures such as the cerebellum and, even more so the monaminergic systems of the pons, are not direct elements in a chain of neuronal stations leading from the perception of specific stimulus patterns to particular behavioral responses, and that it will therefore be very difficult to find simple stimulus-response paradigms to assess the functional significance of these structures. If the function of a given brain structure were to modulate motivational states, perception, and attention or to modify and adapt responses of other parts of the brain, it might indeed not reveal itself in simple behavioral tests and under the restricted conditions of standard laboratories. But a wider exploration of the animal's behavioral repertoire and ecology, paired with sufficiently subtle tests of its internal states and behavioral and physiological responses, could well yield insight into the functional significance of those brain structures that operate, so to speak, in the background of behavioral organization. Irreversible lesions of such structures may not always be a suitable tool for their exploration, but small perturbations and monitoring of their electrical and biochemical activity under a multitude of behavioral and environmental conditions may ultimately reveal their role. As will become apparent from Chapter 17 by Bloom, the tools for such approaches are now well within reach.

We may wonder whether a theory of brain and behavior ultimately can be formulated at the level of individual neuronal activity or whether a higher-level language, detached from single neurons, might be more economical, if not at all the only one possible. Much as a gas can be described conveniently in terms of its volume, temperature, and pressure and not at all in terms of the locations and momenta of its individual molecules, the properties of large assemblies of neurons may have to be described by general state variables of whole systems. The determination and exploration of appropriate variables of this kind may be one of the most challenging tasks in the study of brain and behavior.

Chapter Seventeen

CELLULAR MECHANISMS UNDERLYING BEHAVIOR

FLOYD E. BLOOM

Division of Preclinical Neuroscience and Endocrinology
Scripps Clinic and Research Foundation
La Jolla, California

My concern over the past several years has been the nature of the information encoded into chemical neurotransmitters and the way in which this information can be assembled into systems regulating behavior. The complexity of this concern has been intensified by three recent series of major advances in mammalian cellular neurobiology. First, countless new interneuronal connections have been revealed by the new cellular tracing methods. Second, new methods of electrophysiological analysis, particularly the *in vitro* preparations (see Llinas, Chapter 16 of this volume) have revealed a whole new range of ionic conductance mechanisms which can be influenced by neurotransmitters to enrich their signaling capability. Third, modern methods of chemical analysis have provided an ever-increasing list of new neurotransmitter candidates (Bloom, 1983b). In fact, there is a bewildering onslaught of new, transmitterlike, chemicals being offered to us at rates far faster than their sites and mechanisms of operation can be determined.

The molecular diversity among the many molecular messengers now available for consideration as transmitters, each having apparent classes of specific mechanisms of action at selected synaptic connections, suggests to me that there may be underlying organizational principles associated with neurotransmission by which these properties may be analyzed and compared for functional insight into the operations they control. As other contributions to this book make clear to open-minded readers, it is quite obvious that these organizational principles and operational control mechanisms operate widely throughout many nervous systems, not just those of vertebrates. In this chapter I present some new musings on this complex situation, looking for means to make it simpler so that some integrative actions of the mammalian nervous system may be better appreciated. Other more extensive coverages have been presented elsewhere (Bloom, 1978, 1979, 1983b).

HYPOTHESES OF TRANSMITTER DIVERSITY

One of many possible hypothetical views of transmitter diversity is that the same transmitter is released by all those cells needed to execute the individual steps leading to fundamental biobehavioral operations such as feeding, sleeping, mating, defending, or learning, or even more detailed macrofunctions such as regulation of body temperature or blood glucose. Execution of this type of function might then require connections that coordinate the outputs of several chemically coded nuclei. Theoretical proposals to explain the functions of specific peptides on this operational level of analysis (see Hoebel, 1983; Hokfelt et al., 1980; Iversen, 1983; Smith, 1983) frequently employ this argument: the presence of the same peptides in multiple sites (e.g., brain, gut, and endocrines) implies that the peptide can purposefully integrate all three sets of tissues leading to a concerted multicellular action. Despite the appeal of the "one transmitter–one behavioral function" concept, pharmacological approaches directed against neuronal cell populations containing the same monoamine transmitter have yet to provide compelling evidence that any complex behavioral function is so simply encoded. On the other hand, recent work on peptides related to drinking (Fitzsimons and Writh, 1978) and pain perception (Simon and Hiller, 1978; Snyder, 1980) may prove more successful. However, one must still question whether any single behavioral function could be subserved by cholecystokininlike peptides in the cerebral cortex (see Iversen, 1983) and cholecystokinin-secreting cells in gut (Rehfeld et al., 1979) to account for the broad spatial distribution of this peptide.

Future experimental work may well serve to strengthen such general hypothetical views of transmitters. More incisive operational definitions of the "purpose" of certain behaviors (see Koob and Bloom, 1982) may eventually provide general functional abstractions that will clarify the integral functions of widely separated cells sharing the same peptide or amine. Thus cells secreting angiotensin or vasopressin could be seen not as signals for drinking or for blood pressure elevation but rather as signals to control extracellular fluid volume or movement. At the same time, alternative views of peptide and amines may still be needed to provide different frameworks for interpretation of the forthcoming data.

DOMAINS OF NEUROTRANSMITTER DIVERSITY

A more simple interpretation of neurotransmitter diversity is that the chemical nature of a neurotransmitter (i.e., amino acid, amine, or peptide) does not per se provide insight into the functional role mediated by the neurons that secrete it. Rather, we must focus on the mechanisms by which that transmitter regulates the cells that can respond to it. If that generalization is tenable, then, what other discriminative features of the system may be more useful for organizing the diverse range of chemical messengers? The operations of all neurons can be charted on two domains, space and time, for comparative analysis. The spatial domain of a neuron is the total target cell area to which that neuron sends information. Similarly, the temporal domain is the time course of the neuron's effects on its targets. Let us now ask whether the spatial and temporal domains of chemically characterized neuronal circuits provide any hints to the nature of the operations such circuits may perform.

Present concepts of the structural organization of the brain rely heavily on two principles of connectivity. In the primary sensory and motor pathways, the prevailing principle has been the classical concept of hierarchical relationships in which the transmission of information is highly sequential and specific (Schmitt et al., 1976). Under this concept, primary receptors transmit to primary relay neurons and on upward to the ultimate sensors in the cortex; for motor output systems the reverse would be true, progressing downward from the cortex to the final common motor output. Although parallel processing routes may be added into these hierarchies for internal refinement of control, the classical concept of the hierarchical or "throughput" system is that destruction of one link incapacitates the chain. No transmitter molecule for any such throughput system, up or down, has as yet been identified, with the exception of some of the first input links (Nicoll et al., 1980a) and some of the final common output links (Hokfelt et al., 1980).

The second widely applied principle of neuronal connectivity is that of the local circuit neuron (Rakic, 1975): typically a small, frequently unipolar neuron whose efferent processes bear the morphology of dendrites. The basic feature of a local circuit neuron is that its connections are established exclusively within the local vicinity in which the cell body is found. Such small interneurons can exert significant control over information flowing through that locale, and may do so through "dendritic" release of their transmitter and without action potentials. In some cases of inhibitory local circuits, the amino acids gamma-aminobutyrate (GABA) or glycine (GLY) have been implicated as the transmitter (Werman, 1972), and some of the peptides would seem to fit this class as well (Barker et al., 1980; Nicoll et al., 1980b; Okamoto et al., 1983; Pittman and Siggins, 1981; Siggins and Gruol, 1984; Siggins and Zieglgansberger, 1981).

A third form of spatial domain is that illustrated by the major monaminergic systems of the pons, which I view as a "single-source, divergent" morphology (Bloom, 1979, 1983b; Moore and Bloom, 1979) in which a few neurons diverge to a large number of widely distributed, specific targets. These target cells do not exhibit any obvious direct functional relationships to each other. This suggests that monoamine systems may perform similar general tasks on a large number of circuits. In fact, the noradrenergic system on which my colleagues and I have centered our obsessions appears to project generally to the hierarchical output neurons of the regions it innervates (see Foote et al., 1983).

TEMPORAL DOMAINS

On the scale of time, nerve circuits can be discriminated on the basis of how long they influence their targets after a single activation. If separated from the distances involved, such timing events generally depend on the time course of action of the transmitter on its receptor. Of the known systems, amino acids would be at the fast end of the scale (acting in tens of milliseconds) and monoamines at the slow end (actions lasting hundreds of milliseconds or more) (see Bloom, 1979).

DIVERSITY SPANS THE DOMAINS

However, this two-dimensional approach is clearly too much of an oversimplification, as can easily be seen from only two examples of intratransmitter diversity. For example, the monoaminergic neurons as a class do not operate over a single spatial domain or temporal epoch. The dopamine-containing cell systems alone cover an extremely wide range from the ultrashort systems of the retina and olfactory bulbs to the longer and far more highly arborized mesocortical systems (Moore and Bloom, 1978, 1979). Within the spatial domain, the serotonergic systems seem to be as divergent and extensive as the noradrenergic systems but are perhaps somewhat more succinct in the temporal display of their synaptic actions (Wang and Aghajanian, 1977). Cholinergic systems may also cover a broad spatial domain, but neither their circuitry nor their synaptic time spans of action within the CNS are as yet specifiable (Fibiger, 1982; Levey et al., 1983).

Those neurons that contain biologically active peptides, as determined immunohistochemically (see Bloom and McGinty, 1981), do not fit at all onto a two-dimensional domain map as if they were functionally equivalent units of single coherent operational class: some presumptive peptidergic cells are small interneurons, such as the enkephalin-containing cells, whereas others cover significantly broader spatial domains, such as those cells in the CNS and peripheral nervous system that contain vasopressin (Swanson and Mogenson, 1981), somatostatin (Morrison et al.,

1982, 1983a), substance P (Nicoll et al., 1980a), or vasoactive intestinal polypeptide (Iverson, 1983; Morrison et al., 1983c). Unfortunately, there are as yet no data with which to specify for any of these central peptidergic systems the duration of the synaptic actions they may mediate. Data from studies applying the peptides exogenously to test systems suggests that their durations of action under these conditions may be longer than the effects of simple amino acids (see Nicoll et al., 1980a).

A THIRD DOMAIN OF FUNCTION

Neuronal systems differ from each other in more ways than can be expressed simply in terms of the spatial and temporal domains over which they operate. At least one additional domain can be approached in this analysis, and that domain I have termed tentatively, as "energy" (Bloom, 1975, 1979, 1983a). This third domain of neuronal operation defines the functional properties, that is, mechanisms and consequences of the synaptic operations of that given neuron. The dimensions in strict energy terms of the putative shifts associated with any specifiable synaptic action are still to be calculated. Indeed "energy" as such may not even be the correct quality in which to quantitate this domain. However, I use this term to distinguish transmitters producing passive membrane responses (i.e., permitting responses to preexisting electrochemical gradients) from transmitters producing active responses on the membrane and other segments of the target cell (e.g., the activation or inactivation of an enzyme or an ion-exchange carrier process). Along this energy "dimension" we can obtain further separation between the passive response operations mediated by simple amino acids and some amines and those mediated by some noradrenergic and dopaminergic actions in which one consequence is the activation of adenylate cyclase (Bloom, 1975; Magistretti et al., 1981; Greengard, 1978; Reuter, 1983; Walaas et al., 1983). Some peptide hormones can also act on their central targets through activation of adenylate cyclase (Magistretti et al., 1981; Morrison and Magistretti, 1983). However, this does not seem to be a general feature for most centrally active peptides. Again, such features know no species barriers (see Strumwasser, Chapter 10 of this book; Walters et al. Chapter 11).

ELECTROPHYSIOLOGICAL SIGNS OF FUNCTIONAL DIVERSITY

When GABA or GLY have been examined, their receptors appear coupled to mechanisms that increase membrane conductance to ions whose equilibrium potentials lie near or below resting membrane potentials and frequently produce hyperpolarizations (Werman, 1972). At other sites in the brain, other presumptive amino acid-mediated excitatory and inhibitory synapses also appear to operate through increased membrane conductance to specific ions yielding depolarizations or hyperpolarizations, respectively.

Within the group of noradrenergic projections that have been tested physiologically (see Foote et al., 1983 for review), operations in the time domain indicate that the effects have long latencies (30–50 msec or more) and long durations (300–600 msec or more), if those "effects" of the pathway are judged strictly on the basis of changes in target cell firing patterns or transmembrane properties (Siggins and Bloom, 1981). Neurons of the locus coeruleus, like other monoaminergic neurons, appear to exert a potent feedback inhibition (Groves et al., 1979) within their nucleus of origin; this effect would tend to dampen the changes in firing imposed on these neurons by their afferent connections. Although locus coeruleus neurons can fire in short rapid bursts under certain behavioral conditions (Aston-Jones and Bloom, 1981a,b; Foote et al., 1980, 1983), the long-duration, high-frequency stimulus trains generally used to activate some behavioral effects of this nucleus may well have been highly unphysiological.

The effects of coeruleo-cerebellar and coeruleo-hippocampal projections and the iontophoretic simulations of the effects of coeruleo-cortical projections (see Foote et al., 1983; Siggins and Bloom, 1981) all adhere closely to the interpretation that the action of neurally released noradrenaline is mediated by beta adrenergic receptors coupled to adenylate cyclase, and hence according to the "second messenger" scheme of Sutherland (see Foote et al., 1983; Siggins and Bloom, 1981). The effects as judged by firing rates and transmembrane effects are overtly inhibitory, with hyperpolarizations accompanied by increased membrane resistance. However, when the effects are examined on other aspects of target cell functioning in different experimental contexts (see Foote et al., 1983 for review), the effects of the locus coeruleus appear to fit better the designation of "biasing" or "enabling" than they do simple "inhibition." The biasing or enabling function means only that in the epoch over which noradrenergic receptors are active, certain chemical messages received through other receptors can be enhanced or weighed. Such enabling effects can be regarded as predictable outcomes of the observed changes in membrane properties.

The electrophysiological analysis of neuropeptide actions, based on the application of synthetic peptides rather than on the ideal comparative index of synaptically released peptides, has presented a very large range of conventional and nonconventional actions. Such actions range from the membrane potential and membrane resistance changes of a classical transmitter, as exhibited by Substance P (Nicoll et al., 1980a) and by opioid peptides (Konishi et al., 1981; Nicoll et al., 1980b; Siggins and Zieglgansberger, 1981) on certain target neurons, to much more complex changes in voltage-dependent conductances (Berker et al., 1980; Pittman and Siggins, 1981; Werz and MacDonald, 1982). The latter effects will perhaps become more commonly observed as the methods of *in vitro* electrophysiological analysis become better perfected to permit detection of more subtle changes, especially those actions that would be considered atypical from an historical perspective (see Siggins and Gruol, 1984).

The voltage-dependent mechanisms might easily appear as "no actions" at all because they often reveal that the application of a neuropeptide produces no observable shift in either membrane potential or membrane resistance. However, when the test cell is moved from basal conditions as by the depolarizing action of a simple amino acid, the resultant depolarization can often be blunted or prevented by the simultaneous application of the same dose of the peptide.

This complex allostericlike interaction, with each transmitter acting at its own receptor site, has lead me to give it the abstract term "disenabling" which in my chemical lexicon of neuronal signals is the antonym for "enabling" (Bloom, 1979). I propose, further, that it may be useful to consider the pairs of coexisting neurotransmitters being recognized with moderate frequency throughout the central and peripheral nervous system (Hokfelt et al., 1980) as acting coordinately, with one transmitter initiating an action, and a coreleased peptide acting to modify that action or to terminate it.

BROADER FUNCTIONAL DOMAINS

Elsewhere (Bloom 1975, 1983b) I have suggested that the combination of electrophysiological and biochemical changes produced in target cells by the noradrenergic fibers should be considered together as a holistic set of responsive changes. Although more easily observed, the electrophysiologic events that result from activation of this system (hyperpolarization, increased membrane impedance) are in this holistic view simply another index of the altered state of the target cells resulting from this form of neurotransmission. For example, hyperpolarizing changes in transmembrane potential accompany the response of many nonneural cells to hormones and neurotransmitters that activate cyclic nucleotide synthesis. In the heart, catecholamines not only increase the force and frequency of cardiac contractions but also activate lipolysis and glycogenolysis to provide the cardiac muscle with increased substrates for energy metabolism. Taken as a whole, the electrophysiologic shifts in the properties of the target cell membrane and the concomitant shifts in intracellular metabolism could provide a cell state specific for an altered mode of information processing. On one hand, these effects could be considered "modulatory" in the sense that adenylate cyclase coupled noradrenergic receptors have altered information transmitted by other afferents for processing by the common target cell. However, the argument rapidly becomes semantic: if the act of modulating is the physiologic function of locus coeruleus neurons, then surely their activity transmits that "modulate" or "enable" message.

Under this holistic view of neurotransmitter function, recent data clarifying when noradrenergic circuits are naturally called into action suggest that at least one purpose of the cellular actions on the target cells is to mediate selective attention to novel events in the external world (see Foote et al., 1983). Based on what is documentable now, however, this overall system effect would appear ideally suited to integrate across both time and space, and to shift the metabolic "gears" of the receptive cells necessary to interpret the novel event. For example, activation of VIP bipolar cells (see Morrison et al., 1983a) by specific thalamic afferents to cortex could then interact within those cortical domains activated by these novel sensory events, further enhancing the interpretive opportunities in combination with the extrinsic monoaminergic inputs (see Magistretti et al., 1981).

COMPARATIVE CORTICAL ORGANIZATION OF THE NORADRENERGIC SYSTEM

The continued analysis of the NA coeruleo-cortical system in the rat has demonstrated two major characteristics: (1) there is a rich network of NA innervation throughout all layers and regions of the dorsal and lateral cortex which is characterized by a relatively uniform laminar pattern, and (2) the major NA fibers are oriented and travel longitudinally through the grey matter and branch widely, furnishing the coeruleo-cortical system with the unique capacity to modulate neuronal activity synchronously throughout a vast expanse of neocortex. Thus the NA innervation of neocortex may be viewed as a tangential afferent system whose organization and pattern of termination is different from the highly localized, radial nature of the thalamo-cortical afferents. A tangential system of this design is relatively easy to accept within the framework of the lissencephalic, largely homogenous structure of rat neocortex; however, how does an afferent system designed for such global influence relate to the highly differentiated, gyrencephalic neocortex of the primate?

In recent years we have turned our attention to the anatomical organization of the NA and 5-HT innervation of primate neocortex (see Morrison et al., 1983c). We have used immunohistochemical methods to investigate these systems in three primate species (squirrel monkey, cynomologous, and rhesus). An antiserum directed against human dopamine-β-hydroxylase (DBH) purified from pheochromocytoma tissue provided by Dr. Dan O'Connor, UCSD, was used to localize NA fibers while an antiserum directed against formaldehyde conjugated 5-HT (kindly provided by Mark Molliver) was used to localize 5-HT fibers. We have found these techniques to be superior to formaldehyde or glyoxylic acid induced fluorescence techniques in terms of sensitivity and permanence. In the course of our studies, we have found that the specific density and pattern of NA (or 5-HT) innervation in a particular cortical locus varies systematically as a function of several factors: the cytoarchitectonic region, the cortical lamina, the species of animal, age of the animal, and the functional interaction of the region with other brain areas. In this report we illustrate how each of these factors influence monoaminergic innervation patterns specifically.

Regional Specificity

The three areas we have studied in greatest detail, thus far, are dorsolateral prefrontal cortex (Brodmann areas 9 and 10), the primary somatosensory cortex (areas 3, 1, and 2), and primary visual cortex (area 17). The NA innervation of these two regions is similar in the following respects: (1) fibers are present in all six layers; (2) the innervation is dense and terminallike in layers IV and V; and (3) layer VI is characterized by fibers oriented parallel to the pial surface which follow the contours of the subcortical white matter. However, these regions differ with respect to specific laminar patterns of fiber distribution and orientation and by virtue of the fact that the primary somatosensory cortex has a very dense NA innervation, whereas the density of innervation of dorsolateral prefrontal cortex is low relative to the postcentral gyrus and the rest of the frontal and parietal lobes. The laminar pattern of NA innervation in primary visual cortex differs fundamentally from both prefrontal and primary somatosensory cortices and is discussed later in terms of laminar specificity and transmitter complementarity.

As in the rat, local damage in the frontal cortex leads to a loss of NA fibers in more caudal regions. In addition, in the primate cortex, long tangential fibers can be seen crossing cytoarchitectural boundaries (e.g., from primary motor to primary somatosensory). Thus the tangential intracortical trajectory that is characteristic of the rat brain is also a dominant feature of the NA innervation of the primate brain. The NA innervation of primate cortex exhibits a far greater degree of regional heterogeneity than the rat: most major cytoarchitectural regions exhibit distinctive patterns of innervation in terms of both density and laminar distribution of fibers. These regional variations in primate NA innervation do not seem to follow any simple geometric or functional organizational principles such as a pronounced preference for primary sensory cortices. By contrast, the 5-HT system (to be reported elsewhere) does appear to innervate primary sensory cortices more densely than secondary sensory or association areas.

Laminar Specificity

A detailed examination of primate primary visual cortex was undertaken to characterize possible laminar specialization of monoaminergic innervation patterns in the most obviously laminated and well-characterized region of neocortex. Initially, we characterized the NA and 5-HT innervation patterns within this cortical area in the squirrel monkey (*Saimiri sciureus*). The primary visual cortex has a relatively low density of NA innervation, whereas the 5-HT innervation, particularly in layer IV, is the densest of all neocortical regions. These two fiber systems exhibited a high degree of laminar specialization, and were, in fact, distributed in a complementary fashion: layers V and VI receive a moderately dense NA projection and a sparse 5-HT projection, whereas layers IVa and IVc receive a very dense 5-HT projection and are largely devoid of NA fibers. In addition, the NA fibers manifest a geometric order that is not so readily apparent in the distribution of 5-HT fibers. These patterns of innervation imply that the two transmitter systems affect different stages of cortical information processing: the raphe-cortical 5-HT projections may preferentially innervate the spiny stellate cells of layers IVa and IVc, whereas the coeruleo-cortical NA projection may be directed predominantly at pyramidal cells.

Species Specificity

Based on cytoarchitectural characteristics, investigators have further subdivided layer IVc of visual cortex into IVc-α and IVc-β, and layer V has been further subdivided into Va and Vb. This extended laminar differentiation of layers IVc and V is very easily discernable in Nissl stained sections of cynomolgous or rhesus (old world) monkey primary visual cortex, whereas in layers IVc and V of squirrel monkey (new world) primary visual cortex these further laminar differentiations are very subtle, if they are present at all. We have recently compared these monoamine innervation patterns in the visual cortex of old and new world monkeys. The old world monkeys receive an even denser 5-HT innervation of primary visual cortex than do the squirrel monkeys, whereas the overall density of the NA innervation is equivalent, or even slightly decreased, in the old as compared to new world monkeys. Indeed, the variations of these 5HT patterns are correlatable with the further cortical specialization in the old world species.

The 5-HT innervation of old world monkey visual cortex reflects this further laminar differentiation in that each of these sublaminae receives its own distinctive pattern and density of innervation. The very dense 5-HT innervation of IVc that is evident in the squirrel monkey is, in the old world monkeys, largely directed at IVc-α, which is far more densely innervated than IVc-β. Furthermore, layer Va in the old world monkeys has a higher density than the superficial portion of layer V in the new world monkey. More specifically, a distinct band of tangential fibers is present at the IVc-β-Va border in the old world monkeys which is not evident in the new world monkey. In addition, the overall density of 5-HT innervation in layers V and VI of old world monkeys is higher than that present in these layers of squirrel monkey primary visual cortex.

Two important points emerge from these data: (1) with the possible exception of layer IVc-β, the various sublaminae of layer IV (the geniculo-recipient layer) receive the densest 5-HT innervation in both old and new world monkeys; (2) the further laminar differentiation of visual cortex that is coincident with the more advanced phylogenetic level of the old world monkeys is reflected in further laminar differentiation of the monoamine projections to this cortical region (see Morrison et al., 1983a, 1982a,b).

These observations suggest that coincident with the extensive phylogenetic development and functional differentiation of neocortex in the primate there is a parallel elaboration and differentiation of the ascending NA and 5-HT projections. The remarkable heterogeneity of NA and 5-HT innervation observed in monkey neocortex is highly ordered, that is, it is systematically related to the cytoarchitectonic region, lamina, age, species, and functional nature of the particular area of neocortex being examined. In addition, there are important differences in the NA and 5-HT innervation in each of these "dimensions." In each case, the different innervation patterns point to functional distinctions between these two transmitter-specific systems. For example, laminar complementarity may result from the two systems terminating on different classes of neocortical neurons, whereas regional specificity may result from preferential participation in different aspects of behavior. Although such hypotheses will require further anatomic and physiologic investigation, the observations already in hand clearly point toward a substantial and independent role for each of these transmitter systems in neocortical information processing in the primate.

CONCLUSIONS

Let us assume, at least temporarily, that many classes of chemically coupled transductive systems exist to express the effects of transmitter receptors, and that these couplings can be ion or substrate specific, that they can be dependent or independent variables of energy production or Ca^{2+} translocations, and that they can operate actively or passively over a wide range of time. I view these probabilities as evidence for the concept that cells employ a variety of specific molecules to communicate with their targets and to achieve operational responses across space and time.

In my concept, cells exhibit a broad but finite array of receptor transductive mechanisms, and the diverse array of messenger molecules share ability to evoke the responses of this finite repertoire. On convergent targets such effects may permit enhanced integration of responding. Obviously such interactions permit a form of conditional chemical logic that offers many experimental challenges for its deduction.

Better data are needed to clarify the evolutionary significance of individual transmitter families and the correlative behavioral functions of the neurons that secrete them. However, the growing list of potential transmitter peptides operating in the brain is far from complete. We have recently begun to employ molecular genetic methods to study this question (see

Sutcliffe et al., 1983). Our initial studies lead us to believe there may be scores of new molecules yet to be characterized. All these factors must eventually be brought into the solution of the chemical cryptogram considered here before we can comprehend the full meaning of what specific chemical signals transmit across their own unique temporal and spatial domains.

REFERENCES

Aston-Jones, G. and F. E. Bloom (1981a) Activity of norepinephrine-containing locus coeruleus neurons in behaving rats anticipates fluctuations in the sleep-waking cycle. *J. Neurosci.* **1**:876–886.

Aston-Jones, G. and F. E. Bloom (1981b) Norepinephrine-containing locus coeruleus neurons in behaving rats exhibit pronounced response to nonnoxious environmental stimuli. *J. Neurosci.* **1**:887–900.

Barker, J. L., D. L. Gruol, L.-Y. M. Huang, J. F. MacDonald, and T. G. Smith (1980) Peptides: Pharmacological evidence for three forms of chemical excitability in cultured mouse spinal neurons. *Neuropeptides* **1**:63–82.

Bloom, F. E. (1975) The role of cyclic nucleotides in central synaptic function. *Rev. Physiol. Biochem. Pharmacol.* **74**:1–103.

Bloom, F. E. (1978) Is there a neurotransmitter code in the brain? In *Advances in Pharmacology and Therapeutics*, Vol. 2 *Neurotransmitters*, P. Simon, ed. Pergamon, New York, pp. 205–213.

Bloom, F. E. (1979) Chemical integrative processes in the central nervous system system. In *Neurosciences: Fourth Intensive Study Program*, F. O. Schmitt and F. G. Worden, eds. M.I.T. Press, Cambridge, pp. 51–58.

Bloom, F. E. (1981) Chemically coded transmitter systems. In *Development and Chemical Specificity of Neurons. Progress in Brain Research*, Vol. 51, M. Cuenod, G. W. Kreutzberg, and F. E. Bloom, eds. Elsevier/North-Holland Biomedical Press, Amsterdam, pp. 125–131.

Bloom, F. E. (1983a) The endorphins: A growing family of pharmacologically pertinent peptides. *Annu. Rev. Pharm. Toxicol.* **23**:151–170.

Bloom, F. E. (1983b) The functional significance of neurotransmitter diversity. *Am. J. Physiol.* (in press).

Bloom, F. E. and J. F. McGinty (1981) Cellular distribution and function of endorphins. In *Endogenous Peptides in Learning and Memory Processes*, J. McGaugh and J. Martinez, eds. Academic Press, New York, pp. 199–230.

Fibiger, H. C. (1982) The organization and some projections of cholinergic neurons of the mammalian forebrain. *Brain Res. Rev.* **4**:327–388.

Fitzsimons, J. T. and J. B. Writh (1978) The renin-angiotensin system and sodium appetite. *J. Physiol. (Lond.)* **274**:63–80.

Foote, S. L., G. Aston-Jones, and F. E. Bloom (1980) Impulse activity of locus coeruleus neurons in awake rats and monkeys is a function of sensory stimulation and arousal. *Proc. Natl. Acad. Sci. U.S.A.* **77**:3033–3037.

Foote, S. L., F. E. Bloom, and G. Aton-Jones (1983) Nucleus locus ceruleus: New evidence of anatomical and physiological specificity. *Physiol. Rev.* **63**:844.

Greengard, P. (1978) *Cyclic Nucleotides Phosphorylated Proteins, and Neuronal Function*. Raven, New York.

Groves, P. M., D. A. Staunton, C. J. Wilson, and S. J. Young (1979) Sites of action of amphetamine intrinsic to catecholaminergic nuclei: catecholaminergic presynaptic dendrites and axons. *Prog. Neuro-Psychopharmol.* **3**:315–335.

Hoebel, B. G. (1983) Neurotransmitters in the control of feeding and its rewards: Monoamines, opiates and brain-gut peptides. In *Eating and Its Disorders*, A. J. Stunkard and E. Stellar, eds. (in press).

Hokfelt, T., O. Johansson, A. Ljungdahl, J. M. Lundberg, and M. Schultzberg, (1980) Peptidergic neurones. *Nature* **284**:515–521.

Iversen, L. L. (1983) Nonopioid neuropeptides in mammalian CNS. *Annu. Rev. Pharmacol. Toxicol.* **23**:1–27.

Konishi, S., A. Tsunoo, and M. Otsuka (1981) Enkephalin as a transmitter for presynaptic inhibition in sympathetic ganglia. *Nature* **294**:80–82.

Koob, G. F. and F. E. Bloom (1982) Behavioral effects of neuropeptides: Endorphins and vasopressin. *Annu. Rev. Physiol.* **44**:571–582.

Levey, A. I., B. H. Wainer, E. J. Mufson, and M. M. Mesulam (1983) Co-localization of acetylcholinesterase and choline acetyltransferase in the rat cerebrum. *Neuroscience* **9**:9–22.

Magistretti, P. J., J. H. Morrison, W. J. Shoemaker, V. Sapin, and F. E. Bloom (1981) Vasoactive intestinal polypeptide induces glycogenolysis in mouse cortical slices: A possible regulatory mechanism for the local control of energy metabolism. *Proc. Natl. Acad. Sci. U.S.A.* **78**:6535–6539.

Moore, R. Y. and F. E. Bloom (1978) Central catecholamine neuron systems: anatomy and physiology of the dopamine systems. *Annu. Rev. Neurosci.* **1**:129–169.

Moore, R. Y. and F. E. Bloom (1979) Central catecholamine neuron systems: anatomy and physiology of the norepinephrine and epinephrine systems. *Annu. Rev. Neurosci.* **2**:113–168.

Morrison, J. H., R. Benoit, P. J. Magistretti, N. Ling, and F. E. Bloom (1982) Immunohistochemical distribution of prosomatostatin related peptides in hippocampus. *Neurosci. Lett.* **34**:137–142.

Morrison, J. H., R. Benoit, P. J. Magistretti, and F. E. Bloom (1983a) Immunohistochemical distribution of prosomatostatin related peptides in cerebral cortex. *Brain Res.* **262**:344–351.

Morrison, J. H., S. L. Foote, M. E. Molliver, F. E. Bloom, and H. G. W. Lidov (1982a) Noradrenergic and serotonergic fibers innervate complementary layers in monkey primary visual cortex: An immunohistochemical study. *Proc. Natl. Acad. Sci. U.S.A.* **79**:2401–2405.

Morrison, J. H., S. L. Foote, D. O'Connor, and F. E. Bloom (1982b) Laminar, tangential and regional organization of the noradrenergic innervation of monkey cortex: dopamine-β-hydroxylase immunohistochemistry. *Brain Res. Bull.* **9**:309–319.

Morrison, J. H., S. L. Foote, and F. E. Bloom (1983b) Laminar, regional, developmental, and functional specificity of monoaminergic innervation patterns in monkey cortex. In *Monoamine Innervation of Cerebral Cortex*, L. Descarries, T. A. Reader, and H. H. Jasper, eds. Alan R. Liss, New York.

Morrison, J. H. and P. J. Magistretti (1983) Monoamines and peptides in cerebral cortex: contrasting principles of cortical organization. *Trends Neurosci.* **6**:146–151.

Morrison, J. H., P. J. Magistretti, R. Benoit, and F. E. Bloom (1983c) The distribution and morphological characteristics of the intracortical VIP-positive cell: An immunohistochemical analysis. *Brain Res.* (in press).

Nicoll, R. A., C. Schenker, and S. E. Leeman (1980a) Substance P as a transmitter candidate. *Annu. Rev. Neurosci.* **3**:227–268.

Nicoll, R. A., B. E. Alger, and C. E. Jahr (1980b) Enkephalin blocks inhibitory pathways in the vertebrate CNS. *Nature* **287**:22–25.

Okamoto, K., H. Kimura, and Y. Sakai (1983) Effects of taurine and GABA on Ca spikes and NA spikes in cerebellar purkinje cells in vitro: intrasomatic study. *Brain Res.* **260:**249–259.

Pittman, Q. J. and G. R. Siggins (1981) Somatostatin hyperpolarizes hippocampal pyramidal cells *in vitro. Brain Res.* **221:**402–408.

Rakic, P. (1975) Local circuit neurons. *Neurosci. Res. Progr. Bull.* **13:**293–446.

Rehfeld, J. F., N. R. Goltermann, L.-I., Larsson, P. M. Emson, and C. M. Lee (1979) Gastrin and cholecystokinin in central and peripheral neurones. *Fed. Proc.* **38:**2325–2329.

Reuter, H. (1983) Calcium channel modulation by neurotransmitters, enzymes, and drugs. *Nature* **301:**569–574.

Schmitt, F. O., P. Dev, and B. H. Smith (1976) Electronic processing of information by brain cells. *Science* **193:**114–119.

Siggins, G. R. and D. L. Gruol (1984) Synaptic Mechanisms in the vertebrate central nervous system. In *Intrinsic Regulatory Systems of the Brain, Handbook of Physiology*, S. R. Geiger, ed. (in press).

Siggins, G. R. and F. E. Bloom (1981) Modulation of unit activity by chemically coded neurons. In *Brain Mechanisms and Perceptual Awareness*, O. Pompeiano and C. Ajmone Marsan, eds. Raven, New York, pp. 431–448.

Siggins, G. R. and W. Zieglgansberger (1981) Morphine and opioid peptides reduce inhibitory synaptic potentials in hippocampal pyramidal cells *in vitro* without alteration of membrane potential. *Proc. Natl. Acad. Sci. U.S.A.* **78:**5235–5239.

Simon, E. J. and J. M. Hiller (1978) The opiate receptors. *Annu. Rev. Pharmacol. Toxicol.* **18:**371–394.

Smith, G. P. (1983) The role of the gut in the control of food intake. *Viewpoints Dig. Dis.* **15:**1.

Snyder, S. H. (1980) Brain peptides as neurotransmitters. *Science* **290:**976–983.

Sutcliffe, J. G., R. J. Milner, T. M. Shinnick, and F. E. Bloom (1983) Identifying the protein products of brain-specific genes with antibodies to chemically synthesized peptides. *Cell* **33:**671–682.

Swanson, L. Y. and G. J. Mogenson (1981) Neural mechanisms for the functional coupling of autonomic, endocrine and somatory motor responses in adaptive behavior. *Brain Res. Rev.* **3:**1–34.

Walaas, S. I., D. W. Aswad, and P. Greengard (1983) A dopamine- and cyclic AMP-regulated phosphoprotein enriched in dopamine-innervated brain regions. *Nature* **301:**69–71.

Wang, R. K. and G. K. Aghajanian (1977) Inhibition of neurons in the amygdala by dorsal raphe stimulation: mediation through a direct serotonergic pathway. *Brain Res.* **120:**85–102.

Werman, R. (1972) Amino acids as central neurotransmitters. In *Proceedings of Association for Research on Nervous and Mental Disorders*, Vol. 1, *Neurotransmitters*, pp. 147–180.

Werz, M. A. and R. L. MacDonald (1982) Opioid peptides decrease calcium-dependent action potential duration of mouse dorsal root ganglion neurons in cell culture. *Brain Res.* **239:**315–321.

Chapter Eighteen

MULTIMODAL INTEGRATION IN THE BRAIN

Combining Dissimilar Views of the World

PETER H. HARTLINE

Eye Research Institute of Retina Foundation
Boston, Massachusetts

. . . .

"I see," quoth he, "the elephant
is very like a snake"

. . . .

"Tis clear enough the elephant is
very like a tree."

. . . .

THE BLIND MEN AND THE ELEPHANT
JOHN GODFREY SAXE

In a series of papers begun in the early 1950s, T. H. Bullock and his associates demonstrated that the trigeminal nerve fibers innervating the pit organs of pit vipers were responsive to very slight warming of their endings in the pit membrane (Fig. 1). They further showed that these warm fibers were sensitive enough to respond to an increase of infrared radiation incident on the pit membrane of a magnitude similar to that given off by a

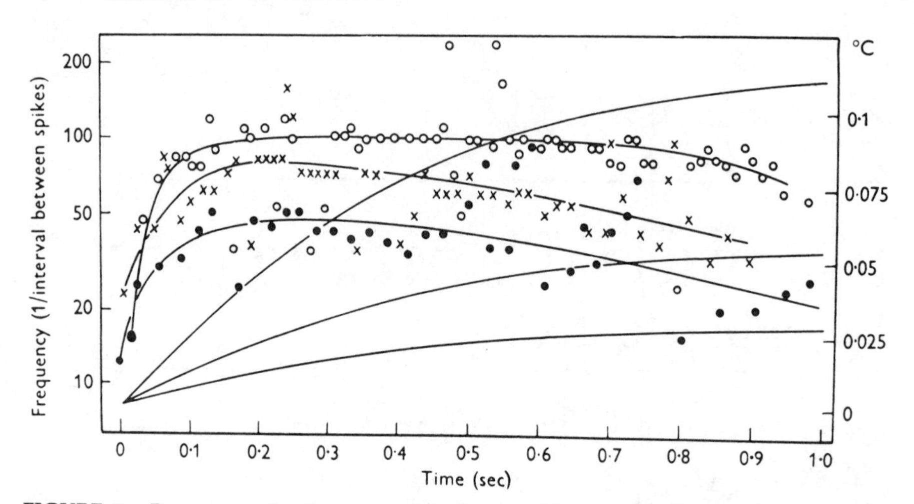

FIGURE 1. Responses of primary trigeminal nerve fiber to warming the pit membrane. Water flowing over membrane was slightly warmed beginning at time $t = 0$. Curves fitted to measured water temperature (scale at right) for three rates of heating are monotonically increasing. Upper (open circles), middle (× crosses), and lower (solid circles) curves show frequency of firing of the nerve fiber, and correspond to the upper, middle, and lower temperature curves. Note the sensitivity to rate of rise of temperature and the small temperature changes required to evoke responses. By 0.06 sec the solid circle curve showed a doubling of response frequency; this corresponded to a measured temperature rise of 0.003°C (Bullock and Diecke, 1956). Calculations based on radiative heating of the pit membrane yielded a temperature change of 0.0005°C after 0.2-sec exposure of a hypothetical membrane to a rat-sized object 10°C above ambient and 50 cm distant from the membrane (Hartline, 1974). [Figure reprinted with permission of author and publisher from Bullock, T. H. and F. P. J. Diecke (1956) Properties of an infra-red receptor. *J. Physiol.* **134**:47–87.]

snake's warm-blooded prey (Bullock and Cowles, 1952; Bullock, 1953; Bullock and Diecke, 1956). The pit's sensory membrane (Fig. 2) is innervated in a very orderly manner by about 7000–8000 trigeminal axons (Bullock and Fox, 1957), presenting a heat-sensitive screen a few millimeters behind the aperture of the pit organ (Fig. 3). No lens is present. Instead, the pit aperture limits the region of the membrane illuminated by any warm source. Of the function of this organ, Bullock and Diecke (1956) conjectured that

> It is thus possible, with the [pit's] structure and the richness of [nerve] supply, to analyse the shadows and obtain information about direction, distance . . . and movements of small objects.

Bullock's initial investigations inspired an international effort to explore the novel infrared sensory system; over the intervening decades, research has largely substantiated the foregoing conjecture. Now, 30 years later, we are approaching an understanding of the processes by which the brain translates the patterns of warmth and their movements on the pit mem-

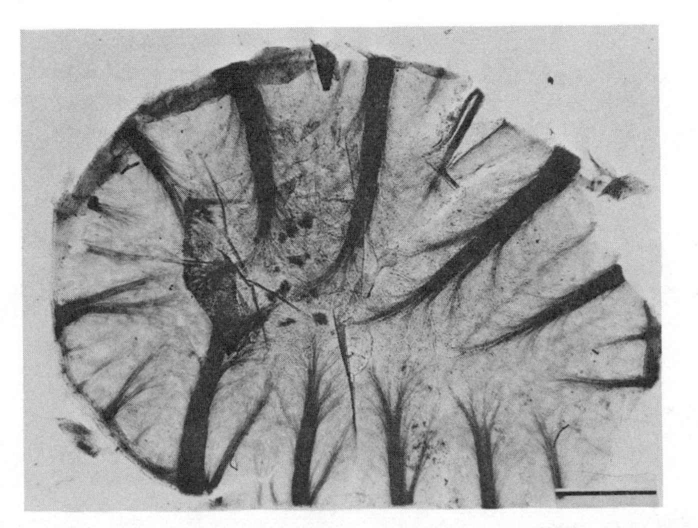

FIGURE 2. Whole mount of pit membrane; branchlets of the trigeminal nerve enter the pit membrane all around its margin and ramify in a very orderly fashion to innervate the entire membrane. Terminal mass endings, described by Terashima et al. (1970), are formed from a tuft of branches from each nerve fiber that further ramify within the confines of a 30–60-μm diameter region, in asociation with a Schwann cell. These endings occupy most of the pit membrane's volume and do not overlap each other substantially. [Printed with permission of author and publisher from Bullock, T. H. and W. Fox (1957) *Q. J. Microsc. Sci.* **98**:219–234.]

brane into information that allows these snakes to strike their prey in the dark with deadly accuracy. We are also discovering how information from this unusual special sense is integrated with visual information, thus giving the snakes the benefit of two senses cooperatively analyzing the patterns of visual and infrared radiation in their surroundings.

The optic tectum (which is referred to as superior colliculus, SC, in mammals), has a prominent visual input. It is thought to be an important brain center for initiating and controlling orienting behaviors and spatial attention processes (reviews: Sprague et al., 1973; Goldberg and Robinson, 1978). Information about infrared sources in the snake's surroundings reaches the optic tectum through a trigeminal nucleus, (Molenaar, 1974; Schroeder and Loop, 1976; Terashima and Goris, 1977; Stanford and Hartline, 1980, 1984) by way of a relay nucleus in the reticular formation, (Gruberg et al., 1979; Newman et al., 1980; Stanford et al., 1981). As with nonvisual neurons in the optic tecta and SC of other animals, infrared sensitive neurons in snake tectum are confined to the lower layers of the tectum (Kass et al., 1978). Tectal IR neurons have many properties similar to those of tectal visual neurons (Goris and Terashima, 1973; Hartline, 1974; Terashima and Goris, 1975; Hartline et al., 1978).

The prominence of both infrared and visual modalities in the optic tectum invites a search for principles of multimodal integration that are well

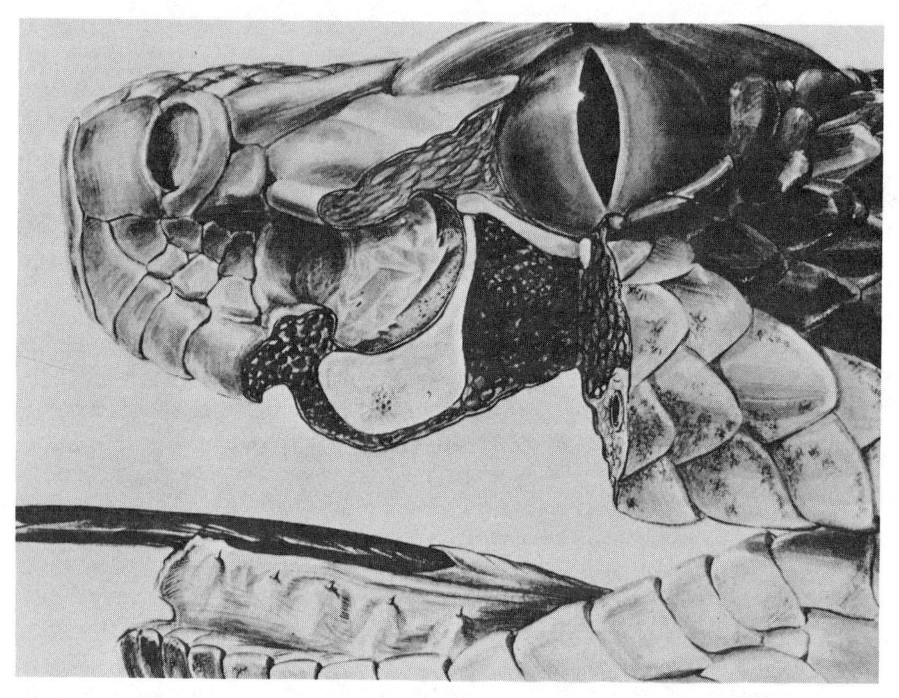

FIGURE 3. Cutaway drawing of pit organ. Note that the heat-sensitive membrane is two to three times the diameter of the pit's opening. A large region of the pit membrane would be illuminated by a remote, small infrared source. Sources located at different angular positions around the snake's head would warm different but overlapping regions. [Printed with permission of author and publisher from Bullock, T. H. and F. P. J. Diecke (1956) *J. Physiol. (Lond.)* **134**:47–87.]

illustrated by these unusual animals but are relevant to understanding the general principles of how nervous systems combine information from many dissimilar sensory modalities. No two senses bring the same information to the host; each sense has its own "point of view," dictated in part by physical processes and in part by sensory and neural specializations. Yet each sense brings to the brain some important part of the composite sensory experience upon which the animal's behaviors must be based. Some information, for instance spatial location, may be provided by several senses. Out of all of the sensory messages it receives, the brain must identify, sort, and synthesize what is relevant, and must avoid delay and indecision if sensory messages conflict. Several candidate functions (but not an exhaustive list) of modality combination follow. I discuss snake multimodal integration with reference to the candidate functions and point out the potential extension of these functions to other species and other senses.

1. *Sensory Substitution.* Any of several senses may provide a message that can signal the appropriateness of a behavior or set of behaviors; any such sense may substitute for another to evoke or modify the behavior. At some point, all such senses must directly or indirectly impinge upon appropriate motor elements that generate the behavior, thus leading to a sharing of output circuitry. To the extent that evolution need not "invent" parallel control circuits for the separate sense, an economy is realized if the senses access higher elements in the circuitry that controls behavior.

2. *Cue Sequence.* Objects or events may be characterized by the sequence of messages that they evoke by way of several senses, and the temporal pattern of messages, analyzed across modalities, may carry important information. Communication sequences involving olfactory, visual, auditory, and tactile stimuli are well known to ethologists. Analysis of cue sequences probably helps humans follow conversation in a noisy environment by watching lips or facial expressions of the people they are listening to.

3. *Division of Labor.* Several modalities may provide information about the same physical attributes or variables that characterize an object or event; each modality, however, may be specialized for a different subset or range of the descriptive space. An obvious example, considering locations in space, is that an animal with frontally placed eyes localizes objects visually in the frontal hemifield, but must rely upon other senses such as hearing or touch to detect and localize objects in the posterior hemifield. Different modalities are thus assigned primacy in different regions of space. Division of labor might occur in other ways. In many animals, for example, hearing becomes important for visionlike tasks at night; similarly, in weakly electric fish, electrolocation undoubtedly takes on a key role when water is turbid.

4. *Color and Texture.* Many attributes that are sensed by one modality differ qualitatively from those sensed by other modalities, yet the object's identity or significance is embodied in the composite sensory experience it generates. Odor, taste, hardness, and temperature belong to various objects as surely as do their shapes and motions. Such qualities maintain the same sort of relationship to an object as does color to a visual form, or texture to a three-dimensional surface. Modality interactions that identify such relationships might be very important, since an object may be recognized by the very combinations of senses that it stimulates. Communication sequences might have multimodality properties, as in a visual signal whose significance is different depending on an accompanying auditory, tactile, or odor signal.

5. *Signal Improvement.* The relationship between two modes of stimulation caused by an object or event may be sufficiently rigid that the animal can improve the certainty of detection or accuracy of analysis by summing input over several modalities, thus improving the signal-to-noise ratio in a

composite sensory channel. A barely audible rustle accompaning a dimly perceived motion may be localized more readily than either stimulus alone.

6. *Cross-Modality Attention.* A stimulus in one modality may not provide enough information to identify or localize an object or event fully, but may prime or alert the nervous system for arrival of complementary information by way of another modality. A message in one modality may enable a subset of circuits of a second modality to gain control of the entire set of output circuits. As example, an odor or a sound might direct an animal's attention toward a potential source of food or danger.

There seem to be qualitative differences between the modality interactions that are needed for some of the foregoing functions as opposed to others. In *sensory substitution* and *division of labor* the primary need for interaction is to prevent two modalities from interfering with each other, as would occur if they evoked conflicting behaviors. For the remaining functions, there must be cooperation between modalities, since it is the spatial and/or temporal correlation of activity in several sensory systems that carries the significant information. One intuitively expects to find convergence of sensory coding mechanisms to achieve compatibility of information represented in convergent modalities. One similarly expects to find nonlinear modality interactions as a neural implementation of the cross-correlation operation.

MULTIMODAL INTEGRATION IN RATTLESNAKES

Sensory Substitution

The merging of infrared and visual systems in snake optic tectum illustrates the sensory substitution function very well. The output circuitry to be shared is, in this example, the tectal neurons involved in initiating movements to orient toward and probably strike at targets sensed by the visual or infrared systems. The role of snake tectum in such behaviors has not been proven, but from fish and amphibians through rodents and primates, ablation and microstimulation studies have implicated the optic tectum as an important spatial orientation center. Both the infrared sense (Noble and Schmidt, 1937; Dullemeijer, 1961) and vision (Cock Buning et al., 1978) can mediate orientation and striking. A preliminary unpublished experiment by L. Stanford, M. Loop, and I, (Fig. 4; see also Newman and Hartline, 1982) showed that strikes based on the infrared sense are usually accurate to within 2–4 degrees.

The spatial organization of the visual system in snake tectum follows the same overall plan as it does in other vertebrates. It can be represented by a spatiotopic map relating average receptive field center coordinates to tectal location (Fig. 5*A*). Largely because of microstimulation studies, it has been proposed that deep tectal neurons having premotor function are arranged

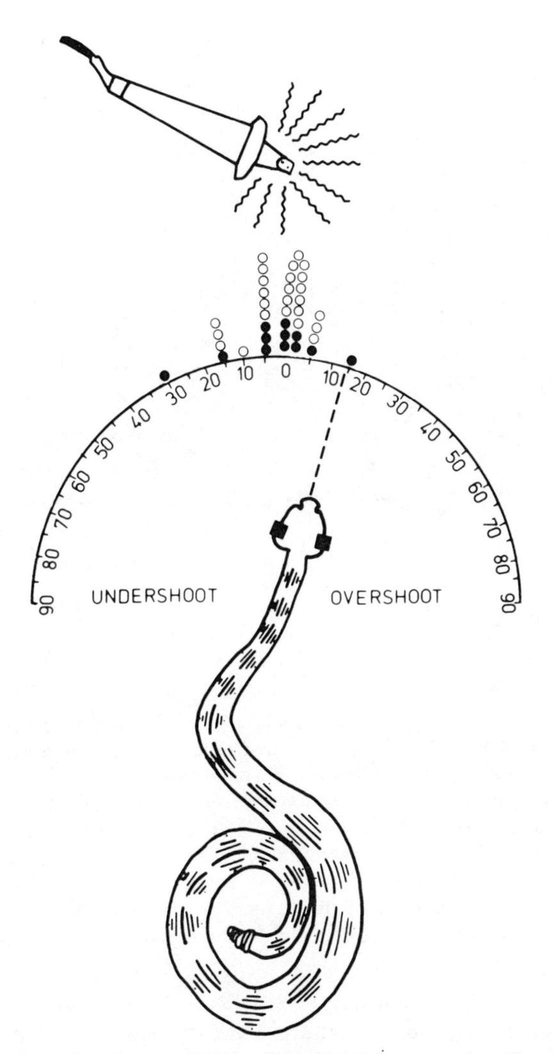

FIGURE 4. Accuracy of strike orientation of blindfolded rattlesnake. The snake's eyes were occluded with aluminum foil and a vaseline–lampblack mixture. A warm soldering iron was introduced at a distance of about 17 cm from the snake's head; it was near the horizontal plane and to the left or right of the extended midline by 0–90 degrees. If the soldering iron was presented concurrently with a mild shock to the tail, the snake might strike (toward the iron) as it was videotaped. The angular error of the direction of the strike was measured from stopped-frame analysis of the tape. A circular histogram (5 degrees bin size) of errors made for left presentation (○) and right presentation (●) shows that this blindfolded snake usually oriented correctly to within ± 2.5–5 degrees (average magnitude of error, 3°). (Unpublished data of L. R. Stanford, M. S. Loop, and P. H. Hartline.)

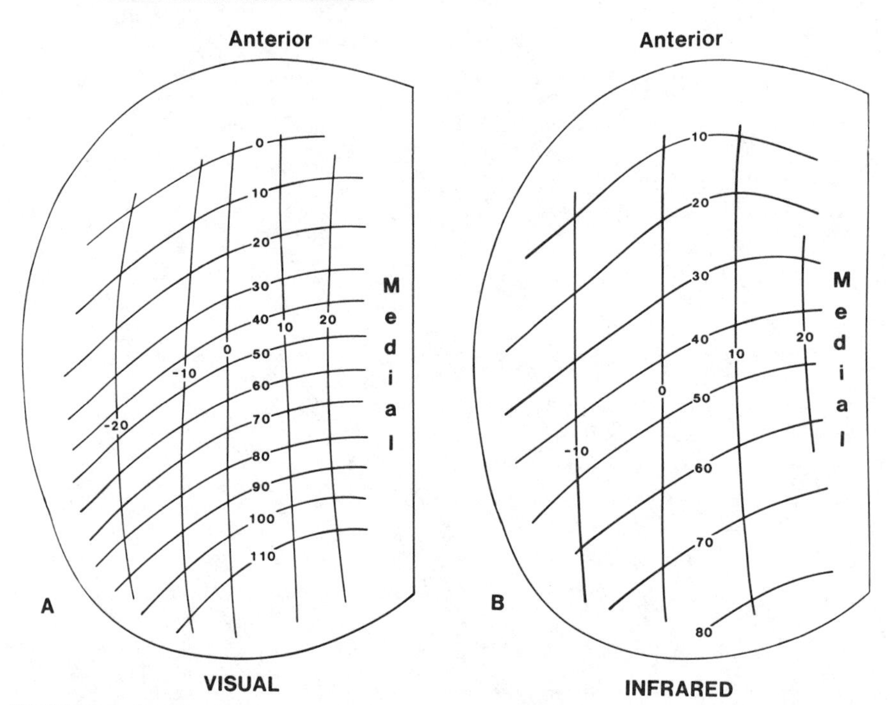

FIGURE 5. Idealized spatiotopic maps of visual (*A*) and infrared (*B*) systems in rattlesnake optic tectum based on four relatively complete maps plus several fragments. For each map, IR and visual receptive field center coordinates were determined at a grid of points on the tectum with a multiunit electrode. Latitude and longitude projection lines were drawn to reflect features that were consistent among the maps of individual snakes. Features to be noted are (1) rostral and caudal regions of high magnification in the infrared system; (2) rostral region of high visual magnification; (3) caudal region of low visual magnification; (4) alignment of visual and infrared axes; (5) greater infrared than visual magnification (a ratio of 1.7 for overall rostro-caudal magnification is shown; this is at the high end of the 1.4–1.7 range).

in a motor map that is approximately congruent to the overlying visual map (Schiller and Stryker, 1972). Thus excitation of a localized population of deep tectal neurons initiates an orientation movement toward the region of space that corresponds to the visual receptive fields of more superficial neurons. If one assumes that excitation of visual neurons by an interesting object causes excitation of the underlying premotor neurons, one has a large part of a qualitative model for the machinery of sensorimotor integration in orienting behaviors.

Against this conceptual framework, the infrared system is seen, at first glance, as very efficiently arranged. Measurements of infrared neuron receptive field center coordinates at a grid of tectal sites yield a spatiotopic map (Fig. 5*B*) whose axes are oriented similarly to the visual axes (Hartline, 1974; Terashima and Goris, 1975; Hartline et al., 1978). However, close examination of the maps reveals a prominent dissimilarity between the two

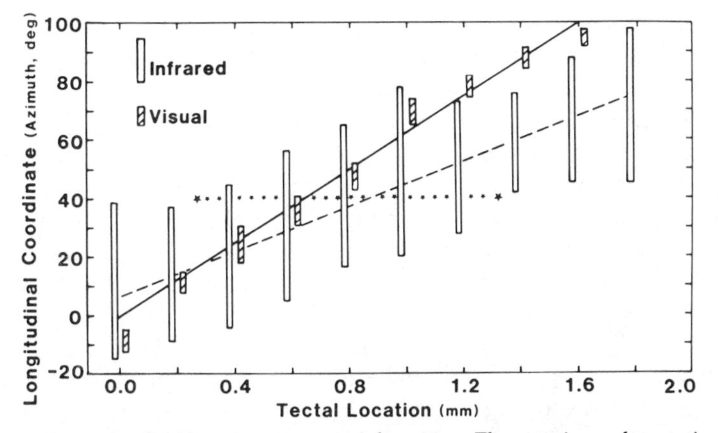

FIGURE 6. Receptive field location versus tectal position. The anterior and posterior borders of receptive fields were mapped along a rostro-caudal transect of the tectum. Each pair of bars represents the field location and size encountered by an electrode placed superficially (visual) and deep (infrared) in the same penetration. The reciprocal of the slope of each hand-drawn line gives the approximate overall rostro-caudal magnification factor for the corresponding sensory system. The visual magnification is 0.016 mm/degree; the infrared magnification is 0.026 mm/degree, yielding a ratio of infrared to visual magnification of 1.6. Compare also the bar sizes for visual and infrared systems. The rostro-caudal spread of the infrared tectal image of an object located at 40 degrees is shown by the dotted line and stars to be about 1.1 mm. A target located at 40 degrees would be at the posterior border of a receptive field between the 0.2- and 0.4-mm tectal positions and would be at the anterior border of a receptive field near the 1.4-mm location: All intermediate tectal locations would be excited to some extent. A similar image for a visual object is not shown because the visual fields are based on superficial neurons rather than the more appropriate, deep visual neurons, which have broader fields.

modalities. As Hartline et al. (1978) emphasized, the average spatial magnification factor, defined for two tectal loci as (tectal separation in millimeters)/(angular separation of receptive field centers mapped to the two loci, in degrees), is larger for the infrared map than for the visual map. The ratio of average infrared to visual magnification ranges from 1.4 to 1.7. This is illustrated directly as the ratio between the mean slopes for visual and infrared system in Figure 6; it can also be seen as the broader separation of map latitude (elevation) and longitude (azimuth) lines in the infrared map as compared to the visual map (Fig. 5).

The different magnifications do not necessarily imply that an object that stimulates both modalities will cause conflicting orientation commands as if it were perceived in two different locations at once. This is because the receptive field center maps do not tell the whole story about spatial sensory organization of the tectum, and the motor organization may be even less well represented by such maps. A small stimulus of either modality does not excite only the neurons at the map coordinates corresponding to its location. Rather, as was pointed out by McIlwain (1975) for visual stimuli, neurons will be excited if they lie within an image region surrounding the map coordinates of the stimulus. The size and shape of the tectal image is determined by the map, by the sizes and shapes of the receptive fields, and by the scatter of receptive field locations of neurons at each place in the map. McIlwain (1982) has suggested that the activity in a tectal image excites a set of outputs whose sum generates the appropriate orientation behaviors much as the sum of many vectors yields a resultant vector. One might imagine a set of *nasal* and *temporal* command units distributed such that the density of *nasal* units is high rostrally and low caudally, whereas *temporal* command units have high density caudally and low density rostrally. *Superior* and *inferior* command units must be arranged with the appropriate mediolateral density distributions. The final oriented output can then be thought of as determined by the balance of activity evoked in *nasal, temporal, superior,* and *inferior* units that fall within the tectal image.

Because the receptive field sizes of infrared neurons tend to be large (40–60 degrees for most) compared to visual neurons (15–30 degrees for the deeper ones), a very different and much larger tectal image must be generated by a small infrared stimulus than by a small visual stimulus. This point is illustrated by Figure 7, which allows comparison of tectal images between modalities (see figure legend for simplifying assumptions). The population of tectal output neurons excited by a visual or infrared stimulus must reflect the difference between tectal image sizes. Considering the spatiotopic organization and receptive field sizes of neurons of nonvisual modalities in the optic tectum of lizards (Stein and Gaither, 1981), owls (Knudsen, 1982), mice (Drager and Hubel, 1975), and cats (Stein et al., 1976), the large size of the tectal images of nonvisual stimuli must be the general rule, not something unique to rattlesnakes.

Because infrared responsive neurons are less sensitive toward the edges of their receptive fields, those that are more peripheral within the tectal image of an infrared stimulus must be less strongly excited than those that are relatively near the map coordinates of the stimulus. Figure 7*B* can be interpreted as dividing the infrared image into regions of high, intermediate, and low excitation, with the added assumptions that response amplitude is determined by distance from the field center, and the function has circular symmetry. The spatial distribution of activity evoked by an infrared stimulus in the population of premotor output neurons must depend on the magnitude of excitatory input that the premotor neurons receive from infrared neurons within the image (it is unknown whether infrared neurons are themselves output neurons). The foregoing analysis makes it clear that the field center maps of two modalities may be rather indirectly related to the output patterns engendered by stimuli in two modalities.

The concept of images has direct bearing on the interpretation of the difference between the magnification of the infrared and visual maps. If one imagines a map that relates the stimulus location in space to the tectal location of the centroid of its infrared image, then much of the intermodal-

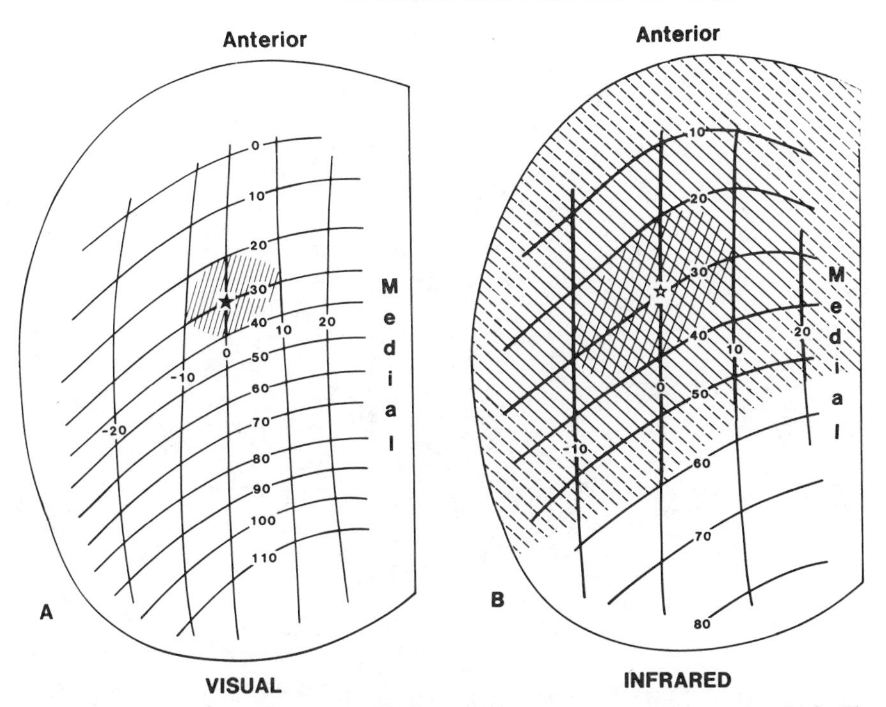

FIGURE 7. Visual (*A*) and infrared (*B*) maps with corresponding images (hatched). The visual image is based on the assumption that all deep visual neurons have round receptive fields 20 degrees in diameter. The infrared image shown with broken hatching is based on a round, 60-degrees-diameter receptive field. The solid and cross hatching can be thought of as images comprised of cells with 40 degrees and 20 degrees receptive field diameters. Alternatively, they can be thought of as indicating areas of greater excitation of neurons with 60-degree fields (where response magnitude versus stimulus location is a single-peaked, radially symmetric function). The hypothetical stimulus objects for image construction are located at the spot in space corresponding to the intersection of the equator with the 30-degree azimuth (longitude) line, which corresponds to the star in each map.

ity magnification difference disappears. This is because the rostral border of the tectal image of an anteriorly located stimulus is cut off by the rostral edge of the tectum (Fig. 8). Consider an anteriorly placed stimulus that is moved anteriorly by 10 degrees; the posterior border of its image moves rostrally in proportion to the field center map's magnification factor, but the rostral border remains fixed. The center-of-excitation (in analogy to center of gravity) of the image moves as little as half the distance moved by the caudal edge of the image. A similar argument holds for a posteriorly placed stimulus and the caudal displacement of the center-of-excitation of its image as the stimulus moves posteriorly. Figure 9 shows the rostrocaudal locations of the centers-of-excitation estimated for an anterior-posterior series of stimulus locations and compares them to receptive field center locations. The overall infrared magnification based on center-of-

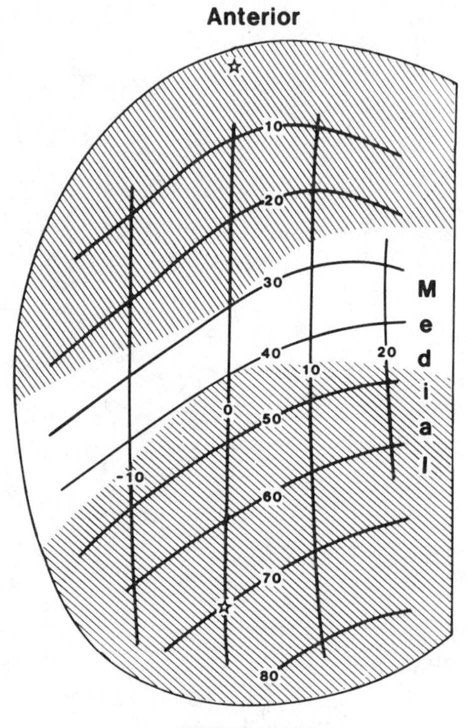

FIGURE 8. Images of infrared objects located on the equator at 0 degrees and 70 degrees azimuth, (based on round 60-degree receptive fields). Note that the images are limited by the rostral and caudal margins of the tectum, respectively.

excitation is close to what is found for the visual system. Thus, although there are important dissimilarities between neural representations of visual and infrared stimuli, there are compensating features of the representations that probably promote efficient sharing of the tectal output circuitry and allow substitution of one modality for the other where orientation is concerned.

Cue Sequence

The nature of neural circuitry for analyzing cue sequence information will almost certainly depend on the time scale of the significant temporal sequences that occur in the animal's environment. In the experiments that I have done, I have not found evidence for special detection of intermodality temporal patterns of visual and infrared stimuli. I have looked only in a relatively unsophisticated way, with intermodality stimulus delay of less than one-half second. Simultaneous or nearly simultaneous presentations of stimuli are most favorable for demonstrating either synergistic or inhibi-

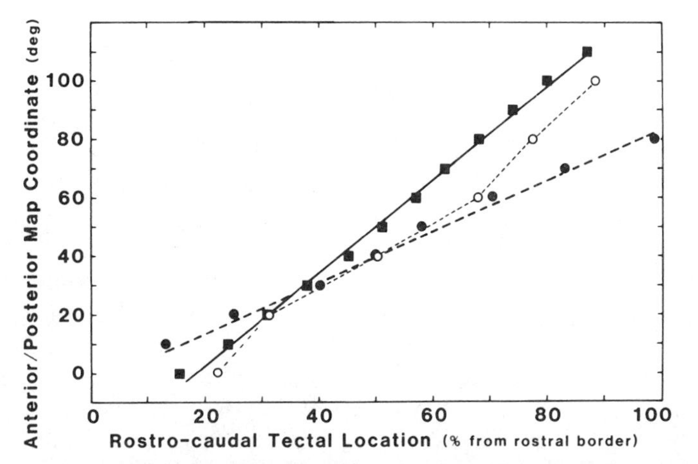

FIGURE 9. Rostro-caudal transect relating infrared and visual receptive field center maps of Figure 5 to the centers of gravity of the infrared tectal images. Squares and solid line relate the visual field centers to rostro-caudal tectal locations near the map's equator; filled circles and heavy broken line indicates the corresponding function for the infrared map, taken along the same transect. Open circles connected by light broken lines show the rostro-caudal coordinates of the centroids of images of infrared objects located (in space) on the equator at various azimuths. The images used for this construction were based on round, 60-degree receptive fields. Although the magnification based on infrared field centers is 1.7 times that based on visual field centers, magnification based on infrared image centroids is close to the visual magnification. This is due to the influence of tectal borders on the infrared images. Tectal curvature was ignored, and the exact location of the tectal borders was uncertain for each of the four spatiotopic maps on which this idealization is based. These factors, plus the assumptions behind image construction, compromise the accuracy of this graph, but not the main qualitative result.

tory modality interactions. One might expect that, since vision is effective at greater distances than is the infrared sense, visual excitation that preceeds infrared excitation by one or a few seconds might signal approach of a warm object. Neurons that carry out analysis appropriate for such a function remain to be found.

Dullemeijer (1961) and Cock Buning et al. (1978) have suggested that, in the feeding sequence, olfactory and visual cues are important for alerting pit vipers to the presence of prey, whereas infrared is more important for initiation of a strike. The time scale of modality sequences such as these is very long, perhaps spanning many minutes. This level of modality sequence analysis may not be occurring in the tectum, nor have my experiments been appropriate to uncover modulatory interactions between modalities that occur over a period of several minutes.

Division of Labor

The nature of the infrared and visual systems invites consideration of several forms of division of labor. One is light-dark specialization. As long as there is sufficient ambient light, the visual system has much greater spatial resolving power, probably has better temporal resolution, and consequently, has better ability to detect movement. If the snake visual system is like other visual systems (which is true in many respects: Hartline, 1984), all of these advantages degenerate as light level decreases. The infrared system, on the other hand, probably functions better under the conditions that accompany darkness. Bullock and Diecke (1956) and others have emphasized that a large temperature difference between a warm object and its surroundings is what makes a strong stimulus. The temperature of the snake's environment becomes lower and more uniform when the sun goes down, which must make mammalian prey or enemies stand out in greater thermal contrast.

Two other ways in which there is an evident division of labor fall along spatial lines. First, the eyes cover almost 360 degrees around the snake's head, whereas the two infrared pits cover a bit more than 180 degrees (Bullock and Diecke, 1956; Newman and Hartline, 1982). Second, the infrared sense is specialized for objects relatively near the snake. In room temperature surroundings, a warm object the size and temperature of a rat evokes a clear response from primary neurons if it is within about 0.5 m of the pit; it excites some tectal neurons if it is within 1–1.5 m (a larger target, such as a person, is detected at 2–3 m). Visual stimuli of appropriate size are effective regardless of distance. Different behavioral repertoires are called for by remote and nearby stimuli, thus it may be useful if multimodal analysis helps to provide distance information.

A class of modality interaction that we have documented in snakes may play a role in division of labor. When both modalities are active at once, it may be desirable to suppress the input arising from one of them. Such suppression might reduce distraction or switch output circuitry so that one modality of input is weighted preferentially over the other. The IR-DEPRESSED VISUAL and VISUAL-DEPRESSED IR neurons that Newman and Hartline (1981) reported (Fig. 10) are clearly candidates for such a function. The fact that both of these types coexist under the same conditions might mean that the two modalities of input are emphasized in two separate subpopulations of tectal neurons. DEPRESSED neurons may also be important for switching between modalities in a way that depends on ambient light. If so, one would expect IR-DEPRESSED VISUAL neurons to lose their responsiveness to visual objects under dim light conditions that still permit visual responses of OR neurons (those excited by either modality: Fig. 10). The appropriate experiments remain to be done.

Color and Texture

Anecdotal accounts and laboratory experiments (Cock Buning et al., 1978) indicate that pit vipers are unlikely to strike toward an appropriate visual target (i.e., a mouse) if the infrared system is deprived of its adequate stimulus. The nervous system apparently recognizes prey as such partly or

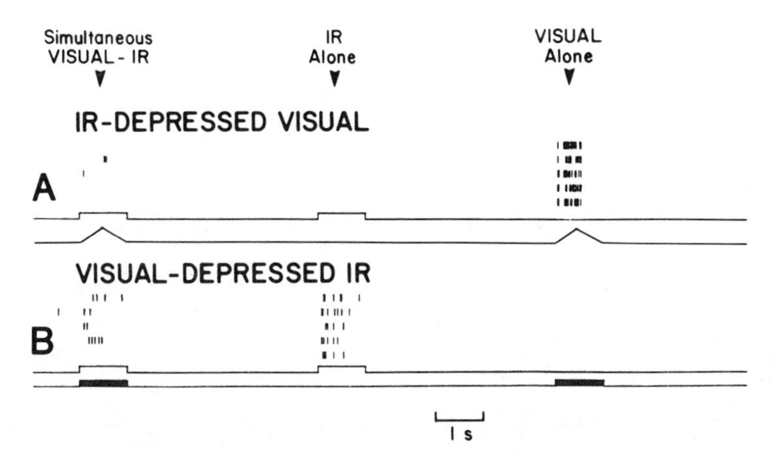

FIGURE 10. Nonlinear modality interaction in DEPRESSED neurons. The raster displays show the responses to separate and combined visual and infrared stimulation of bimodal neurons in *C. viridis*. Each vertical bar represents the time of occurrence of a nerve impulse. Each raster line represents a 15-sec trial (the trials were contiguous, but each is plotted below the previous one). Combining the stimulus of the ineffective modality with the effective one causes reduction in the response compared to what would have been generated by the effective stimulus alone. This kind of interaction would be expected for *division of labor* between the visual and infrared senses. In some IR-DEPRESSED VISUAL cells, offset of the infrared stimulus can cause enhancement. This kind of interaction could extract *color and texture information,* or could contribute to *cross-modality attention* or *signal improvement.* The top stimulus marker trace indicates opening and closure (1-sec duration) of shutter that blocks infrared source; the bottom marker indicates movement of a bright spot stimulus on a tangent screen (in *A*) or off flash (dark bar) of a spot on tangent screen.

mostly by its infrared attribute. From the infrared sense, the snake gains information about an aspect of the stimulus object that is qualitatively different from what is available by way of vision. To a pit viper, the mouse's warmth connotes something analogous to what the red color of a ripe strawberry connotes to a bird or a person.

The class of neurons that has the most appropriate modality interactions to recognize an object by its dual modality nature is that of the AND neurons (Fig. 11; see also Hartline et al., 1978; Newman and Hartline, 1981). We cannot rule out the possibility that some complex single-modality stimulus configuration might also excite such neurons, but their reliable and easily evoked responses to warm, visible objects makes the proposed dual modality recognition function seem plausible. The IR-ENHANCED VISUAL and VISUAL ENHANCED IR (Fig. 12) neurons (Newman and Hartline, 1981) may also help to recognize warmth of a visible target, and to evoke the appropriate behavior; both ENHANCED types are much more responsive if a stimulus is both visible and warm than if it is warm but invisible, or visible but thermoneutral.

Absence as well as presence of a sensory attribute may contribute to the proper identification of an object. It is possible, for instance, that the IR-

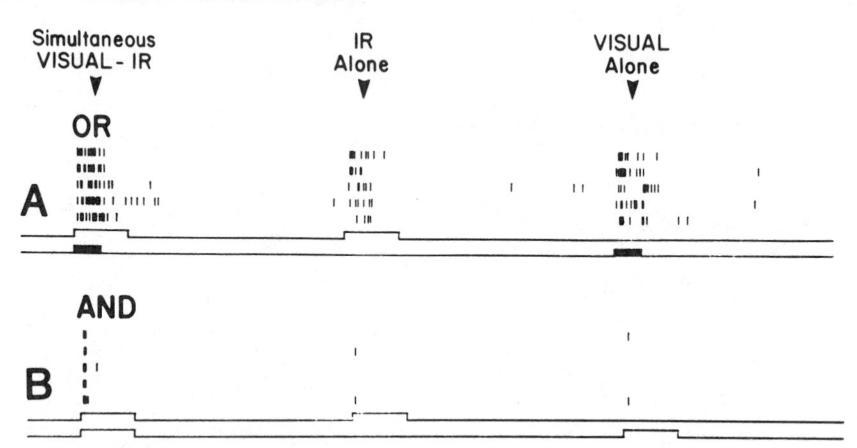

FIGURE 11. Raster displays of responses of OR and AND neurons. Note that the OR neuron's response to the combined stimulation is greater than the response to either modality presented alone. The AND neuron did not respond reliably to either stimulus alone, but gave a brief, high-frequency burst of about four spikes if both stimuli were presented simultaneously. OR neurons may contribute to *cross-modality attention, signal improvement,* and *sensory substitution.* AND neurons probably are providing *color and texture* functionality. Raster and stimulus marker conventions as in Figure 10.

DEPRESSED VISUAL neurons (Fig. 10) help to identify an object as thermoneutral (perhaps inanimate) or cool (perhaps, for some species of pit vipers, an edible frog). Experiments to distinguish a *color and texture* function from a *division of labor* function have not been done; a requirement for close spatial and temporal correlation between two stimulus modalities might be expected for the former.

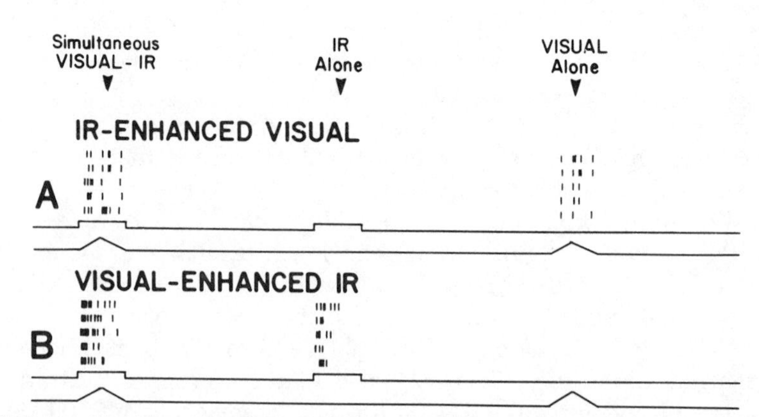

FIGURE 12. Nonlinear modality interaction in ENHANCED neurons. Note that in both cases the enhancing modality alone produced no response but caused the primary modality to evoke a greater response. ENHANCED neurons may be contributing to *cross-modality attention, color and texture,* and *signal improvement* functions. Raster display and stimulus marker conventions as in Figure 10.

Signal Improvement

At dusk, when many rattlesnakes become active and begin hunting, it is quite plausible that the lighting would be poor and the ground temperature would be only slightly different from that of a small mammal. In these conditions, neurons that sum subthreshold excitation from infrared and visual sensors may permit the snake to correctly localize its next meal. Three of the classes of multimodal neurons identified by Newman and Hartline (1981) might be candidates for such a function. The two ENHANCED classes (Fig. 12*B*) have the right modality synergy, but we do not know how the enhancement behaves when the primary or the enhancing modality (or both) are very weakly excited. However, in a few OR neurons (Fig. 11*A*), E. A. Newman and I (unpublished) have documented subthreshold summation and near threshold facilitation. Such neurons could clearly sum the modalities to increase the signal-to-noise ratio of the process by which objects are detected and localized. Linear summation of simultaneous excitation in the two modalities is a sufficient mechanism for signal enhancement. Electric fish have provided another intriguing example in which multimodal integration may have a *signal improvement* function (Bastian, 1982); tectal neurons responsive to an object that disturbs the fish's electric field are more resistant to electrical jamming if visual stimulation accompanies the electric field disturbance.

Cross-Modality Attention

It is known that rattlesnakes can strike accurately at warm targets in the absence of visual cues (Fig. 4). But under natural hunting conditions, at dusk, prey are probably visible, at least part of the time. I suspect that a major advantage of having an infrared sense must be derived from properties of infrared-visual integration that are operational at dusk. Owing to the large receptive fields of tectal infrared neurons, a warm object, as it moves about in the snake's infrared field of view, activates cells in a large tectal image region whose borders shift with the object. I assume that a warm moving object generates peak excitation in neurons whose field it has recently entered, due to the strictly phasic response. Thus the region of maximal excitation must be somewhat inside the advancing edge of the static image diagrammed in Figure 7*B*. I further assume that the tectal distribution of infrared input that is responsible for the enhancing influence of IR-ENHANCED VISUAL neurons is the same as I have found during spatiotopic mapping experiments. Granted these assumptions, IR-ENHANCED VISUAL neurons within the modified infrared image of Figure 13 will be primed for any visual stimulation that occurs while the enhancement persists. For instance, as a mouse moves out of shadow and becomes visible, tectal circuitry primed by infrared radiation would respond particularly strongly; this would presumably cause the mouse to become a more likely target than a competing visual stimulus for an orien-

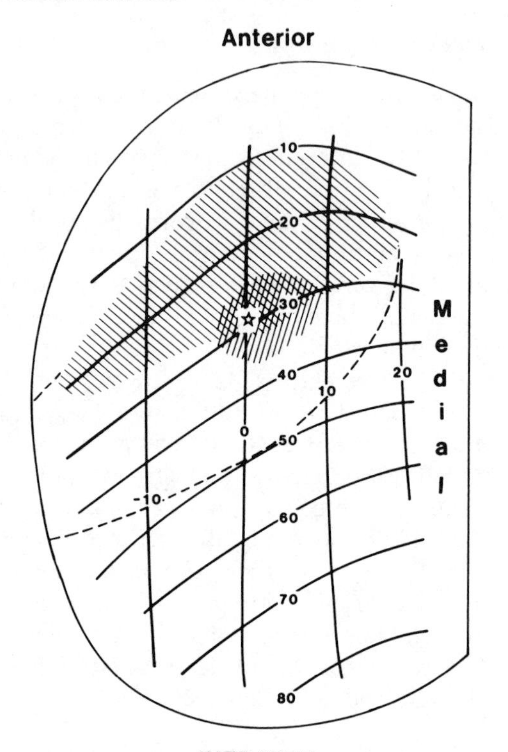

FIGURE 13. Tectal image of a small warm visible object located at 30 degrees azimuth and moving anteriorly along the visual equator. The image construction is based on assumption that infrared excitation becomes significant as the object comes into the central 40 degrees of each neuron's infrared receptive field (large area of diagonal hatching). Excitation is assumed to drop to zero by the time the object has reached the center of the field, owing to the transient response of infrared neurons. (Broken line indicates where the boundary of an infrared image, based on round 40-degrees receptive fields, would fall for a stationary object that suddenly appears at the same location). The visual image (small hatched area) is based on 20-degree round receptive fields. If IR-ENHANCED VISUAL neurons have input corresponding to the respective visual and infrared image neurons, then the IR-ENHANCED VISUAL neurons in the large hatched area are primed or attentive, but not responding. The cross-hatched area contains highly excited enhanced neurons.

tation or striking movement. The result is operationally what one expects from a spatially directed attention mechanism. Key features of an attentional mechanism include nonlinearity of modality summation and temporal persistence of enhancement (so that "attention" would occur even when stimulation is not simultaneous in the two modalities). To determine the extent to which the ENHANCED neurons might be useful in such a role, one needs to determine experimentally whether enhancement has suitable spatial spread, temporal persistence, and fading qualities, as I assumed in the foregoing discussion.

RATTLESNAKE AS A MODEL SYSTEM

In their 1981 paper Newman and Hartline suggested that the modality-combining specializations that they reported for tectal neurons of rattlesnakes might be manifestations of more general vertebrate mechanisms for modality integration. This conjecture has proven to be correct for species of at least two mammalian families.

Mice are known to have a prominent, topographically organized projection of tactile input (fur and vibrissae) in their SC (Drager and Hubel, 1975, 1976). The vibrissae project to about the anterior half of the SC, whereas auditory neurons are found more frequently in the caudal half (this may signal a *division of labor* according to regions of space). The somatotopic map is such that rostral vibrissae project to anterior SC, where nasal visual fields are found. Thus there is a sensible intermodality correspondence. The vibrissae of mice project laterally from the snout to a distance of several head widths. Mice, in exploring even a lighted environment, approach unfamiliar objects and brush them with their vibrissae (as well as sniffing them). Many species of mice have a crepuscular activity cycle, thus the ambient light may be dim to nil when they are carrying out essential activities. Under such conditions, the vibrissae could provide a valuable sense to mediate orientation behavior. They may, for instance, allow a running mouse to avoid colliding with objects along its path. Thus the vibrissae could substitute for vision in a fashion analogous of the substitution of infrared and visual senses that is evident in snake tectum. It is known from the work of Drager and Hubel (1975) that some neurons respond to single-modality presentations of two or three stimulus modalities. These OR neurons certainly could participate in sensory substitution of the type described previously.

S. Wiener and I have begun to search for auditory-visual and somatic-visual interactions in mouse SC. In a few preliminary recordings, we have found that bimodal ENHANCED and DEPRESSED neurons occur in mice as in snakes. Figure 14*A* illustrates a VISUAL-ENHANCED VIBRISSA neuron. We have also found examples of VIBRISSA-ENHANCED VISUAL neurons, the analogs of the IR-ENHANCED VISUAL neuron of Figure 12. Such neurons could have several of the functions listed in the introduction. They could help to identify a particularly important class of object—one that is visible, material, and nearby—illustrating the type of *color and texture* interaction attributed to snake AND or ENHANCED neurons. They could also contribute to a spatially selective *cross-modality attention* mechanism such as the one proposed for IR-ENHANCED VISUAL neurons of snakes. The VISUAL-ENHANCED VIBRISSA neuron of Figure 14*A* may also participate in a *cue sequence* interaction, since an approaching object (or an object approached) would usually stimulate the enhancing visual modality before vibrissal contact is made. We clearly need to test whether

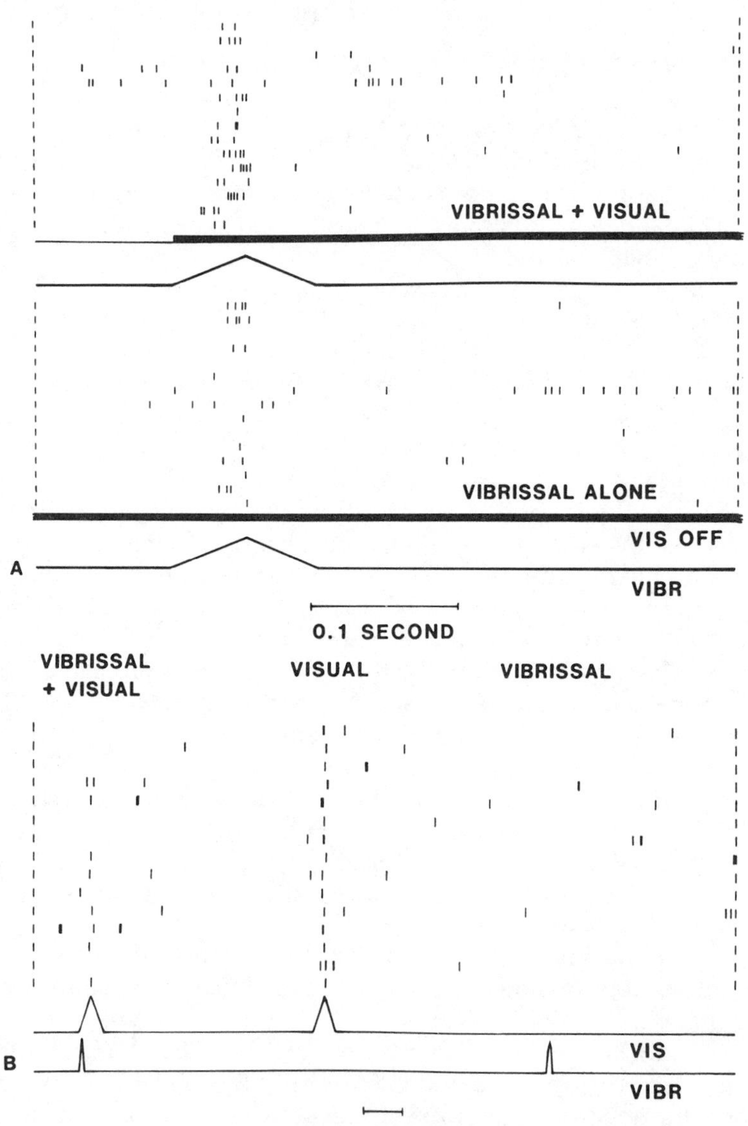

FIGURE 14. (A) Nonlinear modality interactions in VISUAL-ENHANCED VIBRISSAL neuron of mouse. Visual stimulus alone was not effective (not shown). However, approximately twice as many spikes were evoked by combined vibrissal and visual stimulation as by vibrissal stimulation alone. Such neurons could be analyzing *color and texture* information, or could be participating in *modality sequence* or *cross-modality attention* mechanisms. Upper marker indicates offset of a stationary light spot. Lower marker indicates movement of a probe that contacts vibrissae. Stimuli were presented once every 10 sec; the single and dual stimuli were alternated. **(B)** Response of a VIBRISSA-DEPRESSED VISUAL neuron exhibiting inhibition of the vibrissal response by a paired visual stimulus. Visual stimulus marker indicates motion of a light spot. Neurons such as these could be participating in a *division of labor* mechanism, perhaps decreasing the likelihood of orientation toward a nearby object. Raster and timing as in Figure 10.

such neurons are particularly responsive to vibrissal stimulation that follows after (rather than coincides with) visual stimulation as would be expected if cue sequence were important.

Not surprisingly, we have found examples of cross-modality inhibition in mice, such as the VIBRISSA-DEPRESSED VISUAL neuron of Figure 14*B*. This kind of interaction may be involved in a *division of labor* mechanism; if an object is near enough to stimulate the vibrissae, some classes of visual responses must be prevented, and vibrissa-based responses must be allowed to dominate. It would probably not be advantageous, for instance, for a mouse to swing its head to face a dimly seen object, only to smash its nose into a nearby acorn. On the other hand, if VIBRISSA-DEPRESSED VISUAL neurons are highly active, it signals the presence of an object that is beyond immediate reach of the mouse, and one that the mouse will not soon collide with.

We have also seen AUDITORY DEPRESSED examples of vibrissa and visual neurons. Sharp noises frequently cause mice to freeze. Suppression of SC sensory responses that would otherwise evoke orientation movements may be a part of such a freeze response, and AUDITORY DEPRESSED neurons may be a part of the freeze mechanism. In this role, they would be suppressing one class of behaviors in favor of another, and thereby would be performing a *division of labor* function.

The original findings of nonlinear modality interactions in snakes undoubtedly presage similar results in many mammals besides mice. Meredith and Stein (1983), studying neurons of the superior colliculus of cats and hamsters recently reported findings that parallel those in snakes. For example, they found neurons that showed enhancement or depression of visually evoked responses when visual and auditory stimuli were combined. I regard it as likely that similar interactive modes of multimodal integrations are common in the superior colliculi of most mammals and are important for the behavioral function of the colliculus.

ORIGIN OF MULTIMODAL MECHANISMS

How do adaptive cross-modality integrative mechanisms arise? The sensory organs and their messages are not so intimately linked to each other as are the different physical attributes of objects that they report on. As a necessity of laws of nature, the infrared and visible electromagnetic wave patterns that emanate from a warm rat originate from the same location in space, move as a unit, and often appear and disappear nearly simultaneously. They are highly correlated both spatially and temporally. There is no corresponding a priori condition that governs the relationship between the pits and the eyes, or between the spatial organization of infrared input and visual input to the brain.

Several processes may operate to create neural circuitry that extracts information inherent in the interrelationship between sensory messages. One is evolution, which operates over a time scale spanning many generations. A second is adjustment of active anatomical connections; such plasticity may occur during development, or may be needed to compensate for changes in the sense organs during growth or after accidents. The time scale of a plastic change mechanism must be short compared to the life time of the animal, but need not be instantaneous. A third process is a dynamic adjustment of functional connections that occurs while the senses are on line. The time scale for such adjustments must be dictated by changes in the relationship of one sense organ to another, such as those caused by an animal's voluntary actions.

The key feature of a complex multimodal stimulus, as it is represented by the integrative circuitry of the brain, must be that there is a strong tendency for the signals in the separate modalities to have spatiotemporal correlations that are either relatively constant (during biologically important periods) or, if not constant, are predictable according to the circumstances at hand. Spatial variables are usually coded by the populations of responding neurons, that is, by labeled line coding. Thus to extract significant multimodal messages it must be important to create (by evolution or dynamic remapping), develop, or strengthen neural interactions (positive or negative) between appropriate populations of "lines" that bear temporally correlated signals. To the extent that a signal in one line may usually precede, coincide with, or follow that in another, an appropriate temporal window must be built in to the correlation process. To the extent that the relative delay between signals carries biologically significant information, the temporal window must be narrow and appropriately synchronous or phase shifted; to the extent that relative delay is unimportant, the temporal window may be broad.

What adaptations affecting modality integration must be controlled by evolutionary processes? Some can be identified from an unsophisticated consideration of the physics of stimuli and the anatomy and physiology of the sense organs and brain. The pit organs' fields of view overlap the anterior visual field, as the vibrissae of mice overap the nasal visual fields. The overlap results from the physical configuration and location evolved by those sense organs. The adaptive value of such placement is undoubtedly a result of the importance of the anterior hemifield as a location of objects requiring attention and orientation under various lighting conditions. Doubtless, the afferent trigeminal pathways to the tectum and SC are genetically determined and thus specified by an evolutionary process. My guess is that the axes and approximate magnification factor of visual and nonvisual spatiotopic projection maps are likewise determined, within limits, by the genetic blueprint. Thus more anterior regions of the rattlesnake's pit membrane project to more posterior tectal regions. (This is an unusual projection, since the more familiar vertebrate pattern of somatotopic organization involves mapping of more anterior tactile inputs to more rostral tectal regions; see Hartline, 1984.) But the details of infrared and visual spatiotopic maps (e.g., placement of regions of high magnification and exact orientation of map axes) differ greatly from animal to animal, so they may be subject to other mechanisms, for instance, developmental mechanisms involving some degree of experience. My present view is different from that advanced by Hartline, Kass, and Loop in 1978, namely, that the two maps formed independently of each other are completely under genetic control. Supporting an experience hypothesis is the observation that visual and infrared spatiotopic maps in baby rattlesnakes are much more chaotic than they are in adults, though the rostro-caudal and mediolateral axes are similar for the two modalities (unpublished observations). Further evidence for a developmental or experience-related mechanism to achieve spatial correspondence in snakes is lacking. Development of somatic, auditory, and visual connections has been studied in cats, but spatial plasticity was not noted (Stein et al., 1973). However, an experience-dependent spatial calibration of the auditory map to the visual map has recently been reported in the optic tectum of owls (Knudsen et al., 1982; see also Knudsen, Chapter 5, this book).

Grobstein and Comer (1977) found that the degree of plasticity of retinotectal maps needed to achieve and maintain binocular correspondence differs in different species; they suggested that the need for such plasticity depends on how the relationship between the two eyes changes during development or growth. The head of a snake increases in size during the snake's lifetime, and the proportions of the pit organ are different in snakes of different size. This must cause a gradual change in the spatial properties of infrared afferents to the tectum. Applying to a multimodal system the argument made by Grobstein and Comer for a binocular system, I would not be surprised if the interrelationships of infrared and visual connections prove to be somewhat modifiable even in adults.

A clear example of the need for dynamic adjustment of intermodality functional connections involves volitional movements of the eyes relative to the organ subserving another modality (e.g., ears or pit organs). Snakes can move their eyes at least ± 15 degrees from their resting position (unpublished observations), whereas the pit organs are immovable. There is no experimental evidence that there is compensation for the altered intermodality spatial correspondence that results, but I have not done appropriate experiments to uncover such compensation. However, such a mechanism probably does operate in the SC of monkeys: Jay and Sparks (1984) have reported evidence compatible with an active shift of the auditory receptive field of some units (relative to head coordinates) that accompanies a shift of eye position.

SPECULATIONS ON MECHANISMS OF INTERMODALITY PLASTICITY

An experience-related mechanism seems well suited to control the changes in intermodality connections that might be needed after an accident to a sense organ or during development or growth; this is because much information about the proper interrelationships of intermodality sensory correspondence is inherent in the sensory stimuli that are encountered constantly by the animal. But it would seem, at worst, very difficult or impossible (see Grobstein and Chow, 1975) and, at best, unnecessary to code such interrelationships precisely in the genetic blueprint. The hypothesized plasticity mechanisms must alter effective connections to achieve convergence of lines that have high correlations (positive or negative). Several models have been advanced to achieve similar ends. Brindly's (1969) networks that show associative memory, Marr's (1969) model of cerebellar plasticity, and the more recent models of von der Malsburg (1973) and Cooper and his associates (Cooper, 1973; Bienenstock, Cooper, and Munro, 1982) have many of the properties that are needed for adjustment of intermodality connections. Some specific multimodal feature detectors such as AND neurons might be formed through such mechanisms. Consider a mechanism that strenthens the synapses (onto a postsynaptic cell) of inputs of different modalities if their activity is correlated and is in turn correlated with postsynaptic activation. Such a mechanism might suffice to correctly adjust visual and infrared maps to each other, if the maps were approximately in register at the outset. However, more spatial generalization probably is needed to cross-calibrate relatively disparate maps than to create and maintain specific binocular disparity detectors in the visual cortex.

There may be more than a formal similarity between the problem of extracting information from the interrelationships of messages from two independent eyes and the problem of extracting information from two independent senses. The classical finding that, after monocular deprivation, the normal eye attains dominance in driving cells of the visual cortex (Wiesel and Hubel, 1963) might have a multimodality analog. Evolution often applies a successful design principle to many different systems that share a general problem. Based on this corrolary of Occam's principle, I am led to anticipate that in the optic tectum or SC, and more generally in multimodal systems that exhibit developmental (or other) plasticity, sensory deprivation in one modality will lead to an apparent competition in which the deprived modality loses some of its functional connectivity to multimodal neurons.

REFERENCES

Bastian, J. (1982) Vision and electroreception: Integration of sensory information in the optic tectum of the weakly electric fish. *Apteronotus albifrons. J. Comp. Physiol.* **147**:287–297.

Bienenstock, E. L., L. N. Cooper, and P. W. Munro (1982) Theory for the development of neuron selectivity orientation specificity and binocular interaction in visual cortec. *J. Neurosci.* **2**:32–48.

Brindley, G. S. (1969) Nerve net models of plausible size that perform many simple learning tasks. *Proc. R. Soc. (Lond.) Biol.* **174**:173–191.

Bullock, T. H. (1953) Comparative aspects of some biological transducers. *Fed. Proc.* **12**:666–672.

Bullock, T. H. and R. B. Cowles (1952) Physiology of an infrared receptor—the facial pit of pit vipers. *Science* **115**:541–543.

Bullock, T. H. and F. P. J. Diecke (1956) Properties of an infrared receptor. *J. Physiol. (Lond.)* **134**:47–87.

Bullock, T. H. and W. Fox (1957) The anatomy of the infrared sense organ in the facial pit of pit vipers. *Quart. J. Microsc. Sci.* **98**:219–234.

Cock Buning, T. de, R. E. Poelman, and P. Dullmeijer (1978) Feeding behavior and the morphology of the thermoreceptors in *Phython reticulus. Neth. J. Zool.* **28**:62–93.

Cooper, L. N. (1973) A possible organization of animal memory and learning. In *Proceedings of the Nobel Symposium on the Collective Properties of Physical Systems,* Vol. 24, B. Lindquist and S. Lindquist, eds. Academic Press, New York, pp. 252–264.

Dräger, U. C., and D. Hubel (1975) Responses to visual stimulation and relationship between visual, auditory, and somatosensory inputs in mouse superior colliculus. *J. Neurophysiol.* **38**:690–713.

Dräger, U. C. and D. H. Hubel (1976) Topography of visual and somatosensory projections to mouse superior colliculus. *J. Neurophysiol.* **39**:91–101.

Dullemeijer, P. (1961) Some remarks on the feeding behavior of rattlesnakes. *Proc. Akad. Wetench. Amsterdam, Ser. C.* **64**:383–96.

Goldberg, M. E. and D. L. Robinson (1978) Visual system: superior colliculus. In *Handbook of Behavioral Neurobiology,* Masterton, ed., pp. 119–163.

Goris, R. C. and S. Terashima (1973) Central response to infrared stimulation of the pit receptors in a crotaline snake, *Trimeresurus flavoviridis. J. Exp. Biol.* **58**:59–76.

Grobstein, P. and K. L. Chow (1975) Receptive field development and individual experience. *Science* **190**:352–358.

Grobstein, P. and C. Comer (1977) Post-metamorphic eye migration in *Rana* and *Xenopus. Nature* **269**:54–56.

Gruberg, E. R., E. Kicliter, E. A. Newman, K. Kass, and P. H. Hartline (1979) Connections of the tectum of the rattlesnake *Crotalus viridis:* an HRP study. *J. Comp. Neuro.* **188**:31–42.

Hartline, P. H. (1974) Thermoreceptors in snakes. In *Handbook of Sensory Physiology,* Vol. 3, Sect. 3, *Electroreceptors and Other Specialized Receptors in Lower Vertebrates,* A. Fessard, ed. Springer-Verlag, Berlin, pp. 297–312.

Hartline, P. H., L. Kass, and M. S. Loop (1978) Merging of modalities in the optic tectum: Infrared and visual integration in rattlesnakes. *Science* **199**:1225–1229.

Hartline, P. H. (1984) The optic tectum of reptiles: Neurophysiological studies. In *Comparative Neurology of the Optic Tectum,* H. Vanegas, ed. Plenum, New York.

Jay, M. F. and D. L. Sparks (1984) Auditory receptive fields in primate superior colliculus shift with change in eye position. *Nature* **309**:345–347.

Kass, L., M. S. Loop, and P. H. Hartline (1978) Anatomical and physiological localization of visual and infrared cell layers in tectum of pit vipers. *J. Comp. Neurol.* **182:**811–820.

Knudsen, E. I. (1982) Auditory and visual maps of space in the optic tectum of the owl. *J. Neurosci.* **2:**1177–1194.

Knudsen, E. I., P. F. Knudsen, and S. D. Esterly (1982) Early auditory experience modifies sound localization in barn owls. *Nature* **295:**1–3.

Marr, D. (1969) A theory of cerebellar cortex. *J. Physiol. (Lond.)* **202:**437–470.

McIlwain, J. T. (1975) Visual receptive fields and their images in superior colliculus of the cat. *J. Neurophysiol.* **38:**219–230.

McIlwain, J. T. (1982) Lateral spread of neural excitation during microstimulation in intermediate gray layer of cat's superior colliculus. *J. Neurophysiol.* **47:**167–178.

Meredith, M. A. and B. E. Stein (1983) Interactions among converging sensory inputs in the superior colliculus. *Science* **221:**389–391.

Molenaar, G. J. (1974) An additional trigeminal system in certain snakes possessing infrared receptors. *Brain Res.* **78:**340–344.

Newman, E. A., E. R. Gruberg, and P. H. Hartline (1980) The infrared trigemino-tectal pathway in the rattlesnake and in the python. *J. Comp. Neurol.* **191:**465–477.

Newman, E. A. and P. H. Hartline (1981) Integration of visual and infrared information in bimodal neurons of the rattlesnake optic tectum. *Science* **213:**789–791.

Newman, E. A. and P. H. Hartline (1982) The infrared "vision" of snakes. *Science* **246:**116–127.

Noble, G. R. and A. Schmidt (1937) Structure and function of the facial and labial pits of snakes. *Proc. Am. Phil. Soc.* **77:**263–288.

Schiller, P. H. and M. Stryker (1972) Single-unit recording and stimulation in superior colliculus of the alert rhesus monkey. *J. Neurophys.* **35:**915–924.

Schroeder, D. M. and M. S. Loop (1976) Trigeminal projections in snakes possessing infrared sensitivity. *J. Comp. Neurol.* **169:**1–13.

Sprague, J. M., G. Berlucchi, and G. Rizzolatti (1973) The role of the superior colliculus and pretectum in vision and visually guided behavior. In *Handbook of Sensory Physiology,* Vol. 7, Sect. 3, *Central Visual Information* R. Jung, ed. B. Springer Verlag, New York.

Stanford, L. R. and P. H. Hartline (1980) Spatial sharpening by second-order trigeminal neurons in crotaline infrared system. *Brain Res.* **185:**115–123.

Stanford, L. R., D. M. Schroeder, P. H. Hartline (1981) The ascending projection of the nucleus of the lateral descending trigeminal tract: A nucleus in the infrared system of the rattlesnake, *Crotalus viridis. J. Comp. Neurol.* **201:**161–173.

Stein, B. E., E. Labos, and L. Kruger (1973) Sequence of changes in properties of neurons of superior colliculus of the kitten during maturation. *J. Neurophysiol.* **36**(4):667–679.

Stein, B. E., B. Magalhaes-Castro, and L. Kruger (1976) Relationship between visual and tactile representations in cat superior colliculus. *J. Neurophysiol.* **39:**401–419.

Stein, B. E. and N. S. Gaither (1981) Sensory representation in reptilian optic tectum: some comparisons with mammals. *J. Comp. Neurol.* **202:**69–87.

Terashima, S., R. C. Goris, and Y. Katsuki (1970) Structure of warm fiber terminals in the pit membrane of vipers. *J. Ultrastruct. Res.* **31:**494–506.

Terashima, S. and R. C. Goris (1975) Tectal organization of pit viper infrared reception. *Brain Res.* **83:**490–494.

Terashima, S. I. and R. C. Goris (1977) Infrared bulbar units in crotaline snakes. *Proc. Jap. Acad.* **53**(ser. B):292–296.

von der Malsburg, C. (1973) Self-organization of orientation sensitive cells in the striate cortex. *Kybernetik* **14:**85–100.

Wiesel, T. N. and D. H. Hubel (1963) Single-cell responses in striate cortex of kittens deprived of vision in one eye. *J. Neurophysiol.* **26:**1003–1017.

Chapter Nineteen

HOW AUDITORY SPACE IS ENCODED IN THE OWL'S BRAIN

MASAKAZU KONISHI

Division of Biology
California Institute of Technology
Pasadena, California

Nerve cells communicate using many forms of signals: discrete and graded membrane potentials, chemical transmitters, and hormones that originate in one group of neurons and affect the response of other groups of neurons. Action potentials mediate fast, sensory, and motor responses because they are rapid carriers of signals between distant sensory receptors, neurons, and muscles. Encoding information with action potentials is a central issue in integrative neurophysiology. A train of spikes in a single neuron contains such measureable properties as spike number, instantaneous rate of discharge, and interspike intervals. However, which of these properties encodes information is seldom predictable on an a priori basis. Neither the knowledge of neural circuitry nor that of its physiology alone is sufficient for the elucidation of neural codes. An animal's behavior should reveal the nature of its perceptual world, which consists of physical cues defining the characteristics of its sensory space. Analysis of neuronal responses with reference to these cues is indispensable for the study of their neural representations.

INFERENCES FROM BEHAVIOR

The barn owl's auditory space consists of locations defined by differences in the time and intensity of sound perceived by the two ears; interaural time (phase) differences define azimuthal (left-right) locations, and interaural intensity differences define elevational (up-down) locations. The structure of the owl's external ear permits the assignment of a separate cue to each spatial coordinate before the auditory system encodes it (Moiseff and Konishi, unpublished). The barn owl turns its head toward a sound source. This behavior serves as a good measure of sound localization. A dichotic (binaural) signal delivered through earphones can induce this orienting response. The angular distance of head turning is proportional to the magnitude of time shift, and the direction of turning depends on which ear leads. The maximum interaural time difference that the adult owl experiences is about 170 μsec, and the minimum time difference the owl can detect is about 10 μsec.

Of the two types of time, onset (arrival of the first wave) and phase, the owl uses phase for binaural comparison. To derive an interaural time difference from a noise without using onset time, the owl detects a temporal disparity in the signal waveform between the ears. Because this method discriminates between binaurally correlated and uncorrelated noises, it is an effective strategy for extracting the signal buried in noise, just as the use of binocular disparity helps to detect an object in a camouflaged background (Julesz, 1971). The owl distinguishes itself from man and other domestic mammals by its ability to perform binaural phase comparison for such high frequencies as approximately 6–8 kHz, which are also most suitable for binaural intensity comparison.

The owl can use interaural intensity differences for localization in the vertical plane because its ears are vertically displaced from each other; the left ear opening is located higher on the face than the right one. This and other asymmetries in the owl's external ears make the left ear more sensitive to high-frequency sound (6–8 kHz) coming from below the horizon and the right ear to those from above the horizon (Payne, 1971; Knudsen and Konishi, 1978b, 1979). Accordingly, when a noise from earphones is louder in the left ear than in the right ear, the owl turns its face toward a location below eye level. The owl turns its face both in azimuth and elevation simultaneously in response to a dichotic signal that is both binaurally phase shifted and amplitude unbalanced. Thus the owl extracts from a noise signal both binaural time and intensity cues simultaneously and uses them separately for azimuth and elevation, respectively. These findings indicate that a specific combination of time and intensity disparities uniquely defines a location in the owl's auditory space.

AUDITORY RECEPTIVE FIELD AND MAP

A search for neural codes can start from any level within a given neural system. Although tracing certain neuronal response properties from primary sensory neurons to higher-order neurons would seem logical, what logical procedures to use in predicting the response properties of higher-order neurons in comparison to those of lower-order neurons is not obvious. This difficulty is mainly due to the nonlinear operating properties of the nervous system (Capranica, 1972). On the other hand, the response properties of higher-order neurons can be traced "downstream" because they are either relayed to or emerge in those neurons. In keeping with this strategy, we looked for higher-order auditory neurons that are sensitive to variation in sound location (Knudsen et al., 1977). The external nucleus of the owl's inferior colliculus contains neurons that respond selectively to sound emanating from a restricted area in space (see Knudsen, 1983, for the terminology of the owl's auditory midbrain). Such a neuron is called a space-specific neuron, and the area of space where appropriate stimuli excite it is called its receptive field (Fig. 1). Most space-specific neurons

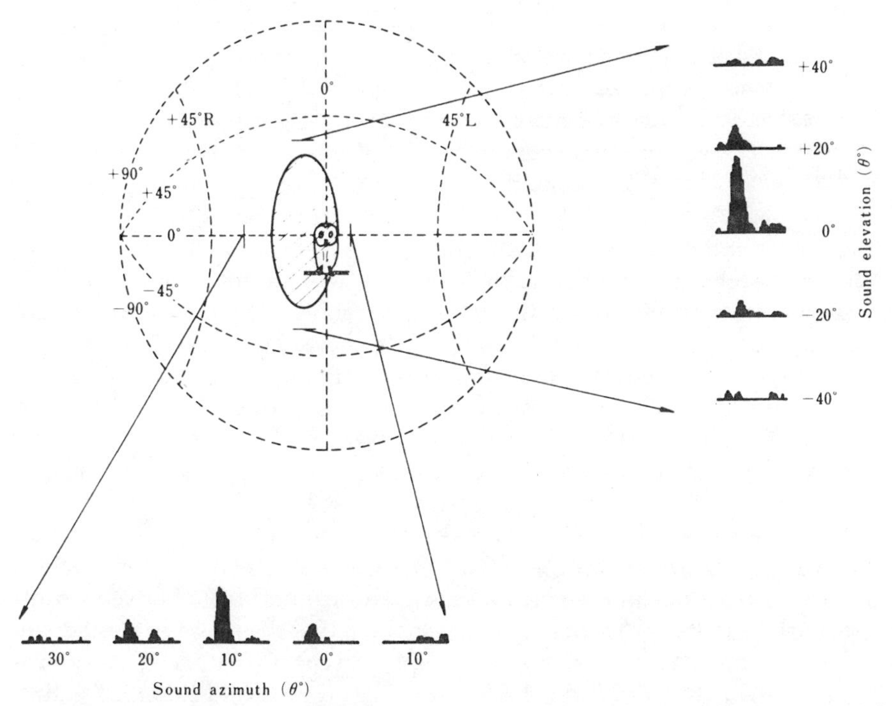

FIGURE 1. The receptive field of an auditory neuron depicted from the observer's point of view. The owl is shown facing out from the center of the stimulus sphere (dashed globe), and the unit's receptive field is projected onto the sphere (diagonally lined area). Below and to the right are shown peristimulus-time histograms of the unit's responses to a sound stimulus presented at different azimuthal and elevational locations within its receptive field.

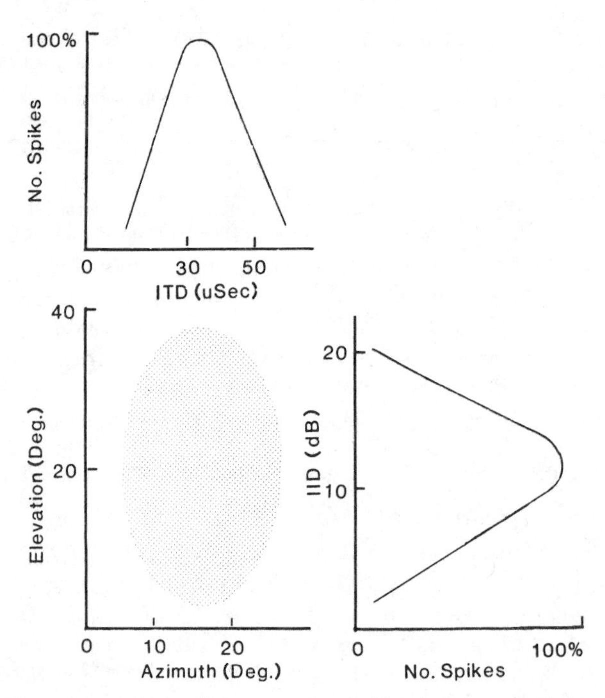

FIGURE 2. Binaural cues for auditory receptive fields. A space-specific neuron requires a binaural stimulus containing a particular combination of interaural time (ITD) and intensity (IID) differences. This diagram is a scale model of a neuron's receptive field (shaded area) and its Δt- and Δt-tuning curves.

have an elliptical receptive field with the long axis oriented vertically. Within a unit's field boundaries there is always an area, called the best area, where the unit shows the lowest threshold. The best area usually corresponds to the azimuthal center of the field, but it need not always correspond to the elevational center. The location of a field's best area is usually immune to changes in the quality (noise, tone, or click) and amplitude of sound (Knudsen and Konishi, 1978b).

Space-specific neurons form a neural map of auditory space (Knudsen and Konishi, 1978a, b): they are arranged so as to project the locations of their best areas systematically onto the contralateral external nucleus. Receptive field azimuths are mapped in the horizontal plane of the nucleus and their elevations in the vertical plane. This isomorphism between auditory space and its neural map is remarkable in that the map is not made by successive projections of the sensory epithelium as in other sensory systems (Konishi and Knudsen, 1982). Many neurons have best areas within the region subtending about 30° from the midpoint of the face. This is the region where the owl localizes sound most accurately (Knudsen and Konishi, 1978c 1979).

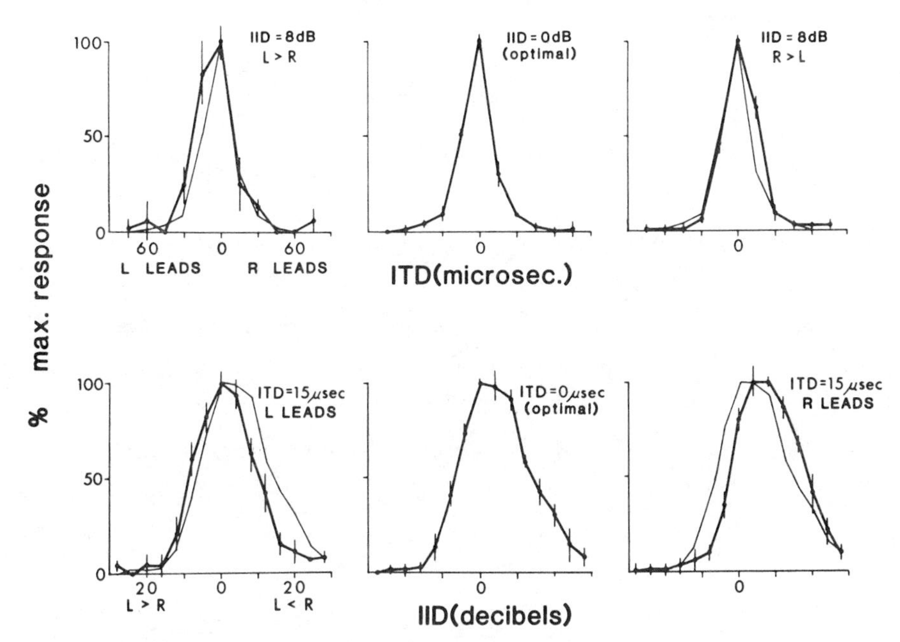

FIGURE 3. The tuning of a space-specific neuron to one interaural cue is not influenced by the value of the other cue. The Δi or Δt setting used to obtain data for each plot is at the upper right. The curves obtained with optimal Δi or Δt values are in the center of each row ("optimal") and are also superimposed on adjacent graphs to facilitate comparison. Curves obtained with nonoptimal Δi or Δt values are shown in the adjacent plots. R and L refer to the sound in right and left ears respectively.

BINAURAL TIME AND INTENSITY DISPARITIES DEFINE RECEPTIVE FIELDS

Space-specific neurons require binaural stimuli that contain a particular combination of interaural time and intensity differences. Binaural stimuli containing inappropriate interaural cues are either ineffective or inhibitory. They are an AND gate of time and intensity cues because either one alone fails to excite them. The best interaural time difference for a space-specific neuron determines the azimuthal center of its receptive field, and the range of interaural time differences to which the neuron is tuned (called hereafter Δt-tuning) determines the azimuthal width of its receptive field. Similarly, the best interaural intensity difference for a space-specific neuron determines the elevational center of its receptive field, and the range of interaural intensity differences to which the neuron is tuned (called hereafter Δi-tuning) determines the height of its elliptical receptive field (Fig. 2; Moiseff and Konishi, 1981).

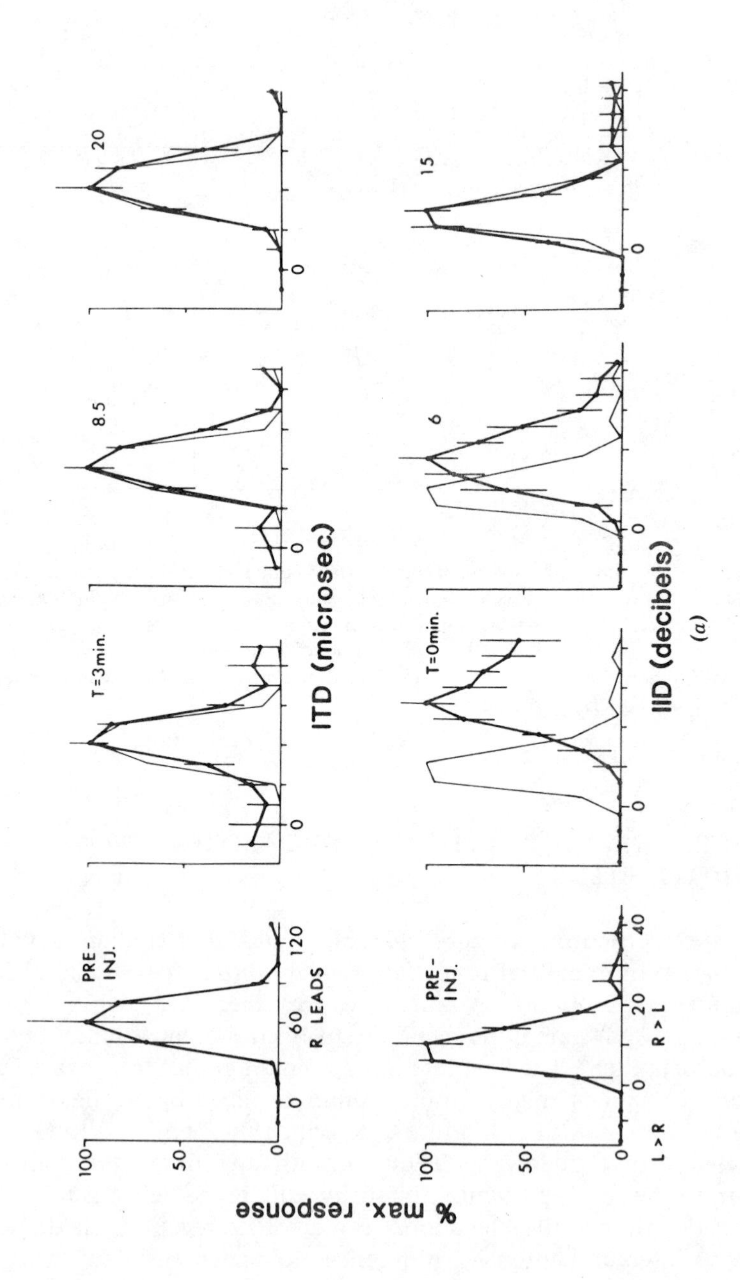
% max. response
100
50
0
PRE-
INJ.
R LEADS
L > R
R > L
T=3min.
8.5
20
T=0min.
6
15
ITD (microsec.)
IID (decibels)
0
60
120
20
40

(*a*)

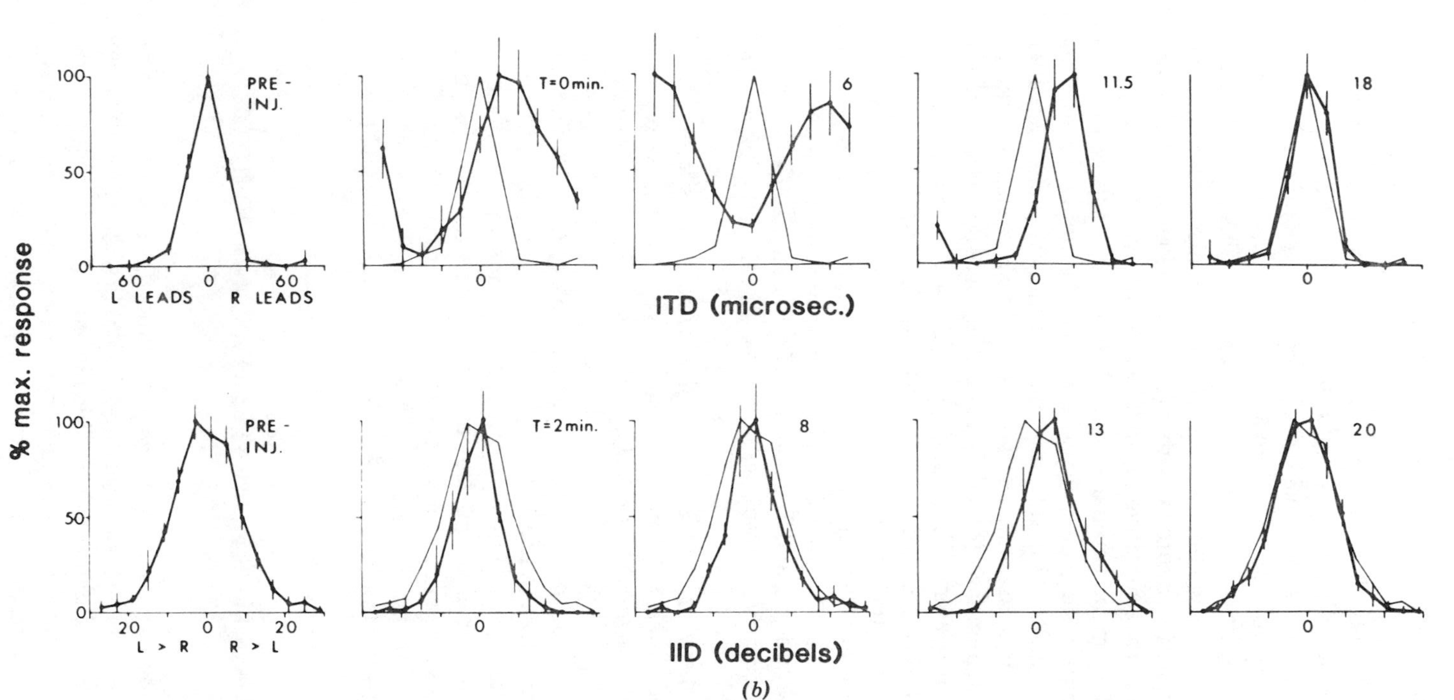

(b)

FIGURE 4. (*a*) The effects of anesthetizing nucleus angularis on a space-specific neuron's Δt-tuning (top row) and Δi-tuning (bottom row). The minutes lapsed since the injection of lidocaine ($T = 0$) are shown at the upper right of each plot. The data points of each plot represent the average number of spikes per stimulus presentation (five repetitions) normalized to the maximum spike count. Vertical bars indicate the standard deviation. The curve obtained prior to lidocaine injection is superimposed on each plot to facilitate comparison. R and L refer to the sound in right and left ears respectively. Note that the Δi-tuning curve broadens and shifts, whereas the Δt-tuning curve remains unchanged. (*b*) The effects of anesthetizing nucleus magnocellularis on a space-specific neuron's Δt-tuning (top row) and Δi-tuning. All abbreviations and conventions are identical to those of Figure 4a. Note that this time the Δt-tuning curve broadens and shifts, whereas the Δi-tuning curve remains unchanged.

SEPARATE PROCESSING OF TIME AND INTENSITY

Where do Δt- and Δi-tuning emerge? An extensive survey of binaural nuclei in the owl's medulla and pons indicates that they fall in two categories: one in which neurons are sensitive to interaural time differences and the other in which neurons are sensitive to interaural intensity differences. No brainstem nucleus below the inferior colliculus contains neurons sensitive to both cues. Of the two binaural nuclei in the lemniscal complex, one nucleus (VLVa) contains neurons with broad Δt-tuning, and the other (VLVp) contains neurons sensitive to Δi. The Δt-sensitive nucleus, VLVa, receives its input from n. laminaris, which is the first site of binaural convergence. Laminaris neurons also possess broad Δt-tuning curves. This nucleus receives its binaural input from n. magnocellularis (called hereafter the magnocellular nucleus), one of the two cochlear nuclei. The other cochlear nucleus, n. angularis (called hereafter the angular nucleus), sends its output to the contralateral VLVp. These anatomical connections and the functional classification mentioned previously suggest that two separate brainstem pathways process time and intensity cues (Takahashi and Konishi, in preparation; Moiseff and Konishi, 1983). It should be pointed out that the separation is not by frequency, as in mammals (e.g., Goldberg and Brown, 1969; Boudreau and Tsuchitani, 1968; Rose *et al*, 1974), because the owl's high-frequency neurons are sensitive to interaural phase differences.

In addition to these lines of evidence, the response of space-specific neurons to Δt and Δi also show that these two cues are separately processed. A space-specific neuron's selectivity for one interaural cue remains the same for all effective values of the other cue, in other words, its Δt-tuning and Δi-tuning do not interfere with each other (Fig. 3). Furthermore, experimental interference with one pathway affects only the processing of the cue assigned to that pathway. A local anesthetic, lidocaine, injected into the magnocellular nucleus reversibly alters the peak position and sharpness of a space-specific neuron's Δt-tuning curve without affecting its Δi-tuning. Conversely, anesthetizing the angular nucleus causes changes only in the neuron's Δi-tuning (Fig. 4*a* and *b*; Takahashi et al., 1984).

THE COCHLEAR NUCLEUS SEPARATES TIME AND INTENSITY CODES

The observations mentioned so far suggest that the separation of the two cues occurs at the level of the cochlear nucleus. The first station in the time pathway is the magnocellular nucleus. This nucleus must provide n. laminaris with monaural temporal codes necessary for their binaural comparison. The detection of Δt at high frequencies (6–9 kHz) by the owl as well as

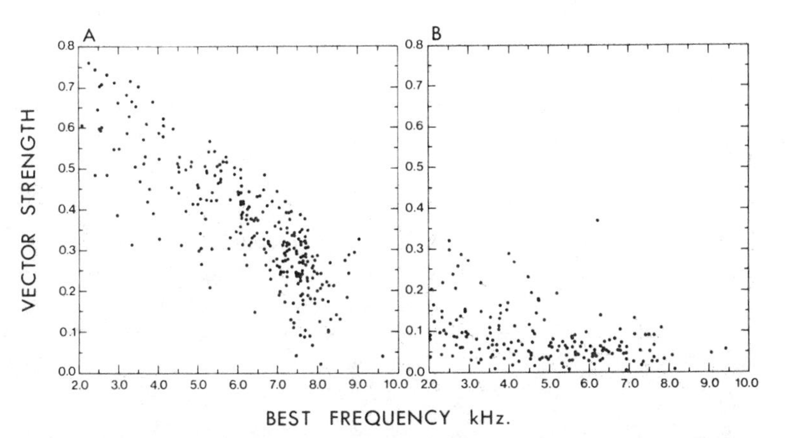

FIGURE 5. Differences in the degree of phase locking between magnocellular (*A*) and angular (*B*) neurons. When a neuron fires at random with reference to stimulus phase, vector strength is 0. Each dot indicates a neuron's best frequency and vector strength. Phase locking occurs in virtually all magnocellular neurons, including those with high best frequencies, whereas it is absent or weak in angular neurons.

by its space-specific neurons indicates that the owl's auditory nerve should be able to encode the phase of high frequencies. The most obvious method of encoding time by a neuron is to use the timing of its spike discharge, although other methods are conceivable. In the magnocellular nucleus virtually all neurons fire at a particular phase of a tonal stimulus. Although an individual neuron cannot fire for every cycle of a high-frequency tone, it prefers a particular phase when it fires. This phenomenon is called "phase locking," which can occur up to about 9 kHz in the owl's magnocellular nucleus in comparison with the highest frequency of approximately 4–5 kHz for neurons in the mammalian anterior ventral cochlear nucleus (Fig. 5*A*; Rose et al., 1974; Lavine, 1971). However, in sharp contrast with the magnocellular nucleus, the angular nucleus lacks phase locking: even low-frequency-sensitive neurons show little or no phase locking (Fig. 5*B*).

The angular nucleus is the first station in the intensity-sensitive pathway. How a train of spikes encodes sound intensity is not immediately obvious, for variables such as latency, spike number, and interspike interval can represent sound intensity. However, spike number as such and latency can be excluded as intensity codes for the following reasons. Because spike number is a function of both stimulus intensity and duration, it cannot uniquely encode intensity. If latency encoded intensity, the owl should turn its face vertically in response to an interaural difference in onset time which causes an interaural difference in latency. The average interspike interval or its reciprocal, the average instantaneous rate of discharge, is sensitive to variation in sound intensity. This rate varies in parallel with the average number of spikes per stimulus. By this criterion, neurons in the owl's angular nucleus show much more finely graded re-

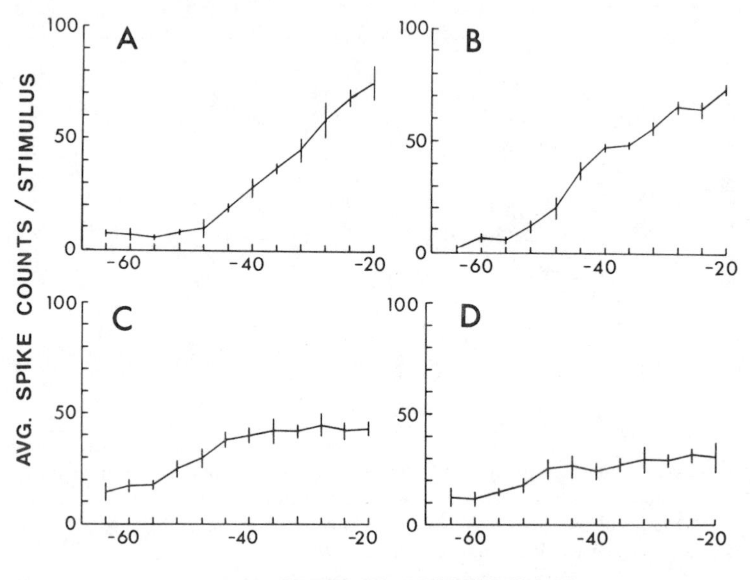

FIGURE 6. Differences in neuronal sensitivity to stimulus intensity between the angular and magnocellular nuclei. (*A* and *B*) The average number of spikes per stimulus as a function of stimulus intensity for typical cells from the angular nucleus. (*C* and *D*) Spike-count functions for typical cells from the magnocellular nucleus.

sponses to a wider range of sound intensity than those in the magnocellular nucleus (Fig. 6). Statistical parameters such as the coefficient of variation in discharge rate and the dynamic range covered by a population of neurons indicate that the properties of angular neurons are much more suitable for intensity coding than those of magnocellular neurons. Thus each nucleus is specialized for processing either intensity or time code, and not both (Sullivan and Konishi, 1984).

The foregoing findings suggest that the cochlear nuclei sort out the time and intensity codes transmitted to them by primary auditory fibers. Preliminary recording from these fibers seems to support this assumption. Each primary auditory fiber bifurcates to send a collateral to each cochlear nucleus (Boord and Rasmussen, 1963). In the magnocellular nucleus the collateral terminates by way of a large calyx on the soma of the postsynaptic cell. In the angular nucleus, the collateral presumably terminates by way of bouton type synapses on the dendrites of the postsynaptic cell (Takahashi and Konishi, unpublished). It would seem that these different types of synapses are responsible for sorting out the two codes. One rarely encounters such a simple mechanism of information processing by the brain. All the preceding facts put together can explain why there are two cochlear nuclei in the owl's brain. This conclusion should encourage further attempts to find a behavioral explanation for similar anatomical and physiological divisions in the mammalian cochlear nucleus (Rose et al., 1974; Lavine, 1971; Goldberg and Brownell, 1974; Osen, 1969; Oertel, 1983).

Finally, the reasons for the separation of the two codes must be considered. There are ultimate and proximate reasons; the former involves reasoning based on the need of the owl, and the latter involves reasoning based on the conditions for neural processing. The use of two independent variables of a signal for two coordinates is perhaps the simplest method in bicoordinate orientation, particularly when they are already separated at the level of the external ear. The main neurophysiological reason for the separation of the two codes is that the processing of one is incompatible with that of the other. The time of spike discharge relative to the stimulus waveform is important for binaural phase comparison, whereas it is unimportant for binaural intensity comparison. The time and intensity codes can coexist as in the auditory nerve, but their binaural comparison requires separate processing. One code for one channel may be a general rule in neural processing.

THE ORIGIN OF Δt-TUNING

The source of Δt-tuning can be traced back from the external nucleus of the inferior colliculus to n. laminaris. Nucleus laminaris holds the key to the elucidation of the mechanism underlying Δt-tuning because it is the first site where this neuronal property appears. It performs Δt-tuning by a method known as coincidence detection (Jeffress, 1948; Colburn, 1973). The spike discharge of a laminaris neuron is a function of the probability of spikes arriving simultaneously from the left and right magnocellular nuclei. Coincident arrival of spikes at a neuron results when a lead in acoustic propagation to one ear cancels out a lag in the transmission of neural signals from that ear. This delay is unique to each neuron. Thus a neuron's transmission" delay equals the interaural time difference that maximally excites the neuron, that is, the peak of its Δt-tuning curve.

The preceding conclusion derives from the following observations on laminaris neurons. They show not only Δt-tuning, which is a binaural response property, but also phase locking to a tonal stimulus delivered to either ear. The phase angle (or time measured from the beginning of each cycle) at which a neuron fires most often when driven by the left ear may be the same as or different from the phase angle at which it fires most often when driven by the right ear. A difference between the left and right phase angles (or times) means a left-right lead or lag in neural transmission. In other words, the neural representation of the stimulus waveform shows an interaural time disparity. For instance, if a neuron shows a difference of 30 μsec, the left side leading the right, the peak of its Δt-tuning curve occurs

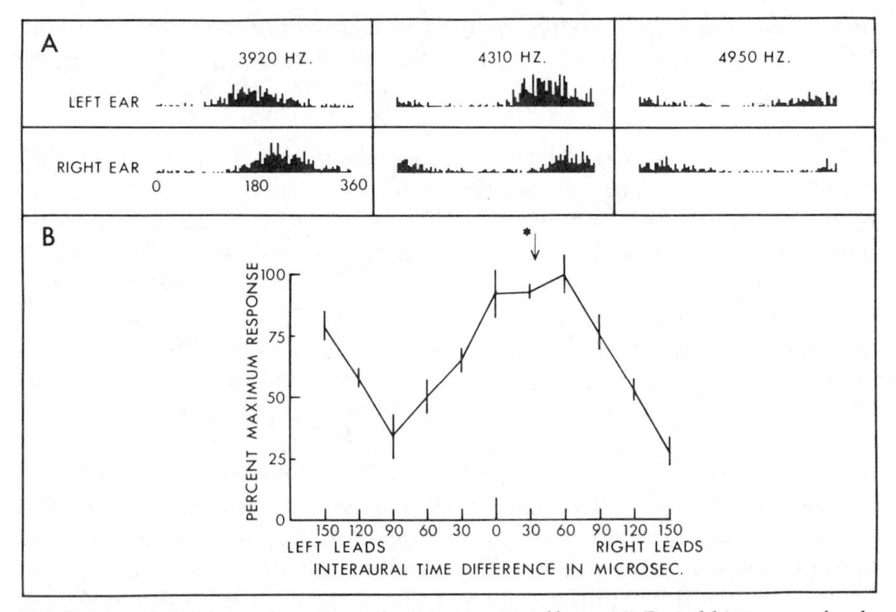

FIGURE 7. Coincidence detection in laminaris output fibers. (*A*) Period histograms for the response of a laminaris unit to monaural stimulation of each ear at three different frequencies. At each frequency the neuron's response to the right ear occurs later in the cycle than does its response to the left ear. This indicates that the input from the right ear has a slightly longer transmission delay. (*B*) Normalized spike counts as a function of interaural time difference for a dichotic stimulus. When the sound delivered to the right ear leads the sound delivered to the left ear by the amount of the monaural transmission time difference (shown by the arrow), the neuron responds maximally. At this interaural time difference, the monaural inputs are in coincidence.

at 30 μsec, the right ear leading the left (Fig. 7) (Sullivan and Konishi, unpublished).

Binaural phase comparison is possible only between inputs of the same frequency or waveform. The left and right inputs to a laminaris neuron are tuned to the same narrow frequency range. When a phase difference is translated into a time difference, this varies with frequency (period length). Therefore, computation of a single Δt from a noise by the delay line method requires that all delays be available in each frequency band. A preliminary study bears out this prediction (Sullivan and Konishi, unpublished).

Further processing in the time pathway involves the sharpening of Δt-tuning. VLVa neurons have sharper Δt-tuning curves than laminaris neurons. Δt-Tuning curves are even sharper in the external nucleus of the inferior colliculus than in VLVa. Unlike laminaris and VLVa neurons, monaural stimulation is either insufficient or inhibitory to space-specific neurons. Furthermore, both Δt- and Δi-tuning curves of space-specific neurons reach below the level of spontaneous discharge (if they fire spontaneously), suggesting that an inhibitory mechanism is responsible for the sharpening of both Δt- and Δi-tuning. This interpretation is consistent with the earlier observation that inhibitory zones surround each auditory receptive field (Knudsen and Konishi, 1978c). Also, the effects of anesthetizing the magnocellular nucleus in shifting and broadening a space-specific neuron's Δt-tuning are perhaps due to lifting of the inhibition.

The processes involved in the genesis of Δi-tuning in space-specific neurons are yet to be determined. Although neuronal sensitivity to Δi is present in VLVp, tuning to Δi does not occur below the level of the inferior colliculus. Convergence of Δt- and Δi-tuning properties on space-specific neurons is another unanswered question.

CONCLUDING REMARKS

Single-unit neurophysiologists believe that recording action potentials from a single neuron at a time should enable them to decipher the brain's codes. This tenet assumes that a neuron's stimulus-response relation contains information about neural coding. According to this view, a higher-order neuron that responds selectively to a complex stimulus can be a rich source of information about neural coding. The antithesis of this view states that the response of a single neuron does not reveal anything about neural coding because it is the distributed properties of a neuronal population that encode information.

A higher-order neuron is not a single channel but a nodal point of convergence of many afferent channels. Its stimulus-response relation, which derives from both its intrinsic properties and afferent connections, is not an epiphenomenon but a representation of the processes underlying the coding of the stimulus. It is not coincidental that the stimulus settings that excite space-specific neurons are the same as those that elicit the owl's sound-localizing behavior.

Space-specific neurons occupy a position higher than at least the fifth-order neuron in the hierarchy of neuronal groups engaged in spatial analysis of sound. They are "complex" neurons by any criterion. The search for the ultimate source of their properties led us to a series of discoveries such as the separation of time and intensity pathways, high-frequency phase coding, and the mechanism of Δt-tuning in n. laminaris. This success contrasts sharply with the view that complex neurons do not reveal anything because they are refractory to logical analysis (Marr, 1982).

Another lesson to be learned from the owl story is the significance of comparative studies. Some properties, whether physiological or anatomical, transcend species, and others are unique to a species. There are similarities and differences, for example, in the auditory systems of the owl and cat. Although the basic mechanism of phase locking and comparison are the same in the two species, the owl performs these functions at much higher frequencies than does the cat. These differences are due to differences in their modes of life. Because sound localization in two dimensions is important for the survival of the owl, its auditory system contains special design features to accommodate that need. Only a comparative approach can distinguish whether or not certain phenomena are universal. The brain is designed to control behavior, and behavior is a product of evolution. Appreciation of this elementary logic is one of the many lessons that we have learned from Ted Bullock (Bullock, 1983).

Acknowledgments

I thank Drs. Eric I. Knudsen, Ted Sullivan, and Terry Takahashi for critically reading and correcting the first draft of the manuscript. This work was supported by NIH Grant NS-14617.

REFERENCES

Boord, R. L. and G. L. Rasmussen (1963) Projection of the cochlear and lagenar nerves on the cochlear nuclei of the pigeon. *J. Comp. Neurol.* **120**:463–475.

Boudreau, J. C. and C. Tsuchitani (1968) Binaural interaction in the cat superior olive S-segment. *J. Neurophysiol.* **31**:442–454.

Bullock, T. H. (1983) Implication for neuroethology from comparative neurophysiology. In *Advances in Vertebrate Neuroethology*, NATO ASI Series, Series A: Life Sciences, Vol. 56, J.-P. Ewert, R. R. Capranica, and D. J. Ingle, eds. Plenum, New York and London, pp. 53–75.

Capranica, R. R. (1972) Why auditory neurophysiologists should be more interested in animal sound communication. *Physiologist* **15**:55–60.

Colburn, H. S. (1973) Theory of binaural interaction based on auditory nerve data. I. General strategy and preliminary results on interaural discrimination. *J. Acoust. Soc. Am.* **54**:1458–1470.

Goldberg, J. M. and P. B. Brown (1969) Responses of binaural neurons of dog superior olivary complex to dichotic tonal stimuli: Some physiological mechanisms of sound localization. *J. Neurophysiol.* **32**:613–636.

Goldberg, J. M. and W. E. Brownell (1974) Discharge characteristics of neurons in anteroventral and dorsal cochlear nuclei of cat. *Brain Res.* **64**:35–54.

Jeffress, L. A. (1948) A place theory of sound localization. *J. Comp. Physiol. Psychol.* **41**:35–39.

Jhaveri, S. and D. K. Morest (1982) Neuronal architecture in nucleus magnocellularis of the chicken auditory system with observations on nucleus laminaris: A light and electron microscope study. *Neurosci.* **7**:809–836.

Julesz, B. (1971) *Foundation of Cyclopean Perception.* University of Chicago Press, Chicago.

Knudsen, E. I. (1983) Subdivisions of the inferior colliculus in the barn owl (*Tyto alba*). *J. Comp. Neurol.* **218**:174–186.

Knudsen, E. I. and M. Konishi (1978a) A neural map of auditory space in the owl. *Science* **200**:795–793.

Knudsen, E. I. and M. Konishi (1978b) Space and frequency are represented separately in auditory midbrain of the owl. *J. Neurophysiol.* **41**:870–884.

Knudsen, E. I. and M. Konishi (1978c) Center-surround organization of auditory receptive fields in the owl. *Science* **202**:778–780.

Knudsen, E. I. and M. Konishi (1979) Mechanisms of sound localization in the barn owl (*Tyto alba*). *J. Comp. Physiol.* **133**:13–21.

Knudsen, E. I., M. Konishi, and J. D. Pettigrew (1977) Receptive fields of auditory neurons in the owl. *Science* **198**:1278–1280.

Konishi, M. and E. I. Knudsen (1982) A theory of neural auditory space: Auditory representation in the owl and its significance. In *Cortical Sensory Organization*, Vol. 3, *Multiple Auditory Areas*, C. N. Woolsey, ed. Humana Press, Clifton, N.J., pp. 219–229.

Lavine, R. A. (1971) Phase-locking in response of single neurons in cochlear nucleus of the cat to low frequency tonal stimuli. *J. Neurophysiol.* **34**:467–483.

Marr, D. (1982) *Vision*. Freeman and Co., San Francisco.

Moiseff, A. and M. Konishi (1981) Neuronal and behavioral sensitivity to binaural time differences in the owl. *J. Neurosci.* **1**:40–48.

Moiseff, A. and M. Konishi (1983) Binaural characteristics of units in the owl's brainstem auditory pathways: Precursors of restricted spatial receptive fields. *J. Neurosci.* **3**:2553–2562.

Oertel, D. (1983) Synaptic responses and electrical properties of cells in brain slices of the mouse anteroventral cochlear nucleus. *J. Neurosci.* **3**:2043–2053.

Osen, K. K. (1969) Cytoarchitecture of the cochlear nuclei in the cat. *J. Comp. Neurol.* **136**:453–484.

Parks, T. N. and E. W. Rubel (1975) Organization and development of brain stem auditory nuclei of the chicken: Organization of projections from N. Magnocellularis to N. Laminaries. *J. Comp. Neurol.* **164**:435–448.

Payne, R. S. (1971) Acoustic location of prey by barn owls (*Tyto alba*). *J. Exp. Biol.* **554**:535–573.

Rose, J. E., L. M. Kitzes, M. M. Gibson, and J. E. Hind (1974) Observations on phase-sensitive neurons of anteroventral cochlear nucleus of the cat: Non-linearity of cochlear output. *J. Neurophysiol.* **37**:218–253.

Sullivan, W. E. and M. Konishi (1984) Segregation of stimulus phase and intensity in the cochlear nuclei of the barn owl. *J. Neurosci.* **4**:1787–1799.

Takahashi, T., A. Moiseff, and M. Konishi (1984) Time and intensity cues are processed independently in the auditory system of the owl. *J. Neurosci.* **4**:1781–1786.

Chapter Twenty

BRAIN PHYLOGENY

Speculations on Pattern and Cause

R. GLENN NORTHCUTT

Division of Biological Sciences
University of Michigan
Ann Arbor, Michigan

The branch of neuroscience called comparative neurology can propose answers. . . . They will often be wrong, or inadequate, after the fashion of answers in science, but then they will be improved or replaced by more adequate answers as better equipped and trained comparative neurologists follow us and correct our mistakes.

T. H. Bullock, 1983

Although vertebrates and their brains have existed for approximately 500 million years, most of our beliefs regarding the changes that have occurred in vertebrate brains, and the processes and mechanisms underlying these changes, are based on the patterns of variation exhibited by the brains of living vertebrates. This is so because brains do not fossilize, and little information beyond relative size and general external configuration of the brains of extinct vertebrates can be gleaned from endocasts.

Despite this limitation, which is not unique to brains but is true also of most soft structures and all functions, much can be deduced about the history (phylogeny) of vertebrate brains. Brain variation in living vertebrates is so extensive, at all levels of organization, that we believe major changes in brain structure and function must have occurred throughout the course of vertebrate evolution. However, it is still unclear when and how various changes occurred.

How do we postulate phylogenetic sequences, deduce their causes, and recognize correlations between structure and function for historical events with a limited to nonexistent fossil record? Most phylogenetic analyses begin with the decision to examine one or more characters (i.e., any definable attribute of an organism) in a number of species. Many characters are invariable within a species (i.e., all members of the species exhibit the character during some period of their life history) but exhibit discontinuous variation among species. Thus varying degrees of character similarities exist in different species, and an observer must decide whether a particular similarity is probably due to a single character transforming through time (homology) or to independent changes of one or more characters by chance or similar selective pressures (homoplasy). Thus a critical second step in phylogenetic analysis involves the resolution of the homology or homoplasy of similar characters. Any statement regarding the homologous or homoplasous relationship of two or more characters is, itself, an hypothesis that can not be proven, only corroborated or falsified in the context of alternate hypotheses. However, further hypotheses regarding the processes of phylogenetic change, and the adaptive significance of such change, are necessarily based on the proposed homology or homoplasy of the characters being examined. If the initial hypotheses are not valid, further interpretations may be erroneous. By the same token, hypotheses regarding the patterns of homologous or homoplasous characters limit the kinds of processes that can be proposed to account for the genesis of these character patterns. Therefore, the evaluation of proposed phylogenetic processes and mechanisms must also include an examination of hypotheses of character similarity.

If phylogenetic analyses are complicated by the formulation of hypotheses based upon hypotheses regarding historical events that are not directly accessible to test and measurement, why bother? The answer is, because phylogenetic analyses are the core of comparative biology, whose tenets state that life is ordered—patterns of characters do exist among different organisms, and these patterns are generated by the process of evolution. The goal of comparative biology is, therefore, to recognize and analyze character patterns and propose explanations for the processes that generate such patterns. Any biologist who examines any aspect of a character in any organism and hopes to compare it to a character in another organism can not avoid phylogenetic analysis.

A complete phylogenetic analysis of vertebrate brains would include all brain characters that vary among species and are believed to have biological significance. Obviously, such an analysis has not been and is still not possible; historically, attention has focused on variation that exists in brain size, connections or networks, and cell morphology. These variables have thus been examined in sufficient detail for hypotheses regarding the genesis of the variation and its biological significance to be generated. Clearly, other parameters of the brain—transmitters, modulators, developmental sequences, physiological properties of cells and networks—are equally important, but for the most part, such characters have been examined in too few species, representing only a limited number of vertebrate radiations, to allow for a meaningful analysis. Therefore, the following discussion focuses on the patterns of variation that can be recognized in brain size, networks, and cell morphology. The analysis begins with definitions and criteria for comparisons, proceeds to an examination of brain size, then suggests corollaries that should follow from these observations. In turn, the corollaries are examined in the context of the variation in networks and its biological significance. Throughout the analysis, attention is also given to earlier hypotheses proposed to account for brain phylogeny.

CHARACTER COMPARISON AND PATTERNS

Two distinctly different types of character comparisons are possible in phylogenetic analyses (Smith, 1967). Characters present in a putative common ancestor can be compared to those in descendent taxa (patristic comparison, Fig. 1*A*); characters among sibling taxa (two or more taxa that have evolved from the same ancestral taxon) can be compared to each other (sibling or cladistic comparison, Fig. 1*A*). The preevolutionary concept of homology (Owen, 1843) held that characters so related were "the same organ in different animals under every variety of form and function." This definition was clearly based on sibling comparisons, as biologists of that period believed that biological species did not change through time. Following the acceptance of Darwin's theory of evolution, morphologists realized that the concept of homology should be based on inheritance due to common ancestry (see Patterson, 1982 for a recent review). As species *do* evolve, hypotheses of homology can involve both patristic and sibling comparisons, and any definition of homology should include statements regarding both types of comparison. Wiley's definition of homology (1981: 121–122) does so and is adopted in the subsequent discussion: "A character of two or more taxa is homologous if this character is found in the common ancestor of these taxa, or, two characters (or a linear sequence of characters) are homologous if one is derived from the other(s)." Given any three taxa, with one the common ancestor (taxon A, Fig. 1*A*) of two descendent taxa (taxa B and C, Fig. 1*A*), there are only two patterns of character distribution in which both patristic and sibling homology can exist. In the first pattern, a character in the ancestral taxon is retained in both descendent taxa; in the second pattern (Fig. 1*B*), a character (character circle) in

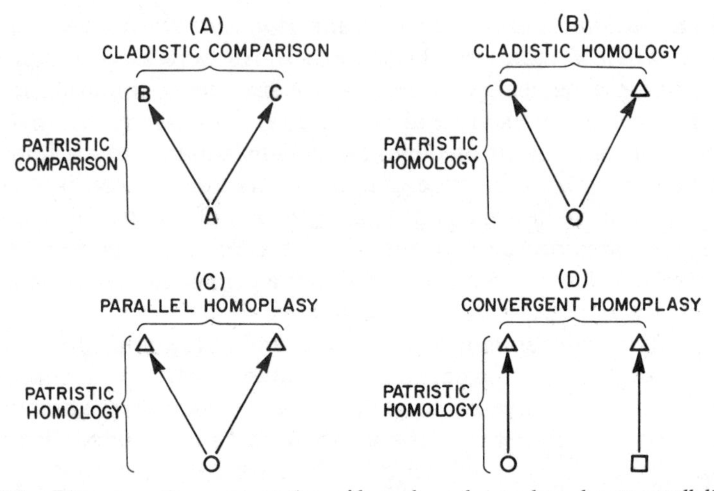

FIGURE 1. Diagrammatic representation of homology, homoplasy due to parallelism, and homoplasy due to convergence. Symbols denote characters, and letters denote taxa. Sibling or cladistic comparisons are enclosed by a horizontal bracket and patristic comparisons by a vertical bracket. (From Northcutt, 1984.)

the ancestral taxon is retained in one descendent taxon but is transformed into a new character (character triangle) in the other descendent taxon.

Definitions of homology clarify the comparisons being made but do not state the criteria necessary to recognize such relationships. Most biologists have emphasized criteria based on phenetic similarity (Remane, 1956; Simpson, 1961; Mayr, 1969; Bock, 1977). It is assumed that a single character changing through time is probably indicated if two or more characters in descendent taxa exhibit similarity in topographical position, high degree of resemblance (i.e., not superficial but detailed), and continuance of similarity throughout intermediate species. More often than not, similarities have been noted, then the foregoing criteria have been applied to support an hypothesis of homology based on the belief that the similarities appear to be greater than could occur by chance. There is clearly an element of circular reasoning inherent in this approach. Rarely have alternate hypotheses been posed and attempts made to falsify them.

In addition, similarity thought to be too detailed to be due to chance does not necessarily indicate a homology. Similarity between characters in sibling taxa may also arise by similar selective pressures acting independently on a character, inherited from a common ancestor, that is then transformed independently in two or more lineages (Fig. 1C). It is important to distinguish this type of character similarity, termed parallelism, from that of homology. Failure to do so leads to the misinterpretation that characters in descent taxa are the same as that in the common ancestor, and it can then lead to the recognition of unnatural taxa and errors in interpreting evolutionary processes.

Similarity among characters in sibling taxa may also arise due to chance. Selective pressures acting on different ancestral characters may independently transform these characters, a phenomenon termed convergence (Fig. 1*D*).

Three separate hypotheses are therefore possible regarding similarity between two or more sibling species: (1) similarity due to homology, (2) similarity due to parallel homoplasy, and (3) similarity due to convergent homoplasy. Parallelism has been distinguished from convergence based on presumed differences in the genetic bases of the characters (Simpson, 1961). Parallel characters are assumed to be similar characters that have arisen independently but are based on the same genes, whereas convergent characters are assumed to be similar characters based on different genes. The genetic bases of most characters are unknown, and a general working "rule" has been that similar characters occurring independently in widely separated taxa are probably due to convergence, whereas similarities occurring independently in closely related taxa are probably due to parallelism. This "rule" is an arbitrary one, as taxonomic "closeness," with respect to convergence and parallelism, is essentially impossible to define. Does parallelism occur only among species of a single genus, among taxa of a single family, order, or class? Wiley (1981, p. 12) has suggested that convergence and parallelism be based on phenotypic criteria, and his definitions are adopted in the subsequent discussion. Convergence (Fig. 1*D*) is "the development of similar characters from different preexisting characters," whereas parallelism (Fig. 1*C*) is "the independent development of similar characters from the same plesiomorphic [primitive] character."

Similarity due to convergent homoplasy should be distinguishable from homology or parallel homoplasy by the degree of phenetic similarity, as two characters that have arisen from different characters should be only superficially similar. Empirically, this assumption appears to be borne out by a large number of examples universally recognized as cases of convergent homoplasy: cephalopod and vertebrate eyes, invertebrate and vertebrate wings (or various structures used as wings in different vertebrate radiations), changes in body shape associated with aquatic locomotion, electroreception in teleosts and other vertebrates, are all such examples.

Similarity due to homology, as opposed to that due to parallel homoplasy, is far more difficult to distinguish, as the characters are presumed to be based on identical, or nearly identical, portions of the genomes. Thus prediction of the degree of phenetic similarity in a sibling homology is impossible.

Traditionally, those who might be called Simpsonian systematists have questioned the existence of a character homology in sibling taxa when (1) the suspected common ancestor is assumed to have had a different character than that of the sibling taxa, (2) the taxa are sufficiently close to assume highly similar genetic and developmental potentials, and (3) the sibling taxa occupy similar habitats and are assumed to be, or have been, subjected to similar selective pressures. However, these conditions do not allow one to falsify an hypothesis of homology; they only indicate that parallelism may have occurred.

Eldredge and Cracraft (1980: pp. 73–74) have argued that parallelism, as a concept, is epistemologically impossible to evaluate, as it requires numerous *ad hoc* assumptions and violates the concept of parsimony. As parallelism requires that derived characters occur independently in two evolutionary lineages, and a properly expressed cladistic hypothesis would necessitate the evolution of the derived condition only once, Eldredge and Cracraft suggest that the concept of parallelism be abandoned and the term convergence be applied to all cases of nonhomology. However, it seems reasonable to assume that both parallelism and convergence (as defined by Wiley) occur as different phenomena in vertebrate evolution, and our inability to distinguish them seems insufficient reason to deny the existence of one or the other.

Le Quesne (1969), in fact, noted a particular test of parallelism. Given two sets of characters, with each set exhibiting a primitive and a derived condition, there are four possible combinations of these characters. If all four combinations occur among taxa being studied, then one of the characters must have undergone two transformations from primitive to derived, or one transformation from primitive to derived and one transformation from derived to primitive. Thus the smallest numbers of species and characters for which this test can detect parallelisms are four and six, respectively (Underwood, 1982). Although the test indicates that parallelism may have occurred, there is no way to determine which character has undergone parallel change. Pair-by-pair comparison of all the characters, however, may reveal those that exhibit a high failure rate, indicating that these have likely undergone parallel evolution.

Cladistic approaches (Hennig, 1966; Eldredge and Cracraft, 1980; Wiley, 1981; Patterson, 1982) also involve methods for distinguishing sibling homology from homoplasy, although they do not provide additional methods for distinguishing convergence from parallelism. According to cladists, shared derived characters (synapomorphs) characterize monophyletic groups, and synapomorphies are sibling homologies; therefore, hypotheses of homology are hypotheses regarding monophyly. Conversely, independently derived characters can not define a monophyletic group and are not cases of sibling homology. Primitive characters (plesiomorphs) and shared primitive characters (symplesiomorphs) are not excluded from consideration as possible sibling homologues, as such characters can be viewed as those whose level of synapomorphy has not been resolved. As additional taxa are included in an analysis, characters initially viewed as symplesiomorphs become synapomorphs defining monophyletic taxa at a higher level. A critical test of a particular hypothesis of homology is possible by testing the congruence of this hypothesis, expressed as a synapomorphy, against other hypotheses of synapomorphy. It is generally assumed that the hypothesis that exhibits the largest number of synapomorphies is the most parsimonious (corroborated hypothesis), whereas other hypotheses that reveal fewer "synapomorphies" (falsified hypotheses) are based on cases of homoplasy.

In analyzing synapomorphies, it is necessary to determine the polarity (i.e., transformation from primitive to derived) of the characters being examined. Two criteria are commonly used: (1) the out-group rule, and (2) ontogenetic character precedence (von Baer's theorem). The out-group rule (Hennig, 1966) proposes that, given two characters that are homologues and found within a monophyletic group, the character that is also found in the sister group is the primitive character, whereas the character found only within the monophyletic group is the derived character. Von Baer's theorem states that two or more closely related taxa will follow the same course of development to the state of their divergence. Thus characters observed to be more general are assumed to be primitive, and those that are less general are assumed to be derived (Nelson, 1978; Eldredge and Cracraft, 1980).

Although various disciplines within systematics have developed methods for distinguishing homology from homoplasy, little attention has been directed to distinguishing parallelism from convergence. This is not surprising, when the object of an analysis is to recognize the most corroborated hypothesis of descent. However, if a phylogenetic analysis focuses on a few characters distributed among widely separated taxa whose phylogeny is highly corroborated (as is frequently the case in neurobiology), tests of alternate phylogenies are of limited use in resolving homoplasy. At present, there are apparently no methods that reveal evidence of parallel evolution when a single character is studied. It can be argued that parallelism should most commonly occur in simple characters that are based on limited portions of the genome in closely related taxa. Given that many, if not most, vertebrate characters are polygenic, and that genetic similarity is greater among closely related taxa (Dene et al., 1976; Goodman, 1976), most cases of parallelism probably occur at ordinal or even lower taxonomic levels. Analysis of some proteins in sets ranging from 8 to 16 species indicates that parallelism may occur in as many as 50% of the species, depending on the protein examined (Guise and Peacock, 1982). Similar studies of more complex characters do not appear to exist; however, even if such studies were to reveal that parallelism occurs more rarely in complex characters, its occurrence could not be ruled out in any particular analysis. For the present, similarities among characters in closely related taxa should be suspected to be due to parallelism if the taxa occupy similar habitats and if the characters cannot be subjected to more rigorous systematic analysis.

VARIATION IN BRAIN SIZE

Brain size, like other organ size, varies a great deal in living animals. Generally, larger species possess larger brains, but a considerable range of variation in brain size is exhibited in animals of the same body size (Fig. 2). Even within a single radiation, brain size in different species can vary for a given body size by a factor of 10 (Jerison, 1973; Northcutt, 1978). Any analysis of brain size must, therefore, isolate body size as one variable, so that measures of brain size are independent of body size, or else the role of body size is recognized. Plots of brain-body weights (Fig. 3) clearly reveal that brain size, for a given vertebrate radiation, does not vary randomly with respect to body size; rather, the plots form orderly arrays. Snell (1891) was the first to demonstrate that this relationship can be described by the power function

$$E = kP^{\alpha}$$

in which E and P are brain and body weights or volumes, respectively, and k and α are constants.

Dubois (1897), using more or less *ad hoc* methods, estimated α to be 0.56 and calculated k for a number of species. He termed k the "index of cephalization" and argued that it was a measure of intelligence.

There are a number of problems with interpreting k. First, it refers to the brain weight of an adult vertebrate weighing one gram, hardly a representative body weight for vertebrates. Second, it is an awkward number mathematically, as it is expressed as length in centimeters rather than being a pure number. These problems have been circumvented by Jerison (1973) and Bauchot and Stephan (1966) who developed indices that relate k to an "average" brain:body size for mammals and "basal insectivores," respectively. A more critical problem is that it has not been possible to relate values of k to any *objective* measure of a meaningful biological parameter such as intelligence or behavioral complexity.

The constant α, termed the exponent of allometry, is a scaling factor that has been empirically determined in a number of studies. Dubois (1897) defined α by assuming that certain pairs of species were equally cephalized and that one could, therefore, determine α by measuring the slope of a straight line passing through the points plotted for the brain weights and body weights of these pairs. Von Bonin (1937), using a more objective approach, fitted a single straight line to the logarithms of brain and body weights of more than 100 mammals, using regression analysis, and found that α was approximately 0.67. Similar values were reported by Jerison (1955) and Bauchot and Stephan (1966).

Generally, relative brain size has been analyzed by the use of minimum convex polygons or by allometric approaches that involve curve-fitting lines to brain:body data converted to logarithms.

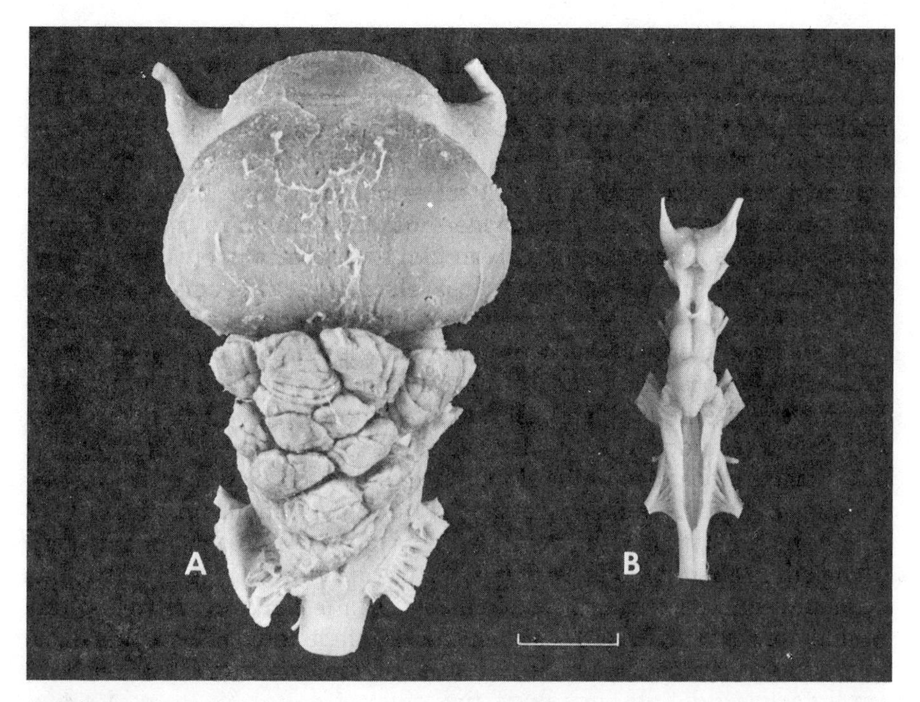

FIGURE 2. Dorsal views of (*A*) the brain of an adult lesser devil ray, *Mobula japanica*, with a body weight of 95 kg and (*B*) the brain of an adult angel shark, *Squatina dumeril*, with a body weight of 80 kg. Bar scale equals 2 cm.

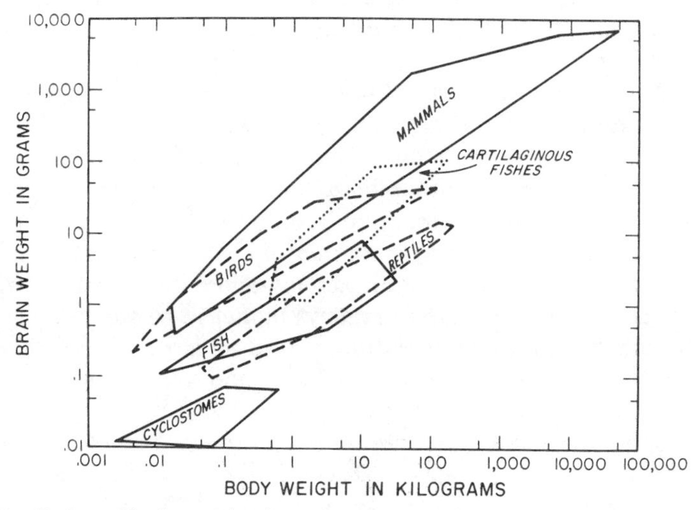

FIGURE 3. Brain and body weights for various species of six vertebrate classes expressed as minimum convex polygons.

Minimum Convex Polygons

Jerison (1970, 1973) devised a semiquantitative method for displaying brain: body data that avoids reliance on empirically determined values of exponents of allometry. Instead, he connected the extreme points for the groups being analyzed, thus encompassing the brain: body data into minimum convex polygons. Figure 3 summarizes Jerison's treatment of data for 198 species of bony fishes, reptiles, birds, and mammals. Data for cartilaginous fishes (Bauchot et al., 1976; Ebbesson and Northcutt, 1976; Northcutt, 1978) and cyclostomes (Ebinger et al., 1984; unpublished observations) have been added to Jerison's treatment of the data originally collected by Crile and Quiring (1940). Data for amphibians are not illustrated in Figure 3, but work cited by Platel (1979) indicates that a polygon for the amphibian data would overlap those for both bony fishes and reptiles.

These minimum convex polygons enclosing interspecific points are elongated so that their major axes exhibit slopes of approximately 0.67, confirming the orderliness of brain: body data and supporting the use of a power function in describing that orderliness.

Equally important, the distribution of data points for a closely related set of taxa (familial or ordinal levels) reveals an orderly distribution within a given polygon, so that brain: body points for more closely related taxa have a more limited distribution along the major axes than do those for less closely related taxa. Thus brain: body points for primates are located within the upper half of the polygon for mammals, whereas brain: body points for insectivores and didelphid marsupials are restricted to the lower portion of the polygon.

The polygons also reveal that vertebrate species of several different radiations have apparently independently evolved very large brains for a given body size. Thus some myliobatiform elasmobranchs (Fig. 2*A*), many birds, and some mormyriform bony fishes (Bass et al., 1981; data for mormyriforms are not plotted in Fig. 3) possess brains as large as those of many mammals. In addition, some primates and cetaceans (Figs. 3 and 4) possess the largest brains, for a given body size, of all vertebrates.

Given the distribution of brain: body points for living vertebrates, what can we deduce about brain size as a character in brain phylogeny? It has been claimed that brain size, relative to body size, has increased through geological time (Lartet, 1868; Marsh, 1874; Dubois, 1897). This process has been described as encephalization, a term that appears to have arisen with studies of relative brain size. Subsequently, encephalization has been used to describe the increase in some forebrain areas (cortices and thalamic nuclei) and their functions in more "progressive" species, as well as the shift of "higher" functions to more rostral brain areas in a linear sequence of phylogeny from "fish to man." Thus the term has been used to refer to several phenomena, but it is generally enmeshed in a *Scala naturae* approach to phylogeny.

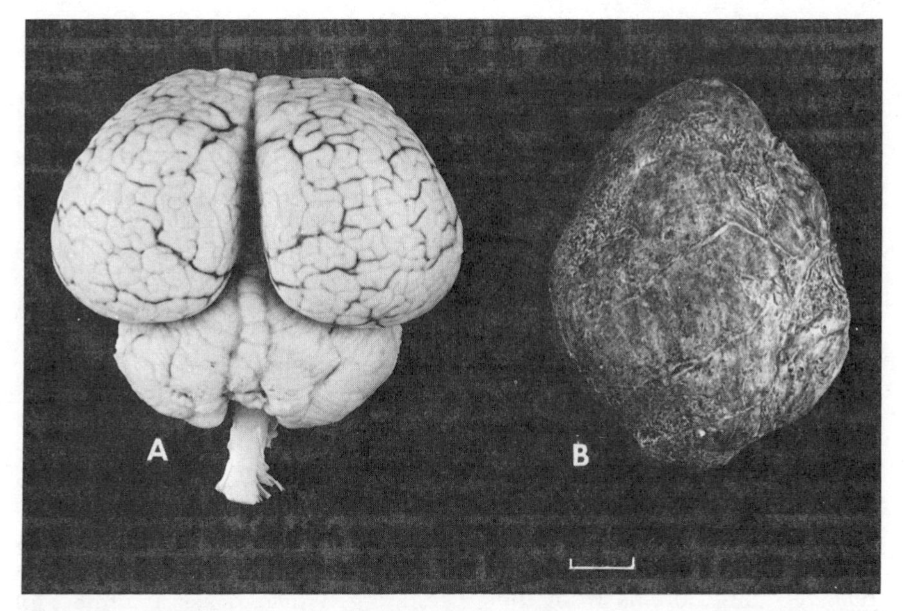

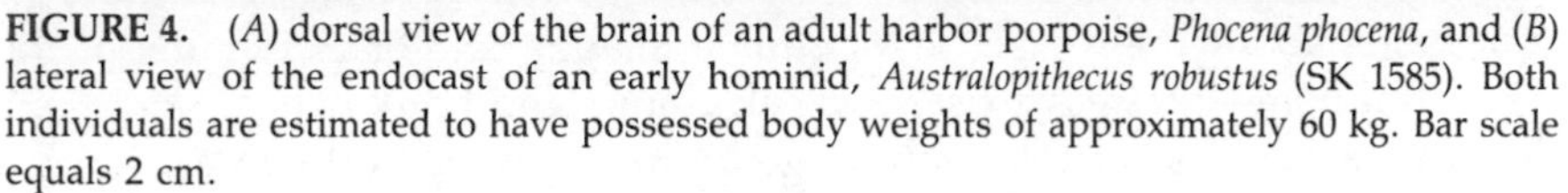
FIGURE 4. (*A*) dorsal view of the brain of an adult harbor porpoise, *Phocena phocena*, and (*B*) lateral view of the endocast of an early hominid, *Australopithecus robustus* (SK 1585). Both individuals are estimated to have possessed body weights of approximately 60 kg. Bar scale equals 2 cm.

Whether or not we refer to the process as encephalization, have brains, relative to bodies, increased in size through time, and if so, have brains become relatively larger and also more complex (i.e., possessing more recognizable cellular aggregates and pathways)? Although the intuitive answer regarding relative size is yes, it is also possible to hypothesize that the existence of relatively small brains in many vertebrate species is not due to the retention of a primitive character but, instead, represents a secondary reduction in brain size and is, thus, a derived character. Analysis of relative brain size in lampreys (the suspected sister radiationof jawed vertebrates) and hagfishes (the suspected sister radiation of lampreys and jawed vertebrates) suggests that relatively small brain size represents the primitive condition of the earliest vertebrates. The alternate hypothesis requires two independent reductions of brain size in living agnathans. Significantly, the hypothesis of small-brained early vertebrates is corroborated by analysis of endocasts of the fossil ostracoderms (Stensio, 1963), the earliest vertebrates for which a fossil record exists.

Analyzing the trends among jawed vertebrates is more difficult. Cartilaginous fishes are the sister radiation of all other jawed vertebrates. If relative brain size in living cartilaginous fishes is considered a primitive character (hypothesis 1), then relatively small brains in most bony fishes, amphibians, and reptiles must be interpreted as a secondary reduction (derived character). Alternatively, relatively small brains in the latter taxa must be a primitive character, and relatively large brains must be a derived character that arose independently in some cartilaginous fishes and bony fishes (some mormyriforms), and in birds and mammals (hypothesis 2). Both hypotheses require approximately equal numbers of evolutionary events. However, relatively large brains in cartilaginous and bony fishes occur only in some members of these groups, and analysis of these individual groups supports the hypothesis that large brains are derived (Northcutt, 1978; Northcutt and Braford, 1980). In regard to amniotes as a group, analysis of the endocasts of fossil reptiles (Jerison, 1973; Hopson, 1979) suggests that brain size in these taxa was similar to that in modern reptiles and that the relatively larger brains of living birds and mammals probably also evolved independently.

Relative brain size in cartilaginous fishes poses an additional problem in that the fossil record, outside of teeth, is not particularly good. However, endocasts of the earliest sharks indicate their skulls may have housed brains comparable in size to those of modern squalomorph sharks (Schaeffer, 1981). If this is the case, cartilaginous fishes must have possessed relatively large brains from their beginning, which raises a number of questions about their exact taxonomic affiliation and/or the brains of the ancestral fishes that gave rise to cartilaginous and bony fishes.

The distribution of relatively large brains in living vertebrates, in conjunction with data from fossil endocasts, is therefore consistent with the hypothesis that brain size, relative to body size, has increased in *many* vertebrate lineages, but this can not be viewed as a linear process. Rather, the data support the contention that relatively large brains have evolved independently in some members of most, if not all, vertebrate radiations. Thus large brain size must be viewed as a homoplasous character in vertebrate phylogeny.

A plot of forebrain: body weights for 26 species of vertebrates (Fig. 5) appears remarkably similar to that for total brain weight (Fig. 3). Although the data base is extremely small, analysis suggests that relative forebrain size, like relative brain size, has probably increased in some members of all vertebrate lineages. Again, however, large forebrain size must have evolved independently and must be viewed as a homoplasous character in vertebrate phylogeny.

It is likely that increases in relative cortical volumes in different mammalian radiations (Hofman, 1982) have also occurred independently. Furthermore, it is not clear what selective pressures and adaptive values are associated with increased brain, or cortical, size (Martin, 1981; Armstrong, 1982, 1983). Similarly, there are few, if any, data that support the concept of a more rostral shift of so-called higher functions (Macphail, 1982).

It is obvious that the large brains of many elasmobranchs, birds, and mammals are not simply enlarged cyclostome brains. Thus, it appears that increase in the number of cellular aggregates and pathways must also have

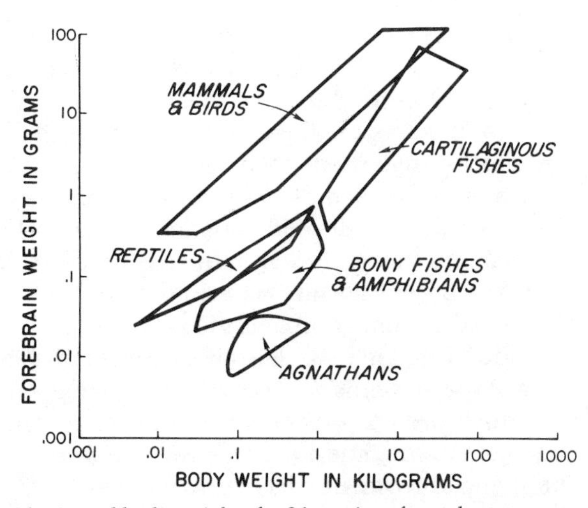

FIGURE 5. Forebrain and body weights for 26 species of vertebrates expressed as minimum convex polygons. (After Northcutt, 1981. Reproduced, with permission, from the *Annual Review of Neuroscience* **4.** © 1981 by Annual Reviews, Inc.)

occurred, yet there are essentially no quantative data on these parameters. A quick and dirty survey of telencephalic cellular aggregates that have been recognized and named in various classes of vertebrates suggests that the number of aggregates in some taxa represent a three- to tenfold increase over that in other taxa. Similar counts of pathways are not possible, as there are too few descriptive and experimental data. Analysis of the distribution of aggregates and pathways is further compounded by very different interpretations of how pathways change in phylogeny; hypotheses range from those that hold new pathways are common (Herrick, 1948; Bishop, 1959; Noback and Shriver, 1969) to those that hold modern vertebrates possess fewer pathways than the earliest vertebrates (Ebbesson, 1980).

Allometry

Although representation of brain : body data as minimum convex polygons allows one to visualize a number of qualitative aspects of relative brain size, it does not allow extensive quantitative measures of data among or within individual polygons. Most studies have empirically determined values of α and k by linear regression techniques applied to brain : body data converted to logarithms. Unfortunately, there has been little consistency in the gathering or analysis of the data comprising this extensive literature. There are several types of scaling inherent in brain : body relationships: (1) ontogenetic scaling, in which brain size, relative to body size, changes in an individual during growth; (2) intraspecific scaling (static adult allometry), in which brain size for *adult* members of a species may alter by sex; and (3) interspecific scaling (static allometry of adult means), in which relative brain size, expressed as means for different species, may differ. Many studies have not separated these three scaling phenomena: data for juveniles have been reported with data for adults, data on sex and the possibility of sexual dimorphism have been omitted, and interspecific analyses have been based on samples of a single individual in one species rather than the mean for the species. Similarly, regression techniques commonly involve least mean square, major axis, or reduced major axis, and many studies do not state which regression method has been used or the rationale for a given regression.

Given these confounding factors, it is not surprising that the reported intraspecific and interspecific exponents of allometry range from −0.04 to 0.50 and 0.23 to 0.94, respectively (Bauchot et al., 1976; Platel and Delfini, 1981; Ebinger et al., 1984). Intraspecific exponents are, however, generally lower than interspecific exponents.

Considerable attention has focused on the biological significance of the exponent of allometry. Many studies have reported a value close to 0.67 (Dubois, 1897; von Bonin, 1937; Bauchot and Stephan, 1966; Jerison, 1973). This value suggests that interspecific scaling of brain size is attributable to the fact that brains integrate information from surface receptors and control surface effectors. Consequently, as bodies expand in weight and size (i.e., as a cube), brains should expand at the same rate as the surface area (i.e., as a square). However, studies of the density of receptors and effectors (Bruesch and Arey, 1942; Carter, 1965; Towe, 1973), relative to body size, do not reveal the scaling that would be predicted if body surface area were determining brain size (Armstrong, 1982).

There appears to be little empirical basis for the widely accepted value of 0.67. Recently, new measures of interspecific exponents of allometry for placental mammals (Bauchot, 1978; Eisenberg and Wilson, 1978; Martin, 1981) have yielded values of 0.73–0.76. Martin noted that basal metabolic rates in most animals (and some plants) scale to body weight with a value of 0.75. To test the possibility that brain size might be directly linked to basal metabolic rates, interspecific exponents of allometry were calculated for birds (180 species) and reptiles (59 species). The avian and reptilian samples yielded exponents of 0.58 and 0.54, respectively. Although exponent values are below what one would expect if basal metabolic rates are directly linked to brain size, these values do not support an exponent of 0.67 for geometric scaling in amniotic vertebrates.

Mammals, however, are viviparous, whereas birds and most reptiles are oviparous. Martin (1981) suggested that brain size may be linked to maternal metabolic turnover in mammals, and to metabolic rate and egg weight in birds and reptiles. Although far more data are needed to evaluate this hypothesis (elasmobranchs offer one fertile area for research, as they exhibit a considerable range of relative brain sizes and reproductive strategies, from oviparity to viviparity), it opens the possibility of relating relative brain size to specific aspects of life-history strategies.

A central issue in all attempts to relate allometry (a power function) to brain : body scaling in vertebrates is the biological significance, if any, of the constants k and α. It should be clear, by now, that the biological meaning of k can be understood only within the context of the exponent of allometry. If this exponent has no single biological basis (geometric, metabolic, or other type of scaling), then k can have no single meaning as a biological predictor. Huxley (1932) believed that the power curve not only described the consequences of growth but also expressed a theoretical basis for the phenomenon of scaling as a consequence of the fact that rates of growth of dependent and independent variables are proportional to the number of cells already present in these variables. However, researchers have noted that ontogenetic growth rarely follows a power curve. Huxley himself (Reeve and Huxley, 1945) questioned whether allometry is itself a process or only describes patterns. An additional problem emerges: if allometry is a real phenomenon, what is the relationship between interspecific and intraspecific scaling, and which represents an evolutionary pathway (Wolpoff, 1985)? If intraspecific allometry represents an evolutionary process, and ancestral-descendent species follow a single intraspecific curve, how can a larger-brained descendent species lie along a higher-sloped interspecific curve when intraspecific curves are usually lower than interspecific curves? Either this is never the case, or intraspecific and interspecific scaling must be the same. On the other hand, the interspecific curve is believed to reflect a general level of brain organization and, in many cases, includes ancestral-descendent sequences (Jerison, 1973; Hopson, 1979). If interspecific curves do not represent evolutionary pathways, what do they represent?

Based on population genetics theory, it has been proposed that both intraspecific and interspecific allometry share a common theoretical basis (Lande, 1979; Wolpoff, 1985). It is believed that intraspecific allometry is a consequence of genetic and environmental sources of covariation. If body size (independent variable) and brain size (dependent variable) are genetically correlated (pleiotropy and linkage), and selection acts on body size, brain size will vary in a linear manner for the log-transformed variables and therefore be allometric. Under stable environmental conditions and continued genetic covariation, low selection levels on one character, or genetic drift, will result in interspecific variation following a curve almost identical to that of an intraspecific curve. This suggests that the selective forces acting to differentiate closely related taxa operate predominantly on body size, and that changes in brain size are largely a correlated response.

Interspecific curves will deviate greatly from intraspecific curves, however, if any significant alteration of the components of covariance occurs. This could happen in one of two ways: (1) significant selection acting on both brain and body sizes, or (2) fundamental changes in the genetic covariation of the variables. It is possible that increase in brain size, due to covariance with increase in body size, is insufficient for an organism to function in a new environment, or that selection on other organ systems alters brain requirements. Under these conditions, selection may act on other parts of the genome and effect brain size independently of body size. It is also possible that selection acts to reduce pleiotropic linkage of brain and body size.

One corollary of this explanation of allometry is that the selective forces that act to differentiate closely related taxa primarily act on overall body size, with brain size largely following body size. Multivariate analysis of Bauchot's (1963) data on brain components in insectivores demonstrates that in species within a taxonomic category as large as an order, most of the total variation in brains is attributable to changes in brain size (Jerison, 1973). Thus many of the differences observed in the brains of closely related taxa may not be due to direct selection but may be a consequence of selection acting on body size, which in turn is known to be highly correlated with metabolic rates, fecundity, competitive strategies, and life histories.

The conditions proposed for differences in interspecific and intraspecific curves provide a second corollary: reorganization of brain components should occur. This would be reflected by changing interspecific curves as a consequence of additional selective forces acting on brain size and changes in the genetic covariation of brain and body sizes.

VARIATION IN NETWORKS: AGGREGATES AND THEIR CONNECTIONS

The population genetics theory of allometry predicts that the brains of closely related taxa should differ primarily in the size of the components constituting the brain, and that these differences are a consequence of selection acting to change body size. In contrast, the brains of more distantly related taxa, constituting diverse higher taxonomic groups, should exhibit reorganization of many brain components due to selective forces having acted directly on both brain and body sizes, and the genetic bases of the covariation of brain and body sizes having altered.

Is there evidence that brain reorganization occurs with the emergence of major taxonomic groups? The answer is definitely yes. As noted earlier, the large brains of elasmobranchs, birds, and mammals are not simply enlarged cyclostome brains. An examination of any level of the central nervous system in vertebrate species representing different taxonomic groups reveals unique derived neural characters. Such characters are particularly evident in the telencephalon, a brain area that has been the object of much study. Figure 6 illustrates transverse sections through one cerebral hemi-

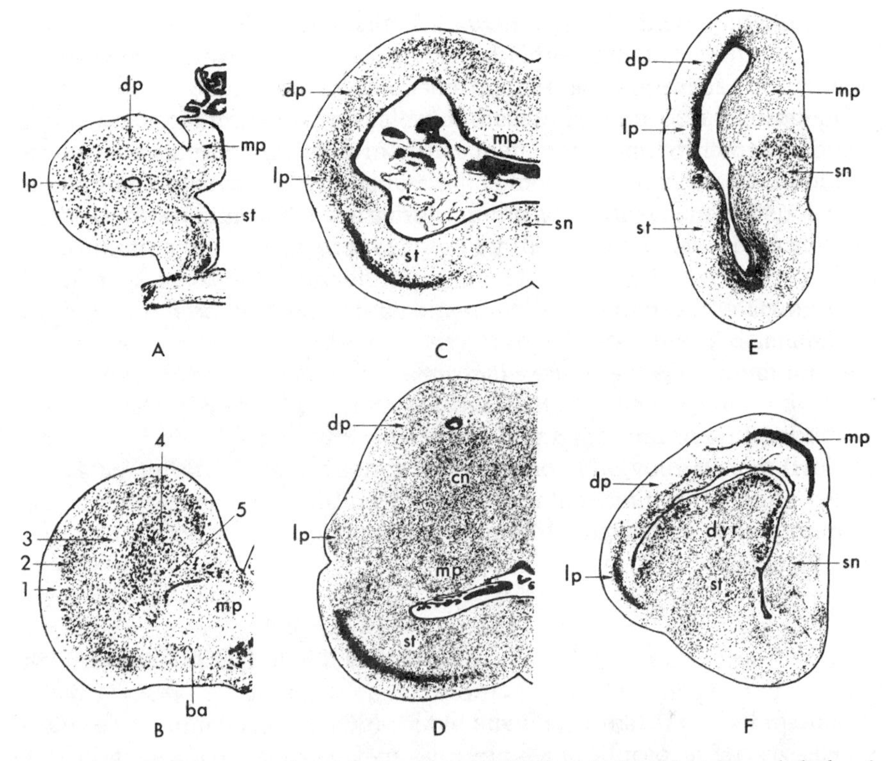

FIGURE 6. Photomicrographs of transverse sections through the right telencephalic hemispheres of (*A*) the marine lamprey, *Petromyzon*, (*B*) the Pacific hagfish, *Eptatretus*, (*C*) the spiny dogfish, *Squalus*, (*D*) the smooth dogfish, *Mustelus*; (*E*) the bullfrog, *Rana*, and (*F*) the tokay gecko, *Gekko*. ba, basal area; cn, central nucleus; dp, dorsal pallium; dvr, dorsal ventricular ridge; lp, lateral pallium; mp, medial pallium; sn, septal nuclei; st, striatum; 1–5, laminae one through five of hagfish pallium.

sphere in six different vertebrate species. These species can be organized into pairs, illustrating differences in relative size of the telencephalon as well as increases in the number of neural aggregates (Northcutt 1978; 1981). Similar species pairs also occur in ray-finned fishes (Northcutt and Braford, 1980).

The telencephalic hemispheres in hagfishes (Fig. 6*B*) are two to three times larger than those in lampreys (Fig. 6*A*) and are characterized by the absence of lateral ventricles, which are present embryonically, and elaboration of the dorsolateral pallium into a number of distinct cellular laminae.

The telencephalic hemispheres in advanced galeomorph sharks (Fig. 6*D*) and rays (Fig. 2*A*) are 4 to 15 times larger than those in squalomorph sharks (Fig. 6C) and skates of comparable body size. The hemispheres in most galeomorph sharks and rays further differ from those in other elasmobranchs by exhibiting marked reduction of the lateral ventricles and elaboration of the deeper layers of the dorsal pallium into a central nucleus.

Thus the telencephalic hemispheres in most galeomorph sharks are not only larger than those in squalomorph sharks, but they also possess a larger number of recognizable cellular aggregates.

The telencephalic hemispheres in reptiles (Fig. 6*F*) and birds are larger than those in amphibians (Fig. 5) of comparable body size, and, again, a larger number of cellular aggregates are present in the hemispheres in reptiles and birds. A particularly notable feature is the large dorsal ventricular ridge (Fig. 6*F*, dvr) which arises embryonically in both reptiles and birds, as an elaboration of the lateral pallial wall, and can be divided into numerous distinct cellular aggregates that receive inputs from several sensory modalities.

Examination of these species pairs reveals several important points: (1) A cladistic analysis reveals that the telencephalic hemispheres in these species pairs represent polarized characters, with hemispheres A, C, and E exhibiting a primitive condition relative to hemispheres B, D, and F. (2) Increase in relative telencephalic size is correlated with increase in number of cellular aggregates. (3) Although the hemispheres do not contain the same number of aggregates, the additional aggregates can be related to previously occurring aggregates on the basis of embryology. (4) The additional aggregates arise embryonically from more than one region of the pallium. There is an important implication of this fourth point: although the same major pallial aggregates (dorsal, lateral, and medial pallia) exist in all six species, elaboration of additional pallial aggregates must be interpreted as independent phylogenetic events (i.e., cases of homoplasy).

If reorganizations in the number of neural aggregates occur with increases in telencephalic size, is the same true of connections? Detailed telencephalic connections have been determined experimentally for only one of the species pairs illustrated (Fig. 7). In ranid amphibians, a portion of the lateral forebrain bundle arises from several different dorsal thalamic nuclei and terminates primarily on the dendrites of cells of the striatum (Gruberg and Ambros, 1974; Kicliter, 1979; Wilczynski and Northcutt, 1983), a subpallial cell group in the ventrolateral wall of the telencephalic hemisphere (Fig. 6*A*). In reptiles and birds, a portion of the lateral forebrain bundle arises from nuclei of the dorsal thalamus and terminates primarily in different subdivisions of the dorsal ventricular ridge (see Ulinski, 1983 for a recent summary), a pallial formation (Fig. 7*B*).

A similar reorganization of ascending pathways may also occur in elasmobranchs. In nurse sharks (galeomorphs), ascending thalamic pathways terminate primarily in the central nucleus (Schroeder and Ebbesson, 1974), a dorsal pallial derived aggregate. However, in *Platyrhinoidis*, a guitarfish characterized by a primitive telencephalon relative to that in other batoids, the ascending thalamic pathways appear to terminate in a subpallial aggregate (R. G. Northcutt and J. Wathey, unpublished observations) comparable to the striatum in amphibians. Clearly, examination of such pathways in species pairs of cartilaginous fishes, ray-finned fishes, and cyclostomes can determine if this type of reorganization is a widespread phenomenon.

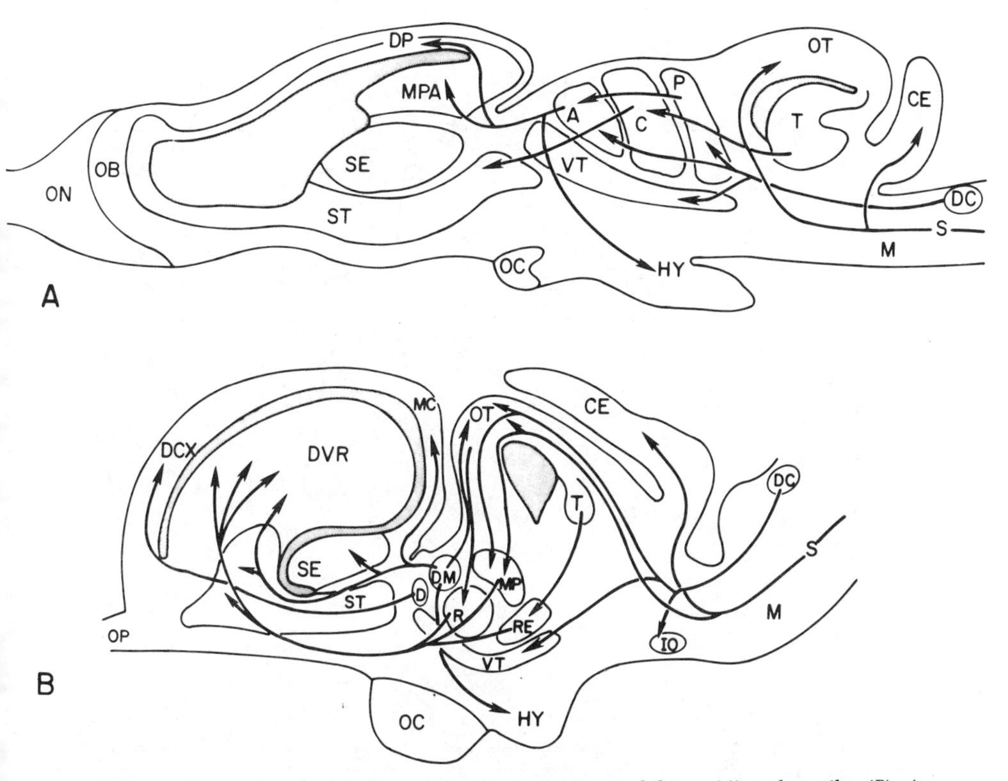

FIGURE 7. Summary of ascending sensory systems in amphibians (*A*) and reptiles (*B*). A, anterior thalamic nucleus; C, central thalamic nucleus; CE, cerebellum; D, dorsal optic nucleus; DC, dorsal column nuclei; DCX, dorsal cortex; DM, dorsomedial thalamic nucleus; DP, dorsal pallium; DVR, dorsal ventricular ridge; HY, hypothalamus; IO, inferior olivary nucleus; M, medulla; MC, medial cortex; MP, medial posterior thalamic nucleus; MPA, medial pallium; OB, olfactory bulb; OC optic chiasm; ON, olfactory nerve; OP, olfactory peduncle; OT optic tectum; P, posterior thalamic nucleus; R, nucleus rotundus; RE, nucleus reuniens; S, ascending spinal pathways; SE, septal nuclei; ST, striatum; T, torus semicircularis; VT, ventral thalamus (after Northcutt, 1981, Reproduced, with permission, from the *Annual Review of Neuroscience* **4**. © 1981 by Annual Reviews, Inc.)

An additional indication of the reorganization of pathways may be provided by the distribution of long ascending and descending pathways in vertebrates. If the distribution pattern of such long pathways indicates that they have arisen independently in a number of vertebrate radiations, this pattern would further suggest that reorganization has occurred. Furthermore, a preponderance of such long pathways in relatively large-brained vertebrates would also support a correlation between increase in brain size and reorganization of pathways.

The distribution of ascending spinal pathways (Fig. 8), as determined experimentally in a number of taxa (see Northcutt, 1984 for a more detailed review) suggests that reorganization has occurred in at least one, if not two, pathways. Spinoreticular pathways occur in all vertebrate species

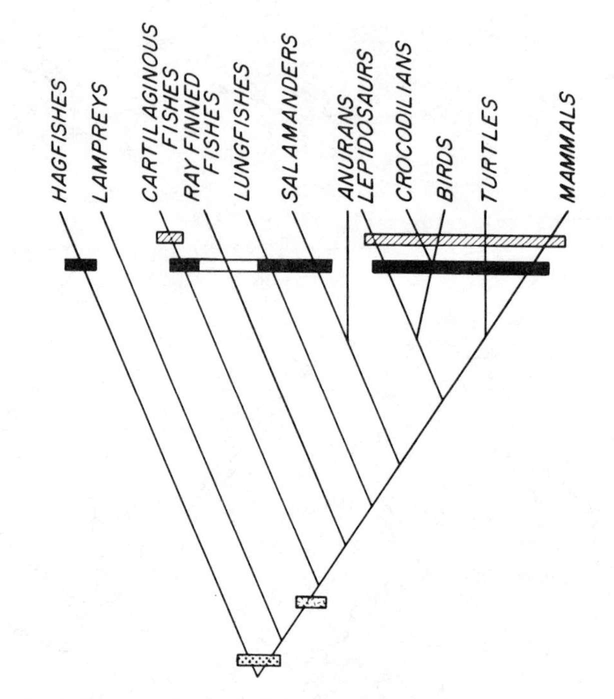

FIGURE 8. A cladogram showing the distribution of various ascending pathways: spinoreticular (stippling); spinocerebellar (random dashes); spinotectal (solid bars); uncertain spinotectal (open bar); spinothalamic (hatching). (From Northcutt, 1984.)

examined and probably represent a shared primitive character that arose with the origin of vertebrates. Alternatively, one must assume that they arose again and again with each vertebrate radiation, an extreme violation of parsimony. Spinocerebellar pathways occur in all jawed vertebrates and appear to be a shared primitive character of gnathostomes. At present, it is not clear whether hagfishes and lampreys possess spinocerebellar pathways, as it is not clear whether these animals possess a cerebellum. If additional studies reveal that cylostomes do not possess a cerebellum, then the most parsimonious hypothesis is that a cerebellum and spinocerebellar pathways arose with jawed fishes. Spinotectal pathways occur in most vertebrates but do not appear to exist in lampreys, rayfinned fishes, and anuran amphibians. A spinotectal pathway was probably present in ancestral vertebrates and independently lost (three evolutionary events). The alternate hypothesis—that spinotectal pathways are homoplasous—would require a minimum of four evolutionary events (three gains and one loss, three gains and two losses, or five gains). A spinothalamic pathway is known to exist in some elasmobranchs and in amniotes. Given this distribution, this pathway probably evolved independently in the two groups (two evolutionary events) and must be considered homoplasous and an indication of reorganization. To consider spinothalamic pathways homologous in those elasmobranchs and the amniotes, one must argue that the pathway was lost five to six times independently.

Similarly, an examination of the distribution of long descending telencephalic pathways (Fig. 9), as determined experimentally in a number of species (see Northcutt, 1984 for a more detailed review) supports the hypothesis that reorganization of pathways has occurred. Telencephalic projections to the diencephalon and midbrain tegmentum occur in all species examined and probably represent a shared primitive character for vertebrates. Striomedullary, or spinal pathways occur in anuran amphibians and lizards. If a similar pathway occurs in other amphibians and in sauropsid reptiles, it is probable that this pathway must at least have arisen with tetrapods and was subsequently lost in birds and mammals. A striomedullary pathway may exist in cartilaginous fishes, as a long descending pathway to the medulla has been described in galeomorph nurse sharks. However, the exact origin of this pathway is unknown, and it may arise in the pallium rather than the striatum. If this pathway is a striomedullary pathway, it may be homologous to the striomedullary pathway of tetrapods, which would require that the pathway arose with the origin of gnathostomes (supporting reorganization at this point in phylogeny) and was lost in lungfishes and ray-finned fishes. An alternate hypothesis—that this pathway arose independently in elasmobranchs and tetrapods—is also equally probable. Telencephalo-tectal pathways exist in many vertebrate taxa, but many of these pathways appear to be homoplasous, as they arise from different telencephalic areas: in elasmobranchs they originate in the dorsal pallium; in teleosts they originate in the striatum; in amphibians they originate in the anterior entopeduncular nucleus; in reptiles they originate in the medial pallium; in birds and mammals they originate in the dorsal pallium. Palliospinal pathways exist in mammals and birds and are probably homoplasous, as such a pathway has not been reported in reptiles. If palliospinal pathways are considered homologous, one must argue for the occurrence of four independent evolutionary losses, as opposed to two evolutionary gains.

The distribution of both long ascending spinal pathways and long descending pathways originating in the telencephalon suggests that many of these pathways are stable and probably arose at the time of the origin of vertebrates. This distribution also suggests that loss of pathways is not an uncommon phenomenon. Both gain and loss of pathways are evidence of reorganization in the central nervous system.

Other phylogenetic hypotheses have also postulated that changes in pathways occur in brain phylogeny. One major hypthesis that resulted from an examination of vertebrate brains during the first half of this century is the invasion hypothesis. This hypothesis proposes that brains change by the addition of new pathways, and it is postulated that this phenomenon occurs when axon collaterals of a neuronal population form connections with other neuronal populations not previously innervated by

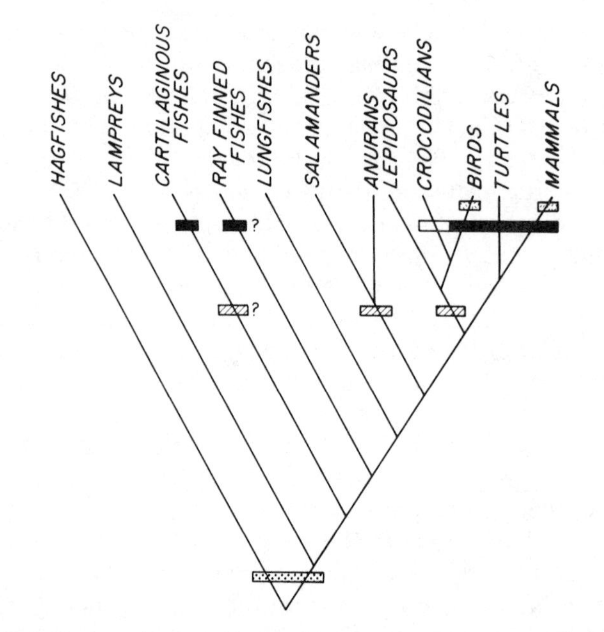

FIGURE 9. A cladogram showing the distribution of various telencephalic efferent pathways: telencephalo-diencephalic and tegmental (stippling); striomedullary or spinal (hatching); palliotectal (solid bars); uncertain palliotectal (open bar); palliospinal (random dashes); question marks indicate telencephalic pathways but pallial or subpallial origin uncertain. (From Northcutt, 1984.)

the first population (Herrick, 1948; Bishop, 1959; Noback and Shriver, 1969). Although the invasion hypothesis, like most hypotheses regarding brain phylogeny, is often interpreted within the context of *Scala naturae,* the distribution of many pathways in vertebrate taxa (Figs. 8, 9) is consistent with this hypothesis. Equally important, if invasion does occur, individual cells of the suspected invading aggregates should possess connections with supposedly "older" targets as well as the "newly" invaded ones. Such appears to be the case with at least a fraction of the cells in the spinal cord of mammals that project to the thalamus. Kevetter and Willis (1983) reported that approximately 11% of the spinal cord cells projecting to the reticular formation also project to thalamic nuclei. Thus variation in the termination of pathways in different vertebrate radiations, and collateralization of the axons of single cells that project to phylogenetically "older" and "newer" targets, both indicate that the phenomenon of invasion has occurred in brain phylogeny.

Recently, however, Ebbesson (1980) argued that there is no evidence for the invasion of new aggregates. He proposed that pathways change through time only by the differential loss of connections and the subsequent segregation (parcellation) of more homogenous neural populations. His hypothesis suggests that brain phylogeny proceeds from brains that are diffuse and undifferentiated to brains that possess more restricted connections and larger numbers of discrete cell groups. As formulated, the hypothesis is illogical, as it would require all existing brain variation to be telescoped into an ancestral population whose brains contained a very limited number of cellular aggregates exhibiting all the connections seen in modern vertebrates. A corollary of the parcellation hypothesis is that brains of vertebrates characterized by earlier grades of organization should possess fewer cell groups with more extensive connections than the brains of vertebrates characterized by later grades of organization. Although the first part of this corollary appears to be generally true, that is, the brains of vertebrates characterized by an earlier grade of organization have fewer cell groups, there is no experimental evidence that the brains of such vertebrates are characterized by cell groups with more extensive connections.

Hypotheses of brain phylogeny that are formulated solely on the bases of increase or decrease in the number of neurons in existing aggregates, formation of new pathways by axon collaterals invading existing neural aggregates, or loss of connections and/or neural aggregates suggest that brains change in a very rigid and mechanical way. In contrast, neurobiologists exploring other aspects of the brain have been struck by the plasticity exhibited by brains during normal development: the widespread effects of single recessive alleles on developmental and organizational patterns, and plasticity and functional recovery following trauma, for example. The population genetics theory of allometry suggests that an additional factor should be considered in phylogenetic hypotheses. If uncoupling or rearrangement of genetic covariation affecting brain and body sizes occurs during the rapid evolutionary events associated with the emergence of higher taxonomic groups, it is possible that these genetic rearrangements profoundly affect other developmental "programs" in addition to those that affect development of brain size and body size. Such altered developmental programs would likely have widespread effects and could cause extensive reorganization.

SUMMARY

Reconstruction of the phylogenetic history of vertebrate brains must be based primarily on interpretation of the patterns of variation exhibited by the brains of living vertebrates, as brains do not fossilize, and little information beyond relative size and external configuration can be recovered from endocasts.

In examining brain variation in vertebrate species, varying degrees of similarity are noted and raise the question whether particular similarities are due to a single character changing through time (homology) or to independent changes of one or more characters due to chance or similar selective pressures (homoplasy). This question can be addressed only by posing multiple hypotheses regarding the cause of the similarity and attempting to falsify them. Similarity due to convergent homoplasy should be distinguishable from homology or parallel homoplasy by the degree of phenetic similarity. Parallel homoplasy may be distinguishable from homology by testing the congruence of an hypothesis of homology, as a synapomorphy, against other hypotheses of synapomorphy. In the event that such tests are impossible, parallelism may be suspected if the taxa under consideration occupy similar habitats and the characters may be subjected to similar selective pressures.

In the examination of all character sets, it is important to determine the direction of character transformation from primitive to derived. Two criteria that frequently allow such determination are the out-group rule and von Baer's theorem.

Brain size and cellular aggregates and their connections are three parameters that have been examined in sufficient detail to establish patterns of variation that suggest certain trends and processes in brain phylogeny.

Cladistic analysis of brain:body data expressed as minimum convex polygons suggests that brain size, relative to body size, has increased in many vertebrate lineages but can not be viewed as a linear process for vertebrates as a group. Similarly, an increase in the number of cellular aggregates and an increase in relative brain size appears to be a valid correlation. A similar correlation of increased numbers of pathways and relative brain size can not be established, as there are insufficient data.

Quantitative examination of brain:body data, converted to logarithms, in which values of α and k have been determined by linear regression techniques generally reveal that intraspecific values of α are smaller than interspecific values. Although certain interspecific values of α have been interpreted as being due to geometric or metabolic scaling, there are insufficient data to establish the determinant(s) of these values.

Disagreement exists regarding whether allometry only describes a pattern or is, itself, a process, and whether interspecific or intraspecific allometry represents an evolutionary process. Recent application of population genetics theory to these problems (Lande, 1979; Wolpoff, 1985) suggests that both types of allometry are processes that share a common theoretical basis as consequences of genetic covariation. It is suggested that brain and body sizes are genetically correlated and that selection acts primarily on body size in closely related taxa, thus in both intraspecific and interspecific allometry brain size varies in a linear manner for the log-transformed variables. However, if significant alteration of the environmental components of covariance occurs by selection acting separately on both brain and body sizes, or if changes occur in the genetic covariation, interspecific allometric curves should deviate greatly from intraspecific curves.

The conditions under which such differences in allometric curves occur suggest that reorganization of brain components should also occur. It is possible that the k values for different higher taxonomic groups of vertebrates are a measure of such reorganization rather than a measure of information-processing capability or intelligence.

Increases in the number of cellular aggregates, and changes in long pathways and increases in their number, are correlated with increased brain size in several different vertebrate radiations. This is consistent with the hypothesis that such reorganization occurs during periods of rapid evolution of new vertebrate groups. It is suggested that changes in genetic "programs" that affect brain size as well as other brain parameters are responsible for such reorganization.

Acknowledgments

Some of the research reported in this paper was supported in part by grants from the National Institutes of Health (NS11006 and EY02485). Mary Sue Northcutt assisted in many phases of the work and in the preparation of the manuscript. I am very grateful to the American Society of Zoologists and Annual Reviews, Inc. for permission to use Figures 1, 8, and 9, and 5 and 7, respectively. I also thank Drs. Mel Cohen and Felix Strumwasser for giving me the opportunity to participate in the symposium. I am highly indebted to Drs. Carl Gans and Milford H. Wolpoff for many stimulating conversations bearing on various concepts discussed in this paper. Last, but not least, I thank Dr. Theodore H. Bullock for innumerable kindnesses, immeasurable inspiration, and unceasing provocation; most of all, I am grateful for his very valued friendship.

REFERENCES

Armstrong, E. (1982) A look at relative brain size in mammals. *Neurosci. Lett.* **34:**101–104.

Armstrong, E. (1983) Relative brain size and metabolism in mammals. *Science* **220:**1302–1304.

Bass, A. H., M. R. Braford, Jr., and C. D. Hopkins (1981) Comparative aspects of brain organization among electric fishes of Africa. *Anat. Rec.* **199:**21A.

Bauchot, R. (1963) L'architectonique comparée qualitative et quantitative du diencéphale des insectivores. *Mammalia* **27,** Suppl. 1:1–400.

Bauchot, R. (1978) Encephalization in vertebrates. A new mode of calculation for allometry coefficients and isoponderal indices. *Brain Behav. Evol.* **15:**1–18.

Bauchot, R. and H. Stephan (1966) Données nouvelles sur l'encéphalisation des Insectivores et des Prosimiens. *Mammalia* **30:**160–196.

Bauchot, R., R. Platel, and J. M. Ridet (1976) Brain-body weight relationships in Selachii. *Copeia* **1976:**305–310.

Bishop, G. H. (1959) The relation between nerve fiber size and sensory modality: Phylogenetic implications of the afferent innervation of cortex. *J. Nerv. Ment. Dis.* **128:**89–114.

Bock, W. J. (1977) Foundations and methods of evolutionary classification. In *Major Patterns in Vertebrate Evolution,* M. K. Hecht, P. C. Goody, and B. M. Hecht, eds. Plenum, New York, pp. 851–895.

Bruesch, S. R. and L. B. Arey (1942) The number of myelinated and unmyelinated fibers in the optic nerve of vertebrates. *J. Comp. Neurol.* **77:**631–665.

Bullock, T. H. (1983) Why study fish brains? Some aims of comparative neurology today. In *Fish Neurobiology*, Vol. 2: *Higher Brain Areas and Functions*, R. E. Davis and R. G. Northcutt, eds. University of Michigan Press, Ann Arbor, pp. 361–368.

Carter, H. B. (1965) Variation in the hair follicle population of the mammalian skin. In *Biology of the Skin and Hair Growth*, A. G. Lyne and B. F. Short, eds. Elsevier, New York, pp. 25–33.

Crile, G. and D. P. Quiring (1940) A record of the body weight and certain organ and gland weights of 3690 animals. *Ohio J. Sci.* **40**:219–259.

Dene, H. T., M. Goodman, and W. Prychodko (1976) Immunodiffusion evidence on the phylogeny of the primates. In *Molecular Anthropology*, M. Goodman and R. E. Tashian, eds. Plenum, New York, pp. 171–195.

Dubois, E. (1897) Sur le rapport du poids de l'encéphale avec la grandeur du corps chez mammifères. *Bull. Soc. Anthropol. Paris* **8**:337–376.

Ebbesson, S. O. E. (1980) The parcellation theory and its relation to interspecific variability in brain organization, evolutionary and ontogenetic development, and neuronal plasticity. *Cell Tissue Res.* **213**:179–212.

Ebbesson, S. O. E. and R. G. Northcutt (1976) Neurology of anamniotic vertebrates. In *Evolution of the Brain and Behavior in Vertebrates*, R. B. Masterton, M. E. Bitterman, C. B. G. Campbell, and N. Hotton, eds. Lawrence Erlbaum Associates, Hillsdale, N.J., pp. 115–146.

Ebinger, P., K. Wächtler, and S. Stähler (1984) Allometrical studies in the brain of cyclostomes. *J. Hirnforsch.* **24**:545–550.

Eisenberg, J. F. and D. Wilson (1978) Relative brain size and feeding strategies in the Chiroptera. *Evolution* **32**:740–751.

Eldredge, N. and J. Cracraft (1980) *Phylogenetic Patterns and the Evolutionary Process.* Columbia University Press, New York.

Goodman, M. (1976) Toward a genealogical description of the primates. In *Molecular Anthropology*, M. Goodman and R. E. Tashian, eds. Plenum, New York, pp. 321–353.

Gruberg, E. R. and V. R. Ambros (1974) A forebrain visual projection in the frog (*Rana pipiens*). *Exp. Neurol.* **44**:187–197.

Guise, A. and D. Peacock (1982) A method for identification of parallelism in discrete character sets. *Zool. J. Linnean Soc.* **74**:293–303.

Hennig, W. (1966) *Phylogenetic Systematics.* University of Illinois Press, Urbana.

Herrick, C. J. (1948) *The Brain of the Tiger Salamander.* University of Chicago Press, Chicago.

Hofman, M. A. (1982) Encephalization in mammals in relation to the size of the cerebral cortex. *Brain Behav. Evol.* **20**:84–96.

Hopson, J. A. (1979) Paleoneurology. In *Biology of the Reptilia*, Vol. 9, Neurology A, C. Gans, R. G. Northcutt, and P. Ulinski, eds. Academic Press, London, pp. 39–146.

Huxley, J. (1932) *Problems of Relative Growth.* Allen & Unwin, London.

Jerison, H. J. (1955) Brain to body ratios and the evolution of intelligence. *Science* **121**:447–449.

Jerison, H. J. (1970) Brain evolution: new light on old principles. *Science* **170**:1224–1225.

Jerison, H. J. (1973) *Evolution of the Brain and Intelligence.* Academic Press, New York.

Kevetter, G. A. and W. D. Willis (1983) Collaterals of spinothalamic cells in the rat. *J. Comp. Neurol.* **215**:453–464.

Kicliter, E. (1979) Some telencephalic connections in the frog, *Rana pipiens. J. Comp. Neurol.* **185**:75–86.

Lande, R. (1979) Quantitative genetic analysis of multivariate evolution, applied to brain: body size allometry. *Evolution* **33**:402–416.

Lartet, E. (1868) De quelques cas de progression organique verifiables dans la succession des temps géologiques sur des mammifères de même famille et de même genre. *C. R. Acad. Sci.* **66**:1119–1122.

Le Quesne, W. J. (1969) A method of selection of characters in numerical taxonomy. *Syst. Zool.* **18**:1–32.

Macphail, E. M. (1982) *Brain and Intelligence in Vertebrates.* Clarendon Press, Oxford.

Marsh, O. C. (1874) Small size of the brain in Tertiary mammals. *Am. J. Sci. Arts* **8**:66–67.

Martin, R. D. (1981) Relative brain size and basal metabolic rate in terrestrial vertebrates. *Nature* **293**:57–60.

Mayr, E. (1969) *Principles of Systematic Zoology.* McGraw-Hill, New York.

Nelson, G. J. (1978) Ontogeny, phylogeny, paleontology, and the biogenetic law. *Syst. Zool.* **27**:324–345.

Noback, C. R. and J. E. Shriver (1969) Encephalization and the lemniscal systems during phylogeny. *Ann. N. Y. Acad. Sci.* **167**:118–128.

Northcutt, R. G. (1978) Brain organization in cartilaginous fishes. In *Sensory Biology of Sharks, Skates, and Rays,* E. S. Hodgson and R. F. Mathewson, eds. Office of Naval Research, Dept. of the Navy, Arlington, Virginia, pp. 117–193.

Northcutt, R. G. (1981) Evolution of the telencephalon in nonmammals. *Annu. Rev. Neurosci.* **4**:301–350.

Northcutt, R. G. (1984) Evolution of the vertebrate central nervous system: patterns and processes. *Am. Zool.* **24**:701–716.

Northcutt, R. G. and M. R. Braford (1980) New observations on the organization and evolution of the telencephalon of actinopterygian fishes. In *Comparative Neurology of the Telencephalon,* S. O. E. Ebbesson, ed. Plenum, New York, pp. 41–98.

Owen, R. (1843) *Lectures on Comparative Anatomy.* Longman, Brown, Green, and Longmans, London.

Patterson, C. (1982) Morphological characters and homology. In *Problems of Phylogenetic Reconstruction* (Systematics Association Special Vol. 21), K. A. Joysey and A. E. Friday, eds. Academic Press, New York, pp. 21–74.

Platel, R. (1979) Brain weight-body weight relationships. In *Biology of the Reptilia,* Vol. 9: Neurology A, C. Gans, R. G. Northcutt, and P. Ulinski, eds. Academic Press, London, pp. 147–171.

Platel, R. and C. Delfini (1981) L'encéphalisation chez la Myxine (*Myxine glutinosa* L.). Analyse quantifiée des principales subdivisions encéphaliques. *Cah. Biol. Mar.* **22**:407–430.

Reeve, E. C. R. and J. S. Huxley (1945) Some problems in the study of allometric growth. In *Essays on Growth and Form,* W. E. Le Gros Clark and P. B. Medawar, eds. Oxford University Press, Oxford, pp. 121–156.

Remane, A. (1956) *Die Grundlagen des naturlichen Systems der vergleichenden Anatomie und Phylogenetik.* Geest und Portig K. G., Leipzig.

Schaeffer, B. (1981) The xenacanth shark neurocranium, with comments on elasmobranch monophyly. *Bull. Mus. Nat. Hist.* **169**:1–66.

Schroeder, D. M. and S. O. E. Ebbesson (1974) Nonolfactory telencephalic afferents in the nurse shark (*Ginglymostoma cirratum*). *Brain Behav. Evol.* **9**:121–155.

Simpson, G. G. (1961) *Principles of Animal Taxonomy.* Columbia University Press, New York.

Smith, H. M. (1967) Biological similarities and homologies. *Syst. Zool.* **16**:101–102.

Snell, O. (1891) Die Abhängigkeit des Hirngewichtes von dem Köpergewicht und den geistigen Fähigkeiten. *Arch. Psychiat. Nervenkr.* **23**:436–446.

Stensio, E. (1963) The brain and the cranial nerves in fossil, lower craniate vertebrates. *Skr. Nor. Videnshaps-Akad. Oslo, Mat. Naturv. Kl.* [N.S.] No. **13**:1–120.

Towe, A. L. (1973) Relative numbers of pyramidal tract neurons in mammals of different sizes. *Brain Behav. Evol.* **7**:1–17.

Ulinski, P. S. (1983) *Dorsal Ventricular Ridge.* Wiley, New York.

Underwood G. (1982) Parallel evolution in the context of character analysis. *Zool. J. Linnean Soc.* **74**:245–266.

Von Bonin, G. (1937) Brain weight and body weight in mammals. *J. Gen. Psychol.* **16**:379–389.

Wilczynski, W. and R. G. Northcutt (1983) Connections of the bullfrog striatum: afferent organization. *J. Comp. Neurol.* **214**:321–332.

Wiley, E. O. (1981) *Phylogenetics.* Wiley, New York.

Wolpoff, M. H. (1985) Tooth size—body size scaling in a human population: theory and practice of an allometric analysis. In *Size and Scaling in Primate Biology,* W. L. Jungers, ed. Plenum, New York, pp. 273–318.

Epilogue

A NEUROBIOLOGICAL APPROACH TO HUMAN BEHAVIOR

C. LADD PROSSER

Department of Physiology and Biophysics
University of Illinois
Urbana, Illinois

One objective of neuroscience is to describe the molecular and cellular mechanisms of conduction in axons, transmission at synapses, and integrative modulation of motor responses by multiple inputs. Another objective is to account for animal behavior of different levels of complexity. A third neurobiological goal is to extend the findings from studies of animal nervous systems to human social behavior. This essay is a brief summary of the last approach to neurobiology.

CLASSIFICATION OF BEHAVIOR

It is not appropriate to speak of lower and higher animals; each species that has survived is adapted behaviorally to its way of life. Evolution has been both horizontal and vertical. The following categories of behavior from simple to complex are not to be considered as distinct but rather to form a continuum. Descriptions of animal behavior are often given in subjective terms.

1. The simplest adaptive behavior is direct response to environmental change or stimuli. A direct response may be positive or negative, often favoring a "preferred" environment, satisfying a need such as food, or avoiding a "harmful" condition. Direct responses are taxes, kineses, tropisms. Direct responses are genetically programmed and stereotyped and may be responses to several convergent stimuli. Direct responses occur in animals without nervous systems as well as in those with nervous systems. Direct responses may be modified according to internal state of an animal.

2. Modification of direct responses by experience may be short term or long term. Examples are habituation, conditioning—classical or operant. Neural mechanisms for modification of direct responses are subject to cellular analysis as presented in this symposium.

3. Many behaviors that depend on communication are, like conditioned direct responses, innate and stereotyped. Communicatory behaviors occur between animals of the same species, rarely with other species. Examples are courtship, warning threat calls of frogs, songbirds, dances of scout honeybees, jamming avoidance responses of electric fish (Marler, 1960, 1976).

4. Some communicatory and motor behaviors are developed by imprinting or other learning processes at certain stages. Examples are imprinted songs of birds, social play and hunting in mammals.

5. The language of communication comes to acquire symbolic meaning in that a sign elicits a sequence of behavior, not a single stereotyped response. Symbols are substitutes for objects of recognized actions and can initiate behavior as well as can the objects. Some examples are the bill and face of a parent gull, parent and sibling recognition by odor in nests of rodents, recognition of symbols by chimpanzees in a learning situation.

6. The most complex level of behavior is cognition as shown by humans. Whether abstract thinking occurs in nonhuman animals has been much debated and is not agreed upon (Sebeok, 1977). Cognition allows origination of behavior and the development of concepts in the absence of symbols. Human language has both phonetic (speech) and psychic (idea) components. Interpretation of animal behavior in terms of self-awareness or consciousness is fraught with anthropomorphic reasoning (Griffin, 1976).

The preceding classification of the continuum of levels of behavior has heuristic value. It is at present only slightly amenable to molecular description. The complex levels of behavior are emergent from interactions of many linked neurons in ways that are not understood.

Communication underlies many aspects of behavior and is amenable to neural and, within constraints, to molecular analysis. Many animals communicate by motions, electric pulses, color displays, and especially by sounds that have meaning for conspecifics, signals that connote aggression, alarm, courtship. These signals are largely genetically coded. The use of verbal language for communication and thought is distinct from use of sounds as symbols. Human speech shows cognition and lability in contrast to the stereotyped calls of many animals. One opinion concerning the origin of human language is that hand gestures, facial expressions, posture, and vocal calls were precursors of words. Another opinion is that the use of words came by the coevolution of association cortex with neural and laryngeal vocal mechanisms and that language is a part of general intelligence. Human language is more than communicative sounds; vocalization and words take on meaning. The cognitive component of language permits thought without articulation in speech. One linguistic position is that humans have a genetically coded innate capacity for language in the sense of syntax. Another view emphasizes the importance of imitation, much as in birds.

Whether chimpanzees are capable of language in the human sense has been much debated. Apes can use sticks as tools, can use symbols for objects and sensations, and can form cross-modal associations, but so can many other animals (including honeybees). Ethological observations of primates in their natural environment record behavioral employment of sounds—22 by gorilla, 23 by chimpanzee. There are marked differences in specificity of calls between normal individual primates and those that have been either deafened after birth or reared in isolation (Marler, 1960, 1976; Searcy and Marler, 1981). Apes lack the vocal apparatus for enunciation, but they can be trained to recognize gestures and objects and to respond by finger movements to a sign language, to match plastic chips, and to press keys of a computer. However, there is no evidence that a chimpanzee uses language in the sense of abstract concepts. The ape cortex has the rudiments of speech association areas. However, the quantitative differences in brain and vocal apparatus between ape and man are so great as to represent a quantum jump in complexity (Steklis and Raleigh, 1979). A conservative view is that, despite biochemical similarities in hemoglobins and enzyme proteins, *Homo* is not derived from or closely related to any other living primate species.

GENERAL CATEGORIES OF CULTURALLY TRANSMITTED CHARACTERS

Genetic programming and cultural transmission of behavior patterns combine to produce the complex behavior of mammals, especially humans. There is genetic specification of reflex patterns of locomotion, feeding, reproduction, and cross-modal sensory interactions. The genotype provides a template on which social (cultural) forces act. Is cultural transmission in humans like the imprinting of young birds and social mammals, or are there general attitudes and specific behavior patterns that are transmitted culturally in all societies and specific patterns for individual societies? It is proposed that some categories of culturally transmitted characters exist in all cultures and that local societies refine and define specific patterns. A person can go from one culture to another and substitute local mores within the general framework.

A list of general categories of culturally transmitted characteristics of humans follows:

1. *Societies.* Social organization is indicated for primitive *Homo* and is common to all cultures. Human social interactions are more varied and labile than those of other social mammals, fishes, or Hymenopterans. Rules of interactive behavior become formulated, and these differ in detail in different cultures. Such rules form the basis of ethical codes.
2. *Language.* In behavior and in brain structure, the most distinctive characteristic of humans is language. This serves two functions—for communication and for formulating concepts and abstract ideas. Linguists postulate that language evolved when specific sounds served to reduce ambiguities of meanings. Each culture has modified the general character of language, but the use of language appears to be a general cultural characteristic based on biological capacity (Gertz, 1964).
3. *Technology.* An early technology was fabrication of stone tools for cutting. Many animals use sticks and stones for obtaining food, sticks and grass for making nests, but primitive man was unique in making stone tools by sharpening them. Invention of wheels, pulleys, and levers occurred relatively recently, but their use spread rapidly in prehistoric cultures. The earliest use of fire for cooking was probably by *Homo erectus.* A general technology was the use of nonhuman sources of energy, wind and water, for doing work. A technology characteristic of recent human cultures is agriculture. Analogies may be drawn to ant-fungus cultures and ant-aphid symbioses; these are stereotyped and genetically coded, not socially transmitted.
4. *Arts.* A cultural activity that developed in human societies during the Pleistocene epoch was decoration of walls, wood carving, and body decoration. Rhythmic sounds, made by instruments, are used in all cultures in entertainment and rituals. The specific forms of art, music, and dance vary locally and are imprinted on children.
5. *Religion and animism.* Religion is used in the restricted sense that whenever man was unable to explain natural phenomena by what he could sense, he attributed the phenomena to gods. There is a general belief in magic, for instance in the curative power of objects. Animism may have evolved in an effort to deal with death.
6. *Taboos.* A taboo found in most cultures is against incest, sibling matings.

7. *Division of labor.* In all cultures there is division of labor between the sexes. This division of labor is based on the biological requirement for females to produce and care for offspring. It is also related to the exceptionally long period required for humans to reach puberty.
8. *Population dispersal.* Human societies are characterized by a tendency to move from one locality to another. Beginning with *Homo erectus,* human populations spread throughout the habitable world. Continuity of gene flow resulted in the present condition that all human populations are one biological species.

NEURAL ADAPTATION OF HUMANS

Some properties of anthropoid brains in general and human brains in particular are distinctive and allow coadaptation of genetic and cultural inheritance. Genetically coded developmental processes make neuronal connections that provide the substrate on which culture is transmitted. Neural differences, often quantitative rather than qualitative, allow for the significant functional differences between pongids and humans and between anthropoids and other primates:

1. Cranial capacity and brain size relative to body size in hominids is large compared to other primates. Average values of the ratio of brain volume to body weight are for man, 0.02, chimpanzee 0.009, gorilla 0.002 (Armstrong and Falk, 1982). The asymmetry of the two sides of the cortex is greater in primates than in non-primates and the asymmetry is marked in *Homo* (Stephan and Andy, 1964, 1969).
2. The most important difference in the brain of humans in comparison with pongids and monkeys is the speech area, which consists of motor, laryngeal, and associative auditory regions. The speech area extends from part of the primary auditory cortex of the temporal lobe forward to a large motor region (Mitra, 1955).
3. The number of small neurons in cerebrum, thalamus, and some other regions is significantly greater in man and in those primates that are capable of symbolic behavior. The percentage in man of stellate cells and granule cells in cortex and in cerebellum is greater than in other mammals (Jacobsson, 1975).
4. With increase in brain size, cortical volume, and quantity of small neurons, the information capacity of the brain goes up exponentially. The nonlinear increase in capacity for information correlates not only with number of interconnected elements, but also with multiple functional states of neurons.
5. Variation is greater in embryonic nervous systems than in adults. In most nervous systems many more neuroblasts are formed than mature as neurons. Neuronal networks are established before they are

used. During development, selective elimination of neuronal collaterals occurs, for example, restriction of collaterals of corpus callosal neurons (Stanfield et al. 1982). It has been postulated that there may be more selection in nervous systems of embryos than of adults (Leary et al., 1981; Northcutt, 1981; Stanfield et al., 1982).

6. Sexual dimorphism, especially in regions concerned with reproductive functions, is highly specified in anthropoid brains.
7. The olfactory bulb is small in advanced primates compared to its size in primitive primates, carnivores, and rodents.
8. Humans are capable of diverse facial expressions; facial muscles and their controlling motor nuclei are more extensive in humans than in other mammals.

GENOTYPE AND CULTURETYPE

In understanding human nature the biochemical measurements commonly used for tracing animal relationships, for example electrophoretic patterns of proteins, are of little utility. Brain structure, especially the speech areas, is most useful. A critical problem is to resolve the coevolution of genetically and culturally transmitted characters. One approach to this problem has been made by sociobiologists, first for social animals in general and then for humans (Wilson, 1975, 1978).

Basic premises of sociobiology are that (1) like all physiological characters, behavior has a genetic basis, and (2) behavioral patterns are subject to natural selection in accordance with their adaptive value. Selection may be of social groups, rather than of individuals. Survival of behavior patterns such as mate selection and reactions to kin and nonkin, may optimize success of a species. How the genotype sets limits on behavioral plasticity is far from completely understood; the sequences of development of neuronal connections are genetically coded and determine adult behavior. In parallel with genic determination by way of neural structures is cultural or social transmission of behavior patterns. Culturetypes are constellations of behavioral characters that are not transmitted genetically but are transmitted culturally. In humans the contribution of culture type is greater than in any other species. How much of human behavior is genetically programmed, how much developmentally determined, and how much culturally transmitted is difficult to estimate.

The units of culturally transmitted characters have been called memes (memory units), by analogy with genes (Dawkins, 1976). It is contended that genes are active agents working for their own survival in a host organism. Memes are self-perpetuating by their own cultural and sociological impact. After death, both genes and memes continue on, with memes perhaps more lasting. Man is biologically a gene machine and culturally a meme machine (Dawkins, 1976). Some of the same terminology can be applied to cultural evolution as to biological evolution, but the mechanisms of transmission and selection are utterly different.

General cultural traits constitute a heritage common to all human societies. Specific cultural traits are restricted to individual societies. Specialized patterns are very labile, general cultural patterns less labile, and genetically coded behavior patterns are stereotyped. Selection of genetic variants of behavior is speeded or is slowed by survival value of the variants in a social environment. In this way cultural inheritance may influence the rate of biological evolution.

The genetic limits for normal function and for survival are wide for molecules, intermediate for cells and tissues, and narrow for integrated whole organisms (Fig. 1). Cultural inheritance is opposite to the biological trend in that adaptiveness of societies is greater than of individuals. A society can function over a wider range of environmental stress than can individuals (Fig. 1).

A number of biological properties of humans correlate with their social inheritance and provide the basis for applying the principles of sociobiology to human societies.

1. The long maturation time of humans has been selected in accordance with the biological value of social determination. Slow development permits extensive imprinting and neural modification.
2. *Homo* is now one species, and gene exchange can take place between all populations. Culturally, however, meme exchange is restricted as long as societies maintain some separation.
3. Cultural transmission is more rapid than genetic; evolution of culture patterns of civilizations occurred in tens of thousands of years, not in the millions of years for genetic separation of hominids.
4. Nonhuman animals tend to show altruism only to individuals in a kin group; humans extend altruistic behavior to nonkin.
5. The most important biologically developed features of man are the language centers in the association areas of the cerebral cortex; these areas serve both speech and abstract aspects of language. Speech areas in the human brain represent a major step in primate evolution. Language has both symbolic and cognitive function, and with language comes ability to think abstractly.
6. A human property also characteristic of some birds and mammals is pair bonding. Sex activity without reproduction functions for bonding; human females have no oestrus and can be sexually receptive at all times.

Whether there is genetic determination for aggressive behavior, grammar, religion is not agreed upon (Wilson, 1978). The essence of extension of sociobiology to humans is the notion that the complex behavior patterns of cultures have a biological basis that is genetically transmitted.

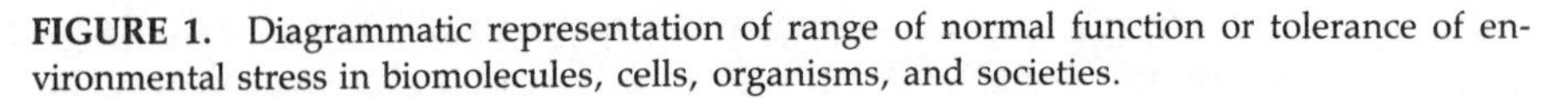

FIGURE 1. Diagrammatic representation of range of normal function or tolerance of environmental stress in biomolecules, cells, organisms, and societies.

Many social scientists have taken a stand against applying to man the principles of sociobiology (Sahlin, 1976). Social scientists note the following:

1. Human reproductive unis are more socially than biologically determined.
2. Mates in human societies are not selected on the basis of strength or weakness in reproductive capacity as in many nonhumans. Mates are usually not close blood relations, and taboos ban incest in most societies.
3. Societal values may determine behavior more than biological values. Culture frees humanity from some of the restrictions of biological emotions and motivations.
4. Thought, beliefs, cultural patterns are epiphenomena, not properties emergent from neural systems.
5. The concepts of social science and philosophy are qualitatively different from the concepts of biology and must be described by a different language. Inquiry into thought and culture are separate from inquiry into the physical properties of life.
6. Language-culture patterns limit social evolution and may eventually lead to the decline and demise of *Homo*; thus, the species may be an evolutionary failure because of cultural, not biological weakness.

A way out of the dilemma between social and biological sciences has been provided by writers such as the psychologist Campbell (Campbell, 1975) and the scientific theologian Burhoe (Burhoe, 1981). They maintain that man is a symbiosis between culturetype and genotype. Neither the genetically coded nor the culturally established man is capable of living as a human being without the other. There is coadaptation of the genotype and culturetype heritages. Interaction of the two heritages provides positive reinforcement and results in social stability. The neural basis for the coadaptation is genetically coded and was probably selected under social pressures. Without the suitable brain substrate the symbiosis between genetic and cultural determination of behavior could not have evolved.

The dual nature of man is reflected in the two types of inheritance and the two types of information. The memes include cultural formulas and symbols, religious rites and myths, codes of values, and interpersonal behavior. The ethical systems and esthetic capacities of humans are dependent on cultural inheritance. Memes persist through generations over the time span of cultural evolution. The genes code for neural structures: association and language areas, synaptic capacities for memory, and endogenous stereotyped behavior. The culturetype requires a genetically determined brain. From this unified, but symbiotic nature of man properties emerge that require the combination of genotype and culturetype. Natural selection of adaptive genes acting within a society is the creative force for human diversity. Such a view of humans is a dualism with a sound biological basis. A proper objective of neuroscience is to contribute to understanding the neural mechanisms of the symbiosis.

Acknowledgment

This epilogue is a condensation of part of a chapter on animal behavior in a forthcoming book, Adaptational Physiology; Molecules to Organisms, by C. Ladd Prosser to be published by Wiley Interscience.

REFERENCES

Armstrong, E. and D. Falk, eds. (1982) *Primate Brain Evolution.* Plenum, New York, p. 332.

Burhoe, R. W. (1981) *Toward a Scientific Theology.* Christian Journals, Ottawa, Ontario.

Campbell, D. T. (1975) On the conflicts between biological and social evolution and between psychology and moral tradition. *Am. Psych.* **30**:1103–1126.

Dawkins, R. (1976) *The Selfish Gene.* Oxford University Press, New York.

Gertz, C. (1964) The transition to humanity. In *Horizons of Anthropology,* Sol Tax, ed. Aldine, Chicago.

Griffin, D. R. (1976) *The Question of Animal Awareness.* Rockefeller University Press, New York.

Jacobsson, M. (1975) Cell types in mammalian brain. In *Golgi Centennial Volume,* M. Santini, ed. Raven, New York, pp. 147–151.

Leary, D. D., B. Stanfield, and W. M. Cowan (1981) Evidence that the early postnatal restriction of the cells of origin of the callosal projection is due to the elimination of axonal collaterals rather than to the death of neurons. *Dev. Brain Res.* **1**:607–617.

Marler, P. R. (1957) Specific distinctiveness in the communication signals of birds. *Behavior* **11**:13–39.

Marler, P. R. (1960) In *Animal Sound and Communication*, U. E. Lanyon and W. N. Tavolga, eds. pp. 348–367.

Marler, P. R. (1976) Sensory templates in species-specific behavior. In Simpler Networks and Behavior, J. Fentriss, ed. Sinauer, Sunderland, Mass., pp. 314–329.

Mitra, N. L. (1955) Quantitative analysis of cell types in mammalian neo-cortex. *J. Anat.* **89**:467–483.

Northcutt, R. G. (1983) personal communication.

Northcutt, R. G. (1981) Evolution of the telencephalon in nonmammals. *Annu. Rev. Neurosci.* **4**:301–350.

Sahlin, M. D. (1976) *Use and Abuse of Biology.* University of Michigan Press, Ann Arbor.

Searcy, N. A. and P. Marler (1981) A test for responsiveness to song structure and programming in female sparrows. *Science* **213**:926–928.

Sebeok, T. A., ed. (1977) *How Animals Communicate.* Indiana University Press, Bloomington.

Stanfield, B. B., D. D. M. O'Leary, and C. Fricks (1982) Selective collateral elimination in early postnatal development restricts cortical distribution of rat pyramidal tract neurones. *Nature* **298**:371–373.

Stephan, H. and O. J. Andy (1964) Quantitative comparisons of brain structures from insectivores to primates. *Am. Zool.* **4**:59–74.

Stephan, H. and O. J. Andy (1969) Quantitative comparative neuroanatomy of primates: an attempt at a phylogenetic interpretation. *Ann. N. Y. Acad. Sci.* **167**:370–387.

Steklis, H. D. and M. J. Raleigh (1979) *Neurobiology of Social Communication in Primates.* Academic Press, New York.

Wilson, E. O. (1975) *Sociobiology.* Harvard University Press, Cambridge, Mass.

Wilson, E. O. (1978) *On Human Nature.* Harvard University Press, Cambridge, Mass.

INDEX